High-Level Synthesis based Methodologies for Hardware Security, Trust and IP Protection

Other related titles:

You may also like

- PBCS080 | Sengupta | Physical Biometrics for Hardware Security of DSP and Machine Learning Coprocessors | 2023
- PBCS076 | Sengupta | Secured Hardware Accelerators for DSP and Image Processing Applications | 2022
- PBCS067 | Sengupta | Frontiers in Securing IP Cores | 2019
- PBCS066 | Sengupta | Frontiers in Hardware Security and Trust: Theory, design and practice | 2020
- PBCS067 | Sengupta and Mohanty | IP Core Protection and Hardware-Assisted Security for Consumer Electronics | 2019

We also publish a wide range of books on the following topics:
Computing and Networks
Control, Robotics and Sensors
Electrical Regulations
Electromagnetics and Radar
Energy Engineering
Healthcare Technologies
History and Management of Technology
IET Codes and Guidance
Materials, Circuits and Devices
Model Forms
Nanomaterials and Nanotechnologies
Optics, Photonics and Lasers
Production, Design and Manufacturing
Security
Telecommunications
Transportation

All books are available in print via https://shop.theiet.org or as eBooks via our Digital Library https://digital-library.theiet.org.

CIRCUITS, DEVICES AND MATERIALS SERIES 84

High-Level Synthesis based Methodologies for Hardware Security, Trust and IP Protection

Anirban Sengupta

The Institution of Engineering and Technology

About the IET

This book is published by the Institution of Engineering and Technology (The IET).

We inspire, inform and influence the global engineering community to engineer a better world. As a diverse home across engineering and technology, we share knowledge that helps make better sense of the world, to accelerate innovation and solve the global challenges that matter.

The IET is a not-for-profit organisation. The surplus we make from our books is used to support activities and products for the engineering community and promote the positive role of science, engineering and technology in the world. This includes education resources and outreach, scholarships and awards, events and courses, publications, professional development and mentoring, and advocacy to governments.

To discover more about the IET please visit https://www.theiet.org/.

About IET Books

The IET publishes books across many engineering and technology disciplines. Our authors and editors offer fresh perspectives from universities and industry. Within our subject areas, we have several book series steered by editorial boards made up of leading subject experts.

We peer review each book at the proposal stage to ensure the quality and relevance of our publications.

Get involved

If you are interested in becoming an author, editor, series advisor, or peer reviewer please visit https://www.theiet.org/publishing/publishing-with-iet-books/ or contact author_support@theiet.org.

Discovering our electronic content

All of our books are available online via the IET's Digital Library. Our Digital Library is the home of technical documents, eBooks, conference publications, real-life case studies and journal articles. To find out more, please visit https://digital-library.theiet.org.

In collaboration with the United Nations and the International Publishers Association, the IET is a Signatory member of the SDG Publishers Compact. The Compact aims to accelerate progress to achieve the Sustainable Development Goals (SDGs) by 2030. Signatories aspire to develop sustainable practices and act as champions of the SDGs during the Decade of Action (2020–2030), publishing books and journals that will help inform, develop, and inspire action in that direction.

In line with our sustainable goals, our UK printing partner has FSC accreditation, which is reducing our environmental impact to the planet. We use a print-on-demand model to further reduce our carbon footprint.

Published by The Institution of Engineering and Technology, London, United Kingdom

The Institution of Engineering and Technology (the "**Publisher**") is registered as a Charity in England & Wales (no. 211014) and Scotland (no. SC038698).

First published 2024

The Institution of Engineering and Technology
Futures Place
Kings Way, Stevenage
Herts, SG1 2UA, United Kingdom

www.theiet.org

British Library Cataloguing in Publication Data
A catalogue record for this product is available from the British Library

ISBN 978-1-83724-117-0 (hardback)
ISBN 978-1-83724-118-7 (PDF)

Typeset in India by MPS Limited

Cover image: Yuichiro Chino/Moment

Contents

Foreword

High-level Synthesis (HLS) is the process of automatically generating a register-level design from a high-level description. Over the last two decades, HLS tools have been employed effectively in the industry to synthesize Very Large-Scale Integration (VLSI) designs. Automated design exploration has been particularly a strong suit of HLS-based design methodologies. Area, power, and performance have been the primary optimization objectives. Over the last decade or so, a new and important design objective has emerged, namely, design security. Today HLS of secure designs is a very active area of research. This book is very timely and presents novel HLS design methodologies to generate secure and trustworthy designs.

In this book, three important research vectors for HLS-based secure hardware have been explored. Firstly, Intellectual Property (IP) protection through watermarking has been thoroughly dealt with. Hardware IP piracy is a major problem in the industry. Watermarking is the process of embedding a signature seamlessly in the IP so that only the IP owner can extract the watermark should there be any litigation about the ownership. The book presents a plethora of watermark insertion techniques based on protein molecular biometric, facial biometric, retinal biometric, mathematical watermark, multi-modal biometric, and crypto chain. Secondly, hardware obfuscation methods have been investigated. Through reverse engineering, attackers can either steal the IP or find security loopholes. Obfuscation is a technique to make the design harder to reverse-engineer. The book proposes structural as well as algorithmic obfuscation approaches. Thirdly, hardware Trojan detection method has been proposed. Hardware Trojans are malicious modifications to the hardware. The author proposes a novel lightweight detection method to counter hardware Trojans.

The book is an excellent resource for diverse audiences. For peer researchers and industry practitioners, the book provides a detailed presentation on the latest advances in the HLS-based secure design topic. For instructors and students, the book can be used as a graduate textbook as each chapter provides exercises to test the understanding.

Srinivas Katkoori
Computer Science and Engineering
University of South Florida
Tampa, FL 33620
7 May 2024

Acknowledgments

I would like to dedicate this book to my beloved grandfather, late Shri Shailendra Kumar Roy, who has been my role model throughout my life. He has been a constant source of inspiration in my career and personal life.

I would like to thank my grandmother (Mrs. Gouri Roy), my parents (Dr. Prabal Kumar Sengupta, Mrs. Aditi Sengupta), my wife (Mrs. Rumpa Sikdar Sengupta), my daughter (Sanvi) and my aunt (Dr. Sashwati Roy) and Uncle (Dr. Himadri Roy) for their support and encouragement. I would also like to thank the Indian Institute of Technology (IIT), Indore, India, for the support in executing this work.

About the author

Anirban Sengupta is a full professor in the Department of Computer Science and Engineering at the Indian Institute of Technology (IIT) Indore, India. He has more than 300 publications and patents, including 6 books, to his credit.

He is the recipient of awards and honors such as fellow of IET, fellow of British Computer Society, fellow of IETE, IEEE Chester Sall Memorial Consumer Electronics Award, IEEE distinguished lecturer, IEEE distinguished visitor, IEEE CESoc Outstanding Editor Award, IEEE CESoc Best Research Award from CEM, Best Research paper Award in IEEE ICCE 2019, IEEE Computer Society TCVLSI Outstanding Editor Award, and IEEE TCVLSI Best Paper Award in IEEE iNIS 2017. He held or holds around 18 editorial positions in IEEE and IET journal boards. He is consistently ranked in Stanford University's Top 2% Scientists globally. Details available at: http://www.anirban-sengupta.com/index.php.

Chapter 1

Introduction to hardware security and trust and high-level synthesis

Anirban Sengupta[1] *and Rahul Chaurasia*[1]

The chapter provides the readers with a formal introduction to achieving hardware security and trust through high-level synthesis (HLS) followed by its motivation and comprehensive exploration of different hardware attack scenarios and countermeasures. An adversary (situated in different design phases) may potentially leverage various attack scenarios to compromise the confidentiality, integrity, and reliability of electronic systems comprising reusable hardware designs. Addressing these threats necessitates a comprehensive understanding of hardware security and a proactive approach to fortify systems against potential threats. This chapter aims to contribute to the development of secure and trustworthy computing systems by providing a basic foundation for security aware HLS-based hardware design methodology, different attack scenarios, and effective countermeasures. This chapter therefore bridges the gap between the existing hardware security challenges and its advanced countermeasure strategies.

The rest of the chapter is structured as follows: Section 1.1 provides the introduction of the chapter including motivation for hardware security and trust; Section 1.2 discusses different attack scenarios; Section 1.3 provides different security countermeasures; Section 1.4 discusses the utility of the HLS framework for integrating security countermeasures; Section 1.5 presents a brief summary of the book; Section 1.6 concludes the chapter by summarizing its significance and implications.

1.1 Introduction

In the rapidly evolving landscape of hardware design, ensuring the security and trustworthiness of reusable hardware designs has become a paramount concern. This chapter delves into the motivation behind the imperative need for robust hardware security and trust, emphasizing the critical role they play in ensuring and maintaining reliable functionality and the integrity of electronic systems (Sengupta *et al.*, 2023d; Sengupta and Chaurasia 2023). The chapter comprehensively

[1]Department of Computer Science and Engineering, Indian Institute of Technology Indore, India

discusses various attack scenarios that pose significant threats to hardware security, ranging from IP piracy, tampering, removal attack, forgery, reverse engineering for covertly inserting hardware Trojans, and supply chain vulnerabilities (Wang *et al.*, 2015; Koushanfar *et al.*, 2012; Syed and Lourde, 2016; Arafin *et al.*, 2017; Colombier and Bossuet, 2015; Rajendran, 2017; Hu *et al.*, 2021; Sengupta *et al.*, 2017; Karmakar *et al.*, 2022; Kuai *et al.*, 2020; Koushanfar *et al.*, 2005; Karmakar and Chattopadhyay, 2020; Sengupta and Rathor; 2019a; Rathor and Sengupta, 2020; Sengupta and Rathor; 2020; Sengupta *et al.*, 2019; Sengupta *et al.*, 2023a; Sengupta *et al.*, 2023b; Chaurasia and Sengupta, 2023a; Chaurasia *et al.*, 2023; Sengupta and Chaurasia 2023; Sengupta *et al.*, 2023b; Chaurasia *et al.* 2022; Sengupta *et al.*, 2023c; Chaurasia and Sengupta, 2021; Chaurasia and Sengupta 2022d; Chaurasia and Sengupta, 2022a; Chaurasia and Sengupta, 2022b; Rathor and Sengupta, 2019; Sengupta, 2016; Sengupta and Rathor, 2020). Recognizing the diversity and sophistication of these attacks, this chapter emphasizes the urgency of developing effective countermeasures to mitigate the inherent risks associated with modern hardware architectures. Moreover, the central focus of the chapter lies in the exploration of leveraging the HLS framework as a strategic hardware design methodology and security countermeasure. HLS enables designers to express hardware functionality at a higher level of abstraction, facilitating rapid development and optimization. By exploiting the inherent flexibility and abstraction provided by HLS, an IP designer fortifies hardware designs against a multitude of potential vulnerabilities. Various approaches (Chaurasia and Sengupta, 2023a; Chaurasia *et al.*, 2023; Sengupta and Chaurasia 2023; Chaurasia *et al.* 2022; Sengupta *et al.*, 2023c; Chaurasia and Sengupta, 2021; Chaurasia and Sengupta, 2022a, 2022b, 2022c, 2022d; Sengupta *et al.*, 2021) involve the integration of security countermeasures directly into the HLS workflow, thereby enhancing the overall resilience of the hardware against diverse attack vectors. Through a detailed analysis of the HLS framework, this chapter further uncovers the advanced methods to embed security measures at the early stages of hardware development (mitigating the risk of post-manufacturing vulnerabilities), ensuring a proactive defense against potential threats ranging from detective countermeasure against IP piracy, nullifying fraudulent IP ownership claim, hindering reverse engineering, etc. The chapter also provides an overview corresponding to trade-offs and challenges associated with incorporating security into the HLS process, aiming to strike a balance between design cost overhead and security considerations.

Ultimately, the chapter contributes to the ongoing discourse on hardware security and trust by offering a comprehensive discussion of diverse attack scenarios and strategies for leveraging the HLS framework to bolster the security of modern electronic systems. This chapter holds significant implications for all the stakeholders of the integrated circuit (IC) supply chain, viz. IP designers/vendors, system-on-chip integrator (SoC), and researchers and scientific community working toward establishing a secure foundation for the future of hardware technologies. The chapter offers insights into the integration of HLS-based security measures as a proactive approach to fortify hardware against evolving threats, fostering a resilient foundation for the digital future.

1.1.1 Motivation for hardware security and trust

In the era of ubiquitous connectivity and increasing dependence on electronic systems, the security and trustworthiness of hardware have emerged as a critical concern. The relentless growth in technology has not only opened new frontiers of innovation but has also exposed hardware vulnerabilities that can be exploited by malicious adversaries that may present in untrustworthy design houses. This has prompted a pressing need for robust hardware security mechanisms to protect intellectual property and ensure the reliability of electronic devices. Hardware security and trust are critical aspects of modern computing systems, and their importance has grown significantly with the increasing integration of technology into various aspects of our daily lives. Several motivations drive the focus on hardware security and trust.

Why designing secure reusable hardware IPs for computationally intensive applications? In digital signal processing (DSP) co-processors, a DSP algorithm is employed to execute tasks corresponding to specific applications. Some commonly utilized DSP algorithms include the discrete cosine transform (DCT), discrete Fourier transform (DFT), fast Fourier transform (FFT), Haar wavelet transform (HWT), discrete wavelet transform (DWT), and inverse discrete cosine transform (IDCT). The DCT is applied when converting an image from the spatial domain to its frequency domain, serving as a fundamental algorithm for image compression-decompression in JPEG-codec co-processors. DFT and FFT are utilized to represent a discrete signal in the frequency domain from its time domain. HWT transforms the waveform of a signal from the time domain to time-frequency, finding extensive use in both lossy and lossless signal and image compression applications. DWT is instrumental in denoising real signals by decomposing them, breaking down digital signals to achieve finer frequency and coarser time resolution through different sub-bands. It is a fundamental algorithm in image compression for JPEG2000. Additionally, digital filters such as finite impulse response (FIR) and infinite impulse response (IIR) filters play a crucial role in various electronic systems, including applications like speech processing, telecommunication, and attenuation removal of selected frequencies (Salivahanan and Vallavaraj, 2001; Sengupta and Mohanty, 2019a).

Further, multimedia processors incorporate algorithms such as JPEG-codec and Moving Picture Experts Group (MPEG) for tasks like image compression. JPEG, widely used in medical imaging and digital camera systems, is particularly notable for image compression. Meanwhile, machine learning IP cores, including convolutional IP, are employed for tasks related to object detection and feature extraction. Therefore, it becomes crucial to generate secured hardware designs for ensuring reliable end system functionality (Chaurasia and Sengupta, 2023b; Chaurasia *et al.*, 2023; Sengupta and Chaurasia, 2023).

Why secure embedded systems? With the rise of the Internet of Things (IoT) and embedded systems, hardware is becoming ubiquitous in everyday objects such as home appliances, medical devices, and industrial equipment. Securing the hardware in these systems is essential to prevent unauthorized access, data breaches, and potential harm (Sengupta and Mohanty, 2019a).

Why secure critical infrastructure? Many critical infrastructures, including power grids, transportation systems, and healthcare facilities, rely heavily on hardware components. Ensuring the security and trustworthiness of the underlying hardware is crucial to safeguard these critical systems from cyber-attacks that could have severe consequences on public safety and national security (Sengupta, 2023).

Why secure cloud computing and data centers? Cloud services and data centers process and store vast amounts of sensitive information. Hardware vulnerabilities could lead to data corruption, leakage of secret information, and compromise of data confidentiality and data integrity. Therefore, securing the hardware in these environments becomes essential for maintaining the trust of users and businesses that rely on cloud services.

Why is trust important in supply chain? The hardware supply chain involves multiple vendors, manufacturers, and distributors. Ensuring the integrity and security of hardware components throughout the supply chain is crucial to prevent the insertion of malicious elements or compromised components. Unauthorized modifications at any stage of the supply chain can lead to serious security vulnerabilities (Sengupta, 2023; Chaurasia and Sengupta, 2023a; Chaurasia *et al.*, 2023; Sengupta *et al.*, 2023).

Why protect against cyber-attacks? Cyber-attacks are becoming increasingly sophisticated, and attackers often target hardware vulnerabilities to gain unauthorized access or control over systems. Securing hardware is a proactive measure to prevent a wide range of attacks, including those targeting firmware, hardware implants, and other low-level components.

Why protect hardware IPs? Companies invest significant resources in the development of proprietary hardware designs and technologies. Protecting these intellectual property assets from theft or reverse engineering is a motivation for implementing hardware security measures (Chaurasia and Sengupta, 2023b; Chaurasia *et al.*, 2023; Sengupta and Chaurasia, 2023).

Why security and trust are important in emerging technologies? As technologies like the Internet of Things (IoT), autonomous vehicles, and artificial intelligence continue to evolve, ensuring trust in the underlying hardware becomes essential. Users need confidence that these technologies are secure and reliable.

In summary, the motivation for hardware security and trust is multifaceted, encompassing concerns related to national security, privacy, data integrity, and the reliable operation of critical systems. As technology continues to advance, addressing hardware security challenges will remain a priority to build and maintain trust in the digital ecosystem.

1.2 Different attack scenarios

In the globalized semiconductor design supply chain, several offshore entities, such as 3PIP vendors, system integrator, and foundry house, collaborate to expedite the design process, aiming for low design costs, reduced design time, and quicker time-to-market. Consequently, hardware IPs may be sourced from different vendors. Depending on design requirements, these IPs are either delivered to an SoC integrator for incorporation into SoC design or directly sent to foundry houses for standalone IC fabrication. Following the integration of IPs at the SoC integrator's facility, the assembled design is then sent to foundry houses for fabrication. Essentially, the data flow in the design cycle is unidirectional, moving from IP vendors to SoC integrator and finally to the foundry house. This one-way flow may involve multiple IP vendors and a foundry house. The engagement of various entities in the IC design chain exposes it to diverse hardware threats, as illustrated in Figure 1.1 (Hu *et al.*, 2021; Sengupta *et al.*, 2017; Karmakar *et al.*, 2022; Kuai *et al.*, 2020; Koushanfar *et al.*, 2005; Sengupta, 2023; Anshul and Sengupta, 2023). Additionally, a brief overview of potential hardware security threats is provided below:

IP piracy: The creation of an IP core for multi-modal Consumer Electronics (CE) designs demands extensive research, investment, validation, and effort. To streamline the design cycle, various offshore entities are enlisted, aiming to reduce overall design costs, complexity, and time-to-market. However, the involvement of offshore design houses or foundries in the design chain introduces a significant challenge of IP piracy. For example, IP piracy may occur when an SoC integrator procures IP cores for integration either directly from an IP vendor or through a broker (acting as an intermediary between the IP designer and the SoC integrator).

Despite the benefits, the integration of offshore entities raises concerns about hardware threats related to IP piracy. Motivated by national interests or the desire for illicit gains, a rogue IP supplier might introduce pirated or counterfeit components (IPs) into the design supply chain. The incorporation of these fake components, disguised as genuine, into CE device SoCs can have adverse effects on both the CE system integrator and end-users. Ensuring security against the threat of IP piracy is of utmost importance for consumers due to the following reasons:

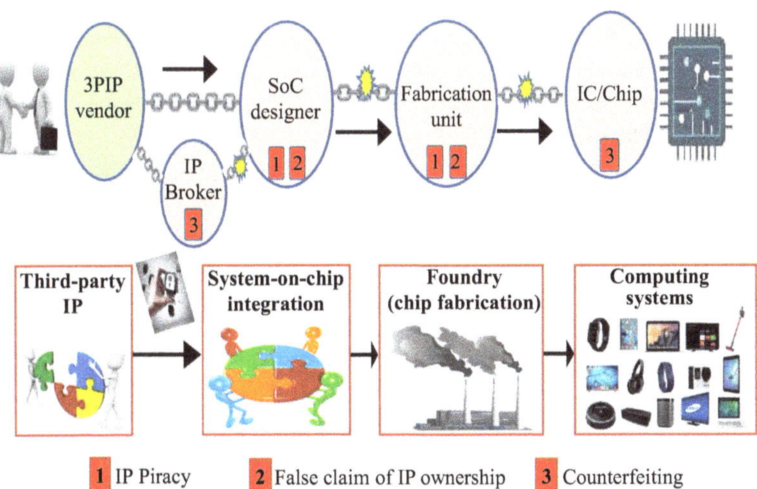

Figure 1.1 Different attack scenarios in the IC supply chain

(i) counterfeited designs lack rigorous testing for reliability and security, and (ii) counterfeit IPs may contain hidden malicious logic, such as hardware Trojans. The integration of infected IPs or ICs into CE systems can result in unreliability and pose safety risks for end consumers. Therefore, the ability to distinguish between authentic and fake IP versions is critical, ensuring the use of only genuine IPs in Consumer Electronics and computing systems.

IP ownership conflict: A deceptive IP buyer or a potential adversary, possibly present in the SoC house or within the foundry, may dishonestly assert ownership rights of the IP by fraud ownership claim. Such fraudulent claims pose a significant risk, potentially resulting in substantial financial losses for the legitimate IP owner. Consequently, the false assertion of ownership is a growing security concern. Traditional IP protection mechanisms like copyright, patent, trademark, and industrial design rights do not readily apply to reusable IP core designs. Therefore, safeguarding the ownership rights of the actual IP owner becomes imperative. In such scenarios, discreetly embedding the designer's signature within the IP core during its design process emerges as an effective and robust strategy (Sengupta *et al.*, 2023d; Chaurasia and Sengupta, 2022a, 2022b, 2022c, 2022d; Sengupta and Mohanty, 2019b; Roy and Sengupta, 2019; Sengupta and Rathor, 2019a). These approaches serve to substantiate the IP vendor's ownership rights and counteract any fraudulent claims of IP ownership by an adversary.

Reverse engineering the design: Reverse engineering of an IP core is a method aimed at uncovering its design, structure, and functionality. Through RE, one can discern the device technology, extract the gate-level netlist, and deduce the IP functionality. Despite the Semiconductor Chip Protection Act of 1984 (SCPA) not considering RE illegal for teaching, analysis, and evaluation purposes, attackers can exploit this process for unauthorized activities such as IP piracy and the insertion of malicious logic. Given the contemporary design supply chain's involvement of offshore design houses, complete security cannot be guaranteed. An adversary within these offshore design houses might manipulate the original register transfer level description or engage in reverse engineering to introduce malicious logic. Consequently, establishing robust security measures against the RE threat is essential before their integration into SoC systems (Sengupta and Rathor, 2020; Sengupta *et al.*, 2017).

IP rights infringement: In the design chain of an IP core, there are two key entities: the seller (IP vendor) and the buyer (IP user). The IP vendor is the creator of the IP, while the IP user is the purchaser. From the buyer's perspective in the supply chain, an untrustworthy IP seller may distribute or sell unauthorized copies of custom IPs, which are designed based on the specifications provided by the IP buyer. This could result in the illicit use of IPs, especially concerning hardware IPs tailored for specific purposes, such as mission-critical applications.

From the seller's perspective, a deceptive IP buyer may falsely claim ownership rights to the IP after receiving it. Therefore, establishing a unique one-to-one mapping between these entities is crucial. A secure IP core should support the detection of unlawfully redistributed or resold duplicates of an IP core by a deceitful IP seller. Additionally, it should safeguard the design in cases where the IP buyer falsely claims ownership of the IP. This dual protection mechanism is essential for maintaining the integrity of the IP core and ensuring a fair and secure transaction between the IP vendor and the IP buyer.

Illegal IP overproduction: It typically refers to the unauthorized manufacturing or distribution of ICs or other hardware components that contain intellectual property (typically occurs at the fabrication stage in the foundry through a rogue designer). This can happen when a company or individual produces hardware that incorporates proprietary designs, algorithms, or other protected IP without obtaining the necessary licenses or permissions from the rightful owner. Such actions can lead to legal disputes, including lawsuits for copyright infringement. Therefore, it is crucial for hardware manufacturers to respect IP rights and obtain proper licenses or permissions for any IP they incorporate into their design/products to avoid legal issues related to IP overproduction.

Further, a detailed description of the complete IC design process, highlighting potential hardware threats and attacks in the globalized semiconductor design supply chain, is presented in Figure 1.2 (Rajendran *et al.*, 2016; Rajendran *et al.*, 2012). Globalization has rendered the design process susceptible to various hardware attacks, which can occur at different IC design stages/divisions. One of the untrustworthy divisions within the hardware design supply chain, from the perspective of SoC integrator, is highlighted using brown color. Further, the untrustworthy division from the viewpoint of an IP vendor is highlighted using red color and the trusted division is highlighted using green color. The dashed lines within Figure 1.2 delineate various potential locations of attacks within the design supply chain process, while the solid lines in blue color represent the comprehensive hardware IC design flow from specification to manufacturing.

The process begins with system specifications, serving as behavioral descriptions of the intended hardware design. These specifications undergo progression through the hardware design process, which involves acquiring IP cores or designs from third-party vendors and integrating them into a single chip by the SoC integrator. Note: *the necessity for involving multiple third-party entities is to expedite the design process, aiming for low design cost, reduced design time, and quicker time-to-market (as discussed earlier)*. Upon integration, a register transfer level (RTL) file is generated, which is then synthesized into a gate-level design file, also known as a netlist file. This netlist file is subsequently transmitted to fabrication and manufacturing facilities/divisions.

As presented in Figure 1.2, hardware attacks are categorized into the following types:

- IP piracy and false claims of IP ownership within the SoC integrator's environment by an adversary.
- Backdoor hardware Trojan insertion through untrustworthy third-party IP cores.
- Netlist level attacks for decryption (deobfuscation) post reverse engineering from layout stage in the foundry (fabrication house) and
- IC level attacks that may potentially occur within the foundry or open market.

The consequences of the following attacks may lead to the following: access to the design netlist file enables an adversary to execute attacks like insertion of

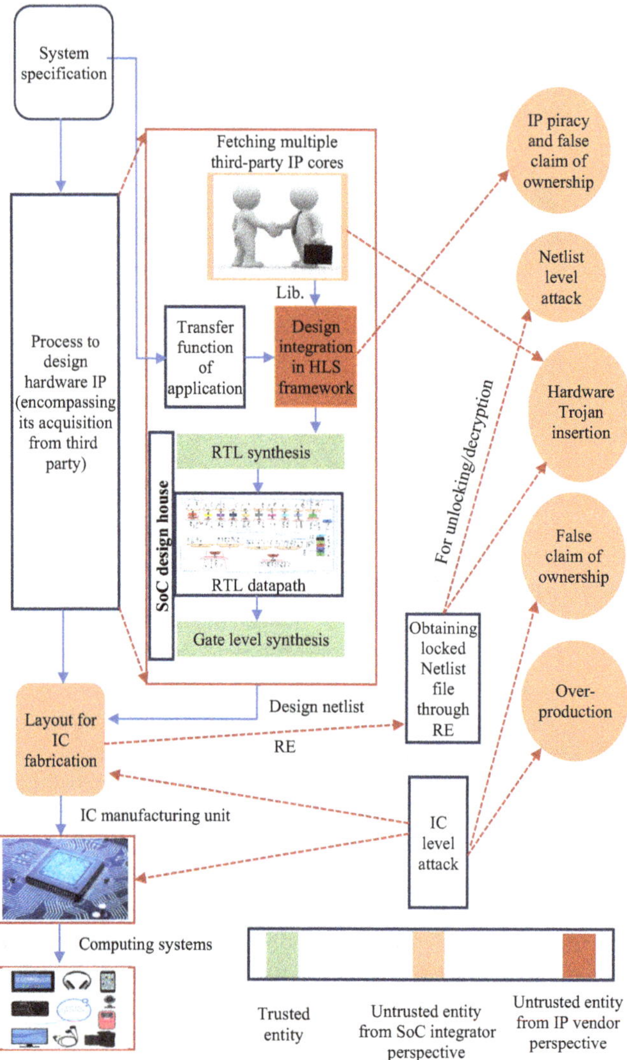

Figure 1.2 Potential hardware threats and attacks in hardware IC design flow

hardware Trojans, piracy, etc. Further, throughout the fabrication process and post-fabrication stage, a rogue adversary within the fabrication facility may initiate attacks such as overproduction and false claims of IP ownership. Thus, the different types of possible attacks include IP piracy (counterfeiting and cloning), reverse engineering, hardware Trojan insertion (insertion of malicious logic), fraudulent claims of IP ownership, and overproduction (producing more than the licensing limit). Where, counterfeited and cloned IPs may contain malicious logic, leading to fatal consequences for end consumers and IP vendors as well as leakage of sensitive information and improper functional output. Additionally, an adversary can covertly insert hardware Trojans into the design after reverse engineering the netlist file, remaining dormant until triggered. Moreover, overproduction of IPs/ICs beyond the original licensing limit poses further risks, highlighting the essential need to secure IP cores against such hardware threats.

1.3 Security countermeasures

In the modern digital age where hardware threats are rampant, securing the end computing systems that comprises reusable components (IPs) has become paramount. To combat vulnerabilities and detection of IP piracy for ensuring reliable functionality and system integrity, various countermeasures have been introduced/developed. The countermeasures offering unique strengths in thwarting external hardware attacks can be broadly categorized into two main types: (a) forensic detective countermeasure and (b) preventive countermeasure as presented in Figure 1.3. Forensic detective countermeasure techniques encompass various methodologies utilized to identify pirated, cloned, or forged hardware IPs within the design chain. These approaches are designed to ascertain fraudulent IP ownership via forensic analysis and identify instances of IP abuse or infringement. Examples include hardware watermarking such as single-phase, multi-phase, quadruple-phase, and multi-level watermarking; hardware steganography; the utilization of digital signatures; and biometric-based security for hardware IP protection. Additionally, structural obfuscation and functional obfuscation stand out as effective preventive countermeasure techniques. Let us delve into each of the above.

1.3.1 IP (hardware) watermarking

In this technique, a watermark signature comprising auxiliary variables is generated based on IP vendor specifications. Subsequently, the watermark is covertly implanted into the design for enabling IP piracy detection and thereby hindering evasion of illegal piracy detection from an adversarial perspective. In the literature, some of the watermarking techniques implant the IP vendor watermark during higher abstraction levels (such as using HLS), and some of them implant at both the higher and low abstraction levels of the design (such as HLS and RTL). There are several watermarking techniques, some of the effective IP watermarking techniques are as follows: (Sengupta and Bhadauria, 2016; Sengupta and Roy, 2017; Koushanfar *et al.*, 2005) presented a low-cost single-phase watermarking methodology. This methodology implants the IP vendor watermark during the register allocation phase of HLS. *Note: A detailed discussion on HLS and its various phases has been discussed in a subsequent section.* Sengupta *et al.* (2018) proposed a triple-phase watermarking technique based on HLS to protect reusable IP cores against piracy. This approach employs a seven-variable signature across three distinct phases of HLS: scheduling, hardware allocation, and register allocation. Rathor *et al.* (2023) introduced a quadruple-phase watermarking strategy to safeguard DSP designs from piracy and ownership infringement. This approach utilizes

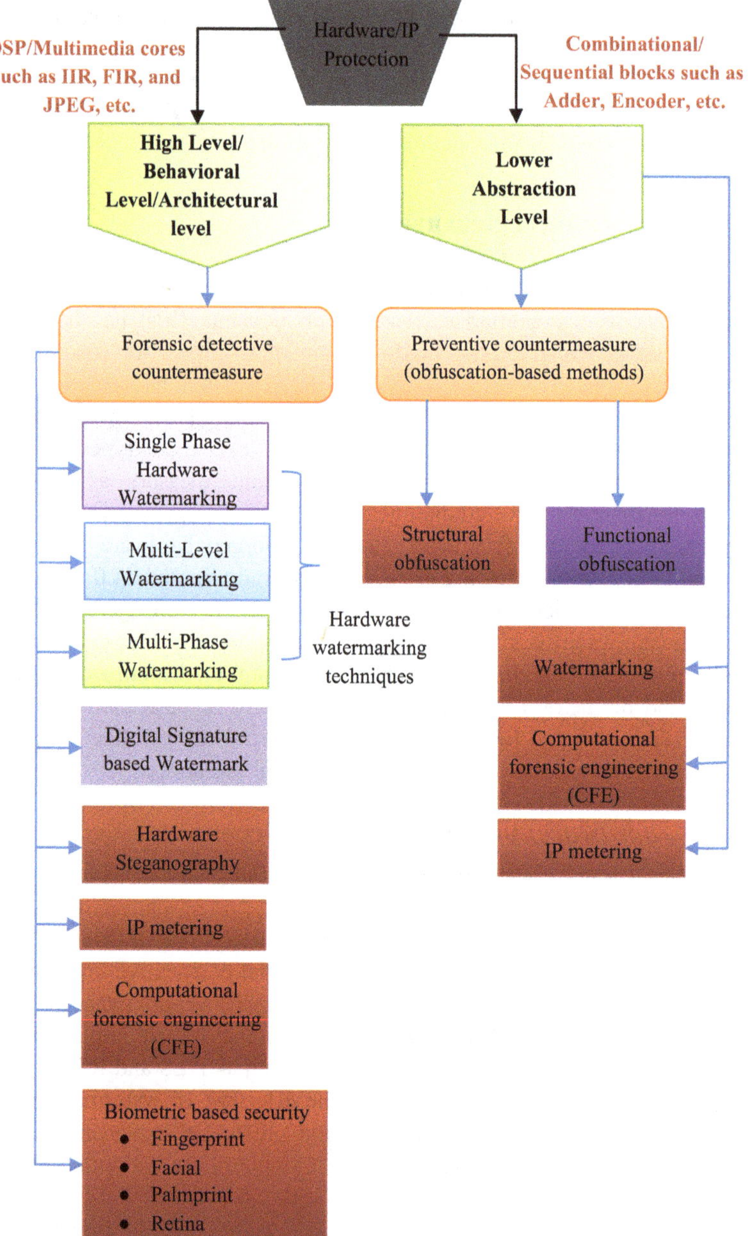

Figure 1.3 Classification of different hardware security countermeasures

graph partitioning, encoding trees, and multifold mapping to generate a resilient hardware watermarking signature, embedded during four phases of behavioral synthesis: scheduling, register binding, functional unit (FU) binding, and interconnect binding. Further, Roy and Sengupta (2019) outlined a multi-level hardware security approach based on watermarking. Here, watermarks are inserted at various higher design abstraction levels, including architectural and register transfer levels, ensuring robust security against IP piracy for DSP cores. Moreover, the implanted watermark is also effective for verification of ownership rights during legal conflict in IP court (Sengupta *et al.*, 2018).

1.3.2 IP steganography

It is a non-signature-based hardware IP security technique. Steganography methodology enables IP piracy detection by implanting the concealed stego mark into the hardware design without using any external signature. The IP steganography technique involves extraction and generation of stego-constraints utilizing design data, secret stego-keys, thresholding parameters, and mapping rules. These stego-constraints are then implanted, along with secret information, into the target hardware (as digital evidence) for enabling IP piracy detection along with seamless ownership verification from a genuine IP vendor's perspective. There are a few hardware steganography techniques in the literature, a detailed description can be found in Sengupta and Rathor (2019a), Sengupta and Rathor (2019b), Rathor and Sengupta (2020), and Sengupta and Rathor (2020).

1.3.3 Digital signature

To enhance the security of the reusable IP core, an encrypted-hash-based digital signature approach was also introduced as an alternative. The encrypted-hash-based digital signature approach provides robust security against counterfeited IP cores through SHA-512 and RSA encryption. The initial step of this security methodology involves encoding the Directed Acyclic Graph (DFG) of the target DSP design into a bitstream. Following this, a bitstream digest is generated using a secure hash algorithm. Next, RSA encryption, utilizing the IP vendor's private key, is employed as the encryption technique for generating encrypted digital signature. Subsequently, the generated digital signature is encoded into secret security constraints using encoding rules specified by the IP vendor. These digital signature-based security constraints are then integrated into the target design during architectural synthesis for enabling IP piracy detection along with ownership verification. The detailed hardware security methodology using digital signature has been presented in Wong *et al.* (2001), Sengupta *et al.* (2019a), and Chaurasia *et al.* (2022).

1.3.4 Hardware metering

Hardware metering encompasses various methods for monitoring any illegal overproduction of designs/ICs, which can include physically identifying serial numbers on chips or storing unique identifiers in memory. This approach can be broadly categorized into two types: passive metering and active metering. The passive metering can be classified into two types: non-functional and functional. In non-functional identification, a unique ID is generated independently of the chip's functionality. However, non-functional identification techniques can further be classified into reproducible and unclonable types. Reproducible identifiers may include unique IDs stored within the chip package, die, or memory, represented as indented or digitally stored serial numbers. Whereas unclonable non-functional identification techniques utilize extrinsic or intrinsic methods. Intrinsic methods exploit IC leakage, power, timing, and path signatures, which are unique due to process variations. This method does not require additional logic and can be applied to existing designs. On the other hand, extrinsic unclonable identifiers, such as physical unclonable functions (PUFs), involve additional logic like ring oscillators (RO), static random access memory (SRAM), etc. capable of generating random values due to process variations. Further, the functional identification-based metering technique links identifiers to the chip's internal functional details during synthesis. All these methods are passive in nature.

However, in the literature, an active metering method exists as well, utilizing external locking techniques to lock ICs. This approach enables the IP owner to control access by using key encryption to secure chip functionality. Activation of each chip requires an external key generated only by the IP owner. A detailed discussion about hardware metering can be found in Alkabani and Koushanfar (2007) and Roy *et al.* (2008).

1.3.5 Hardware fingerprinting

This technique is used to embed a secret mark belonging to an IP buyer into a design. This fingerprinting is created through encoding and cryptographic functions. The primary objective is to address the threat posed by a dishonest IP seller that may attempt to sell unauthorized copies of the IP or illegally redistribute additional copies without the knowledge of the IP buyer. Essentially, the threat model involves combating the unlawful redistribution or resale of IP core copies in the market by a dishonest IP seller. It serves as a forensic detective control to detect and mitigate the presence of such illegal IPs in the supply chain. It is important to note that the threat model of hardware fingerprinting differs significantly from that of hardware watermarking and steganography. While hardware fingerprinting is aimed at protecting the rights of IP buyers, hardware watermarking and steganography are primarily utilized to safeguard the rights of IP vendors. For further insights into hardware fingerprinting, additional details can be found in Roy and Sengupta (2017), Sengupta *et al.* (2019c), Chaurasia and Sengupta (2022c), and Sengupta and Chaurasia (2023).

1.3.6 Computational forensic engineering (CFE)

It is a mechanism to determine the entity responsible for creating a specific intellectual property. It examines various characteristics of the given IP and assesses the probability that a particular entity generated it. The technique operates by distinguishing among different synthesis tools utilized in creating an IP design. The core problem addressed by CFE is as follows: given a solution (Z) to a particular optimization problem (X) and a finite set of algorithms (A_n) available for solving X, the objective is to identify which algorithm (A_i) was employed to solve the problem instance X and produce solution Z. CFE comprises four primary phases, including feature and statistics collection, feature extraction, algorithm clustering, and validation. These phases collectively enable the identification of the algorithm used in solving the optimization problem and thereby infer the entity responsible for creating the IP (Sengupta and Kachave, 2018).

1.3.7 Biometric-based security

Biometric-based security techniques offer more robust security in terms of IP piracy detection and nullifying fraudulent IP ownership claims, as compared to other approaches like hardware watermarking, steganography, digital signature, etc. This is because biometric-based hardware security techniques leveraged naturally unique biometric features of the IP vendor and multi-layer encoding for generating a larger number of unique hardware security constraints. In the literature, there are several physical biometric-based techniques for hardware security such as facial biometric, fingerprint biometric, palmprint biometric, retinal biometric, etc. (Chaurasia and Sengupta, 2022a; Chaurasia and Sengupta, 2022b; Chaurasia and Sengupta, 2022c; Chaurasia and Sengupta, 2021; Sengupta *et al.*, 2023c; Chaurasia *et al.*, 2022; Sengupta and Chaurasia, 2022; Chaurasia *et al.*, 2023; Sengupta and Chaurasia, 2023a). Further, a new class of biometric called "*molecular biometric*"-based approaches for hardware security is also introduced in the literature. The molecular biometric-based approaches exploit DNA information and protein sequence information of IP vendors for generating secret information for hardware security. Additionally, biometric-based approaches are being increasingly used for hardware security and trust. This is because, it offers the following advantages: uniqueness, hard to replicate for an adversary, and also they associate naturally unique identity of original IP vendor for seamless verification of his/her IP rights during ownership conflicts.

1.3.8 Structural obfuscation

Structural obfuscation involves intentionally modifying the design and structure of IP without altering its functionality. This tactic aims to impede attackers, particularly those in untrustworthy foundries, from comprehending and analyzing the underlying functionality of the IP design. By doing so, it enhances hardware security against threats such as reverse engineering, providing a higher level of ambiguity from an adversarial perspective. Structural obfuscation increases the time and resources required for reverse engineering, thereby raising the barrier for attackers and making it more challenging to understand the design logic. Its significance in hardware security is evident in its role in countering reverse engineering, securing intellectual property, and mitigating hardware attacks. Further, a detailed description of structural obfuscation techniques can be found in Lao and Parhi (2014), Vijayakumar *et al.* (2017), Sengupta *et al.* (2018), Sengupta and Chaurasia (2022), Sengupta and Roy (2017), and Sengupta and Rathor (2021).

1.3.9 Functional obfuscation/logic locking

Functional obfuscation, also known as logic locking, is a technique aimed at safeguarding hardware designs from various threats including reverse engineering, intellectual property theft, and tampering. Unlike structural obfuscation, which modifies the design and structure of the hardware without affecting its functionality, functional obfuscation focuses on locking the actual functionality of the hardware design while maintaining its intended operation. This method involves implementing measures such as adding key-gates, introducing misleading control flow, and employing encryption or obfuscated design logic. These locking units render the IP core accessible only to individuals possessing the correct key. The main objective is to lock/encrypt the functionality of the hardware IP, making it difficult for attackers to exploit even if they possess knowledge of the structural complexity and functionality of the design netlist. The most advanced techniques in functional obfuscation are detailed in Yasin *et al.* (2016), Rajendran *et al.* (2012), and Sengupta *et al.* (2019b).

1.4 Exploiting HLS framework for security countermeasure

The high-level synthesis framework involves transforming behavioral description of the application into its corresponding hardware representation, comprising of memory elements, storage units, multiplexers/demultiplexers, and interconnections (known as register transfer level). However, this general process encompasses various complex procedural steps critical to different research aspects of high-level synthesis (Sengupta and Sedaghat, 2011; Sengupta, 2023; Sengupta *et al.*, 2017). This section provides a comprehensive overview of the steps leading to the final RTL design. Section 1.4.1 constructs the framework for the different procedural steps for high-level synthesis by discussing the "General high-level synthesis" process. While Section 1.4.2 discusses the most important component of high-level synthesis called design space exploration (DSE).

1.4.1 Generic high-level synthesis

The generic process for high-level synthesis can be outlined as follows: initially, the process commences with a high-level system specification, encompassing factors such as resource area allocation, clock cycle requirements per operation, power consumption at specified frequencies, and user-defined constraints on area, execution time, and power usage. Subsequently, the desired behavior or application for the system is taken as input and transformed into a data flow graph. Following this, exploration of the design space occurs, considering the user-defined constraints like area, execution time, and power consumption. The data flow graph representing the application is then scheduled, organizing operations into different time slots and grouping similar operations together. Once scheduling is completed accurately, a block diagram of the data path circuit is generated. The subsequent step involves constructing the controller structure to provide necessary synchronization signals. Ultimately, the combined structure comprising the data path and control path represents the resulting system at the RTL. An overview of the generic high-level synthesis process is illustrated in Figure 1.4.

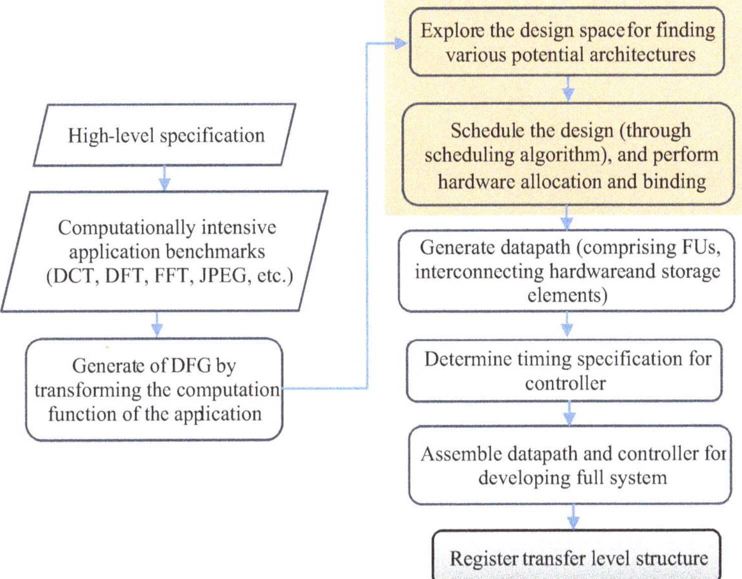

Figure 1.4 Generic high-level synthesis design flow (for transforming high-level specification of a design to register transfer level structure)

1.4.2 Design space exploration in high-level synthesis

Design Space Exploration is the most important step in high-level synthesis. Based on the constraints and specifications, the exploration of the optimal design point is very essential because this solution is to be carried forward in the next steps of high-level synthesis to reach the RTL structure. Also, if the constraints are satisfied while exploring the design space, an optimum result is expected further in the lower levels of abstraction. Based on the research performed till date, design space exploration in high-level synthesis can be broadly classified into two categories. First, the design space exploration of architectures and second, the integrated design space exploration of scheduling, allocation, and binding as shown in Figure 1.5. The design space exploration of architectures is discussed in subsequent Section 1.4.2.1, while the integrated design space exploration of scheduling, allocation, and binding is discussed in subsequent Section 1.4.2.2.

1.4.2.1 Design space exploration of architectures

For modular multi-objective computing systems, fast and precise evaluation of the optimal system architecture is one of the most significant stages in the development process. The assessment and selection of the optimal design point is generally a complex procedure that requires a lot of elaborate analysis. This process of architecture evaluation based on the user-provided objective parameters is done through a sophisticated process called design space exploration of architectures. With the help of this exploration, several aspects are determined like the number of optimum

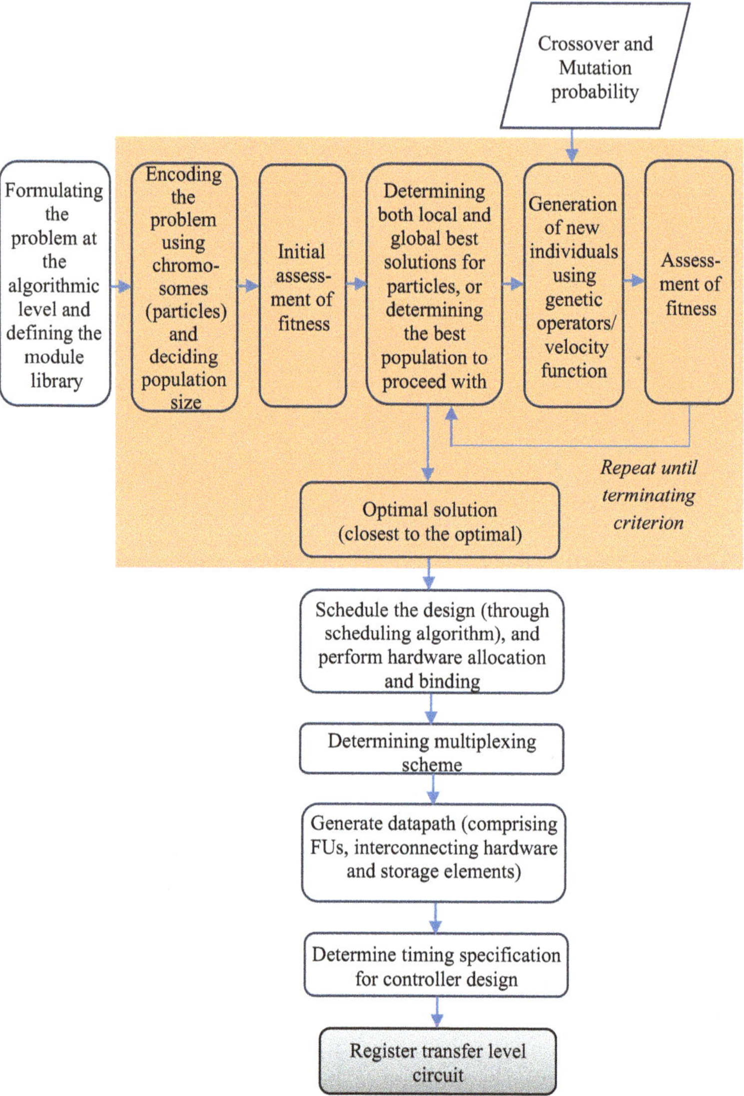

Figure 1.5 Design flow from algorithmic level to RTL integrating design space exploration using evolutionary algorithm

resources, clock frequency, etc. Various authors have introduced different methodologies for efficient design space exploration of architectures. For example, Keinert *et al.* (2009) presented design space exploration of architectures based on area delay tradeoffs for the development of SoC designs. In this technique, parameters such as hardware cost, system throughput, and software execution time of the design have been optimized. Further, during the design space exploration, a single design point is evaluated by simulating highly accurate performance models, which are automatically generated from the behavioral model and behavioral synthesis results. Moreover, the authors presented a promising approach for area delay tradeoffs; however, their method lacks any analysis for power consumption-performance tradeoffs. Power consumption has become of crucial significance for both low-power application-specific processor designs and power-limited portable devices such as laptops and digital audio players. Hence, optimizing power consumption for the present generation of VLSI designs is crucial.

1.4.2.2 Integrated design space exploration of scheduling, allocation, and binding

For the modular VLSI computing architectures, the problem of solving the exploration process in a fast and accurate manner is very important. High-level synthesis is composed of interdependent tasks such as scheduling, allocation, and module selection. For today's VLSI designs, the cost of solving the combined scheduling, allocation, and module selection problem by exhaustive search is prohibitive. However, to meet design objectives, extensive design space exploration is often critical to obtaining superior designs. Integrated design space exploration addresses multiple issues encountered during high-level synthesis such as scheduling, allocation, and binding. These issues are highly critical for the successful functioning of the system based on the user-specified objectives. The characteristic of integrated exploration lies in the fact that it does not only find the optimal architecture for the design but also explores the optimal scheduling and allocation needed to accomplish the task in the provided constraints. Hence the main focus of design space exploration has shifted to the integrated exploration effort made by many researchers. Krishnan and Srinivas Katkoori (2006) presented a framework for efficient design space exploration during high-level synthesis of datapath for data-dominated applications. The framework uses a genetic algorithm (GA) to concurrently perform scheduling and allocation with the aim of finding schedules and module combinations that lead to superior designs while considering user-specified latency and area constraints. The GA uses a multi-chromosome representation to encode datapath schedules and module allocations and efficient heuristics to minimize functional and storage area costs, while minimizing circuit latencies.

1.5 Summary of the book

The book provides a balanced coverage of several state-of-the-art advances in the research community by encompassing some important branches on "security aware HLS," viz. *HLS-based watermarking, HLS-based fingerprinting, HLS-based obfuscation, and HLS-based hardware Trojan and its countermeasure*, through the following topics:

- ✓ High-Level Synthesis-Based Watermarking Using Protein Molecular Biometric with Facial Biometric Encryption

- ✓ High-Level Synthesis-Based Watermarking Using Retinal Biometric
- ✓ HLS-based Mathematical Watermarks for Hardware Security and Trust
- ✓ High-Level Synthesis-Based Watermarking Using Multi-modal Biometric
- ✓ High-Level Synthesis-Based Watermarking Using Crypto-Chain Signature Framework
- ✓ HLS-Based Fingerprinting
- ✓ High-Level Synthesis-Based Structural Obfuscation for Hardware Security and Trust
- ✓ Algorithmic Transformation-Based Obfuscation for Secure Floorplan-Driven High-Level Synthesis
- ✓ Fundamentals on HLS-Based Hardware Trojan
- ✓ Detective Countermeasure against HLS-based Hardware Trojan Attack

1.6 Conclusion

In conclusion, this chapter serves as a foundational resource for understanding and addressing the critical issues surrounding hardware security and trust. By introducing the concept of high-level synthesis (HLS) and delving into various attack scenarios and countermeasures, it highlighted the potential vulnerabilities inherent in electronic systems and emphasized the importance of proactive measures to safeguard confidentiality, integrity, and reliability. Recognizing that adversaries can exploit weaknesses at different design phases underscores the need for a comprehensive approach to hardware security. Through exploration of security-aware HLS-based hardware design methodologies and effective countermeasures, this chapter aimed to empower designers and researchers with the basic knowledge to enhance the security and trustworthiness of computing systems. By bridging the gap between existing challenges and advanced strategies, it paves the way for the development of secure and resilient hardware architectures in the face of evolving threats.

References

Y. Alkabani and F. Koushanfar, (2007) "Active Hardware Metering for Intellectual Property Protection and Security," *USENIX Security Symposium.*

A. Anshul and A. Sengupta, (2023) "A Survey of High Level Synthesis Based Hardware Security Approaches for Reusable IP Cores," *IEEE Circuits and Systems Magazine (CASM)*, vol. 23, no. 4, pp. 44–62.

M. T. Arafin, A. Stanley and P. Sharma, (2017) "Hardware-based Anti-counterfeiting Techniques for Safeguarding Supply Chain Integrity," *2017 IEEE International Symposium on Circuits and Systems (ISCAS)*, pp. 1–4.

R. Chaurasia and A. Sengupta, (2021) "Securing Reusable Hardware IP Cores Using Palmprint Biometric," *2021 IEEE International Symposium on Smart Electronic Systems (iSES)*, pp. 410–413, doi: 10.1109/iSES52644.2021.00099.

R. Chaurasia and A. Sengupta, (2022a) "Protecting Trojan Secured DSP Cores against IP Piracy Using Facial Biometrics," *2022 IEEE 19th India Council International Conference (INDICON)*, India, pp. 1–6, doi:10.1109/INDICON56171.2022.10039864.

R. Chaurasia and A. Sengupta, (2022b) "Security vs Design Cost of Signature Driven Security Methodologies for Reusable Hardware IP Core," *2022 IEEE International Symposium on Smart Electronic Systems (iSES)*, India, pp. 283–288, doi:10.1109/iSES54909.2022.00064.

R. Chaurasia and A. Sengupta, (2022c) "Symmetrical Protection of Ownership Right's for IP Buyer and IP Vendor Using Facial Biometric Pairing," *2022 IEEE International Symposium on Smart Electronic Systems (iSES)*, India, pp. 272–277, doi:10.1109/iSES54909.2022.00062.

R. Chaurasia and A. Sengupta, (2022d) "Crypto-Genome Signature for Securing Hardware Accelerators," *2022 IEEE 19th India Council International Conference (INDICON)*, India, pp. 1–6, doi:10.1109/INDICON56171.2022.10039955.

R. Chaurasia and A. Sengupta, (2023a) "Designing Optimized and Secured Reusable Convolutional Hardware Accelerator against IP Piracy Using Retina Biometrics," *Proceedings of 9th IEEE International Symposium on Smart Electronic Systems (IEEE – iSES)*, India, Accepted Dec. 2023.

R. Chaurasia and A. Sengupta, (2023b) "Retinal Biometric for Securing JPEG-Codec Hardware IP Core for CE Systems," *IEEE Transactions on Consumer Electronics*, vol. 69, no. 3, pp. 441–457, doi:10.1109/TCE.2023.3264669.

R. Chaurasia, A. Anshul, A. Sengupta and S. Gupta, (2022) "Palmprint Biometric Versus Encrypted Hash Based Digital Signature for Securing DSP Cores Used in CE Systems," *IEEE Consumer Electronics Magazine (CEM)*, vol. 11, no. 5, pp. 73–80, doi:10.1109/MCE.2022.3153276.

R. Chaurasia, A. Reddy Asireddy and A. Sengupta, (2023) "Fault Secured JPEG-Codec Hardware Accelerator with Piracy Detective Control Using Secure Fingerprint Template," *2023 IEEE International Symposium on Defect and Fault Tolerance in VLSI and Nanotechnology Systems (DFT)*, Juan-Les-Pins, France, pp. 1–6, doi:10.1109/DFT59622.2023.10313536.

B. Colombier and L. Bossuet, (2015) "Survey of Hardware Protection of Design Data for Integrated Circuits and Intellectual Properties," *IET Computer and Digital Techniques*, vol. 8, no. 6, pp. 274–287, https://doi.org/10.1049/iet-cdt.2014.0028.

W. Hu, C. -H. Chang, A. Sengupta, S. Bhunia, R. Kastner and H. Li, (2021) "An Overview of Hardware Security and Trust: Threats, Countermeasures, and Design Tools," *IEEE Transactions on Computer-Aided Design of Integrated Circuits and Systems*, vol. 40, no. 6, pp. 1010–1038.

R. Karmakar and S. Chattopadhyay, (2020) "Hardware IP Protection Using Logic Encryption and Watermarking," *2020 IEEE International Test Conference (ITC)*, pp. 1–10.

R. Karmakar, S. S. Jana and S. Chattopadhyay, (2022) "A Cellular Automata Guided Finite-State-Machine Watermarking Strategy for IP Protection of

Sequential Circuits," *IEEE Transactions on Emerging Topics in Computing*, vol. 10, no. 2, pp. 806–823.

J. Keinert, M. Streubuhr, T. Schlichter, *et al.*, (2009) "SystemCoDesigner—An Automatic ESL Synthesis Approach by Design Space Exploration and Behavioural Synthesis for Streaming Applications," *ACM Transactions on Design Automation of Electronic Systems (TODAES)*, vol. 14, no. 1 article 1.

F. Koushanfar, S. Fazzari, C. McCants, *et al.*, (2012) "Can EDA Combat the Rise of Electronic Counterfeiting?," *DAC Design Automation Conference 2012*, pp. 133–138.

F. Koushanfar, I. Hong and M. Potkonjak, (2005) "Behavioral Synthesis Techniques for Intellectual Property Protection," *ACM Trans. Des. Autom. Electron. Syst.*, vol. 10, no. 3, pp. 523–545.

V. Krishnan and S. Katkoori, (2006) "A Genetic Algorithm for the Design Space Exploration of Datapaths During High-level Synthesis," *IEEE Transactions on Evolutionary Computation*, vol. 10, no. 3, pp. 213–229.

J. Kuai, J. He, H. Ma, Y. Zhao, Y. Hou and Y. Jin, (2020) "WaLo: Security Primitive Generator for RT-Level Logic Locking and Watermarking," *2020 Asian Hardware Oriented Security and Trust Symposium (AsianHOST)*, pp. 01–06.

Y. Lao and K. K. Parhi, (2014) "Protecting DSP Circuits through Obfuscation," *2014 IEEE International Symposium on Circuits and Systems (ISCAS)*, Melbourne, VIC, Australia, pp. 798–801, doi:10.1109/ISCAS.2014.6865256.

J. Rajendran, Y. Pino, O. Sinanoglu and R. Karri, (2012) "Security Analysis of Logic Obfuscation," in *Proceedings of the 49th Annual Design Automation Conference*, pp. 83–89, ACM.

J. J. Rajendran, O. Sinanoglu and R. Karri, (2016) "Building Trustworthy Systems Using Untrusted Components: A High-Level Synthesis Approach," *IEEE Transactions on Very Large Scale Integration (VLSI) Systems*, vol. 24, no. 9, pp. 2946–2959.

J. J. V. Rajendran, (2017) "An Overview of Hardware Intellectual Property Protection," *2017 IEEE International Symposium on Circuits and Systems (ISCAS)*, pp. 1–4.

M. Rathor and A. Sengupta, (2019) "Robust Logic Locking for Securing Reusable DSP Cores," *IEEE Access*, vol. 7, pp. 120052–120064, doi:10.1109/ACCESS.2019.2936401.

M. Rathor and A. Sengupta, (2020) "IP Core Steganography Using Switch Based Key-driven Hash-chaining and Encoding for Securing DSP Kernels Used in CE Systems," *IEEE Transactions on Consumer Electronics*.

M. Rathor, A. Anshul, K. Bharath, R. Chaurasia and A. Sengupta, (2023) "Quadruple Phase Watermarking during High Level Synthesis for Securing Reusable Hardware IP Cores," *Elsevier Journal on Computers and Electrical Engineering*, vol. 105, 108476.

D. Roy and A. Sengupta, (2017) "Low Overhead Symmetrical Protection of Reusable IP Core Using Robust Fingerprinting and Watermarking during High

Level Synthesis," *Elsevier Journal on Future Generation Computer Systems*, vol. 71, pp. 89–101.

D. Roy and A. Sengupta (2019), "Multilevel Watermark for Protecting DSP Kernel in CE Systems [Hardware Matters]," *IEEE Consumer Electronics Magazine*, vol. 8, no. 2, pp. 100–102, doi: 10.1109/MCE.2018.2880849.

J. A. Roy, F. Koushanfar and I. L. Markov, (2008) "EPIC: Ending Piracy of Integrated Circuits," *2008 Design, Automation and Test in Europe*, Munich, Germany, pp. 1069–1074, doi:10.1109/DATE.2008.4484823.

J. A. Roy, F. Koushanfar and I. L. Markov, (2008) "EPIC: Ending Piracy of Integrated Circuits," *2008 Design, Automation and Test in Europe*, pp. 1069–1074.

S. Salivahanan and A. Vallavaraj, (2001) *Digital Signal Processing*, McGraw-Hill Education (India) Pvt Limited.

A. Sengupta, (2016) "Intellectual Property Cores: Protection Designs for CE Products," *IEEE Consumer Electronics Magazine*, vol. 5, no. 1, pp. 83–88, doi:10.1109/MCE.2015.2484745.

A. Sengupta and S. Bhadauria, (2016) "Exploring Low Cost Optimal Watermark for Reusable IP Cores During High Level Synthesis," *IEEE Access*, vol. 4, pp. 2198–2215, doi:10.1109/ACCESS.2016.2552058.

A. Sengupta and R. Chaurasia, (2022) "Securing IP Cores for DSP Applications Using Structural Obfuscation and Chromosomal DNA Impression," *IEEE Access*, vol. 10, pp. 50903–50913, doi:10.1109/ACCESS.2022.3174349.

A. Sengupta and R. Chaurasia, (2022) "Securing IP Cores for DSP Applications Using Structural Obfuscation and Chromosomal DNA Impression," *IEEE Access*, vol. 10, pp. 50903–50913.

A. Sengupta and R. Chaurasia, (2023) "Handling Symmetrical IP Core Protection and IP Protection (IPP) of Trojan Secured Designs in HLS using Physical Biometrics," *IET Book "Physical Biometrics for Hardware Security of DSP and Machine Learning Coprocessors,"* Chap. 5, pp. 147–198, doi:10.1049/PBCS080E_ch5.

A. Sengupta and R. Chaurasia, (2023) "Securing Fault-Detectable CNN Hardware Accelerator against False Claim of IP Ownership Using Embedded Fingerprint as Countermeasure," *Proceedings of 9th IEEE International Symposium on Smart Electronic Systems (IEEE – iSES)*, India. Accepted Dec. 2023.

A. Sengupta and D. Kachave, (2018) "Forensic Engineering for Resolving Ownership Problem of Reusable IP Core Generated during High Level Synthesis," *Elsevier Journal on Future Generation Computer Systems*, vol. 80, pp. 29–46.

A. Sengupta, S. Mohanty, (2019a), "IP Core and Integrated Circuit Protection using Robust Watermarking," In: *IP Core Protection and Hardware-Assisted Security for Consumer Electronics*, Chap. 4, pp. 123–170, https://digital-library.theiet.org/content/books/10.1049/pbcs060e_ch4.

A. Sengupta and S. P. Mohanty, (2019b) ," *IP Core Protection and Hardware-Assisted Security for Consumer Electronics*. Stevenage: The Institute of Engineering and Technology (IET)

A. Sengupta and M. Rathor, (2019a) "IP Core Steganography for Protecting DSP Kernels Used in CE Systems," *IEEE Transactions on Consumer Electronics*, vol. 65, no. 4, pp. 506–515.

A. Sengupta and M. Rathor (2019b), "Crypto-Based Dual-Phase Hardware Steganography for Securing IP cores," *IEEE Letters of the Computer Society*, vol. 2, no. 4, pp. 32–35, doi: 10.1109/LOCS.2019.2942289.

A. Sengupta and M. Rathor, (2020) "Securing Hardware Accelerators for CE Systems Using Biometric Fingerprinting," *IEEE Transactions on Very Large Scale Integration (VLSI) Systems*, doi:10.1109/TVLSI.2020.2999514.

A. Sengupta and M. Rathor, (2021) "Structural Transformation and Obfuscation Frameworks for Data-intensive IPs," IET Book *Secured Hardware Accelerators for DSP and Image Processing Applications*, Print: 978-1-83953-306-8, eBook: 978-1-83953-307-5.

A. Sengupta and M. Rathor, (2020) "Structural Obfuscation and Crypto-Steganography Based Secured JPEG Compression Hardware for Medical Imaging Systems," *IEEE Access*, vol. 8, no. 1, pp. 6543–6565.

A. Sengupta and D. Roy, (2017) "Automated Low Cost Scheduling Driven Watermarking Methodology for Modern CAD High-Level Synthesis Tools," *Elsevier Journal of Advances in Engineering Software*, vol. 110, pp. 26–33.

A. Sengupta and D. Roy, (2017) "Protecting an Intellectual Property Core during Architectural Synthesis Using High-Level Transformation Based Obfuscation," *IET Electronics Letters*, vol. 53, no. 13, pp. 849–851.

A. Sengupta and R. Sedaghat, (2011) "A High Level Synthesis Design Flow from ESL to RTL with Multi-parametric Optimization Objective," *IETE Journal of Research*, vol. 57, no. 2, pp. 169–186.

A. Sengupta, S. Bhadauria and S. P Mohanty, (2017) "TL-HLS: Methodology for Low Cost Hardware Trojan Security Aware Scheduling with Optimal Loop Unrolling Factor during High Level Synthesis," *IEEE Transactions on Computer Aided Design of Integrated Circuits & Systems (TCAD)*, vol. 36, no. 4, pp. 655–668.

A. Sengupta, D. Roy, S. P. Mohanty and P. Corcoran, (2017) "DSP Design Protection in CE through Algorithmic Transformation Based Structural Obfuscation," *IEEE Transactions on Consumer Electronics*, vol. 63, no. 4, pp. 467–476.

A. Sengupta, S. P. Mohanty, F. Pescador and P. Corcoran, (2018) "Multi-Phase Obfuscation of Fault Secured DSP Designs with Enhanced Security Feature," *IEEE Transactions on Consumer Electronics*, vol. 64, no. 3, pp. 356–364.

A. Sengupta, E. R. Kumar and N. P. Chandra, (2019a) "Embedding Digital Signature Using Encrypted-hashing for Protection of DSP Cores in CE," *IEEE Transactions on Consumer Electronics*, vol. 65, no. 3, pp. 398–407.

A. Sengupta, D. Kachave and D. Roy, (2019b) "Low Cost Functional Obfuscation of Reusable IP Cores Used in CE Hardware Through Robust Locking," *IEEE Transactions on Computer-Aided Design of Integrated Circuits and Systems*, vol. 38, no. 4, pp. 604–616, doi:10.1109/TCAD.2018.2818720.

A. Sengupta, U. Singh and P. Kalkute, (2019c) "Crypto Based Multi-Variable Fingerprinting for Protecting DSP Cores," *Proceedings of 9th IEEE International*

Conference on Consumer Electronics (ICCE) – Berlin, Berlin, pp. 1–6, doi:10.1109/ICCE-Berlin47944.2019.9127235.

A. Sengupta, R. Chaurasia and T. Reddy, (2021) "Contact-Less Palmprint Biometric for Securing DSP Coprocessors Used in CE Systems," *IEEE Transactions on Consumer Electronics*, vol. 67, no. 3, pp. 202–213, doi:10.1109/TCE.2021.3105113.

A. Sengupta, M. Rathor, and R. Chaurasia, (2023) "Biometrics for Hardware Security and Trust: Discussion and Analysis," *IEEE IT Professionals (ITPro)*, vol. 25, no. 4, pp. 36–44.

A. Sengupta, R. Chaurasia and A. Anshul, (2023a) "Robust Security of Hardware Accelerators Using Protein Molecular Biometric Signature and Facial Biometric Encryption Key," *IEEE Transactions on Very Large-Scale Integration (VLSI) Systems*, vol. 31, no. 6, pp. 826–839.

A. Sengupta, R. Chaurasia and K Bharath, (2023b) "Exploring Unified Biometrics with Encoded Dictionary for Hardware Security of Fault Secured IP Core Designs," *Elsevier Journal Computers and Electrical Engineering*, vol. 111, Part A, p. 108928.

A. Sengupta, R. Chaurasia and A. Anshul, (2023b) "Hardware Security of Digital Image Filter IP Cores against Piracy Using IP Seller's Fingerprint Encrypted Amino Acid Biometric Sample," *Proceedings of 8th Asian Hardware Oriented Security and Trust Symposium (AsianHOST)*, China, Accepted Sept. 2023b.

A. Sengupta, M. Rathor and R. Chaurasia, (2023c) "Biometrics for Hardware Security and Trust: Discussion and Analysis," *IT Professional*, vol. 25, no. 4, pp. 36–44, doi:10.1109/MITP.2023.3277594.

A. Sengupta, R. Chaurasia and M. Rathor, (2023d) "HLS based Swarm Intelligence Driven Optimized Hardware IP Core for Linear Regression Based Machine Learning," *IET Journal of Engineering*, vol. 2023, no. 8, e12299.

A. Sengupta and M. Rathor, (2020) "Enhanced Security of DSP Circuits Using Multi-Key Based Structural Obfuscation and Physical-level Watermarking for Consumer Electronics Systems," *IEEE Transactions on Consumer Electronics (TCE)*, vol. 66, no. 2, pp. 163–172.

A. Sengupta, D. Roy, S. Mohanty and P. Corcoran, (2017) "DSP Design Protection in CE through Algorithmic Transformation Based Structural Obfuscation," *IEEE Transactions on Consumer Electronics*, vol. 63, no. 4, pp. 467–476.

A. Sengupta, (2023) "Physical Biometrics for Hardware Security of DSP and Machine Learning Coprocessors," *The Institute of Engineering and Technology (IET)*, Book DOI:10.1049/PBCS080E

A. Syed and R. M. Lourde, (2016) "Hardware Security Threats to DSP Applications in an IoT Network," *2016 IEEE International Symposium on Nanoelectronic and Information Systems (iNIS)*, pp. 62–66.

A. Vijayakumar, V. C. Patil, D. E. Holcomb, C. Paar and S. Kundu, (2017) "Physical Design Obfuscation of Hardware: A Comprehensive Investigation of Device and Logic-Level Techniques," *IEEE Transactions on Information Forensics and Security*, vol. 12, no. 1, pp. 64–77.

X. Wang, Y. Zheng, A. Basak and S. Bhunia, (2015) "IIPS: Infrastructure IP for Secure SoC Design," *IEEE Transactions on Computers*, vol. 64, no. 8, pp. 2226–2238.

J. L. Wong, D. Kirovski and M. Potkonjak, (2001) "Intellectual Property Metering," *Information Hiding*, New York: Springer, pp. 66–81.

M. Yasin, J. J. Rajendran, O. Sinanoglu and R. Karri, (2016) "On Improving the Security of Logic Locking," *IEEE Transactions on Computer-Aided Design of Integrated Circuits and Systems*, vol. 35, no. 9, pp. 1411–1424.

Chapter 2

High-level synthesis-based watermarking using protein molecular biometric with facial biometric encryption

Anirban Sengupta[1] *and Aditya Anshul*[1]

This chapter discusses a watermarking-based hardware security methodology using an encrypted protein molecular biometric signature to secure hardware intellectual property (IP) cores against counterfeiting, IP piracy, and false assertion of IP ownership. In this method, the protein molecular biometric signature (watermark) is generated from the IP seller's protein sample using protein sequencing, followed by robust encryption using facial biometric signature-based encryption key. Subsequently, the obtained watermark (encrypted protein signature) is transformed into covert security constraints using the IP seller's mapping/embedding rule, which are then embedded into the design as an IP seller's digital evidence. Integration of the IP seller's authentic watermark provides a detective countermeasure against IP piracy and nullification of false assertion of IP ownership. The encrypted protein molecular watermarking methodology provides stronger digital evidence against piracy and greater resilience against tampering.

2.1 Introduction

In the modern electronics era, digital signal processing (DSP) and multimedia coprocessors (hardware intellectual property (IP) cores) have become indispensable components of embedded electronic/computing systems such as tablets, laptops, smartphones, digital cameras, health bands, and internet-of-thing (IoT) devices. They serve as the backbone for ensuring optimal performance and efficiency across various applications/functions. They seamlessly execute data and computation-intensive functions/tasks ranging from data compression-decompression, image processing, and digital data filtering to performing complex mathematical calculations. Discrete cosine transform (DCT), discrete wavelet transformation (DWT), finite impulse response (FIR) filter, etc. are some real-world examples of coprocessors (Anshul and Sengupta, 2023a). Designing data-intensive applications as

[1]Department of Computer Science and Engineering, Indian Institute of Technology Indore, India

dedicated reusable hardware IP core provides a compelling solution to meet the growing demand for high-performance, energy-efficient, and scalable computing solutions in today's data-driven world (Schneiderman, 2010). Therefore, these coprocessors designed as dedicated hardware IP cores employ high-level synthesis (HLS) framework. This sophisticated process ensures the seamless integration and optimization of these applications within electronic devices, thereby enhancing their overall performance and functionality (Rathor *et al.*, 2023a; Gal and Bossuet, 2012).

The design supply chain process of hardware computing systems is now global and often includes multiple third-party entities (Rostami *et al.*, 2014; Anshul and Sengupta, 2023a). This widespread involvement of various third-party IP (3PIP) sellers (vendors) introduces greater risks, including IP piracy and false IP ownership assertion. Additionally, pirated/counterfeited IPs present significant dangers to end-users, potentially compromising sensitive information such as user credentials, causing excessive heat dissipation, or impairing the proper functioning of hardware (Rizzo *et al.*, 2019; Anshul and Sengupta, 2023b). An adversary within the system-on-chip (SoC) integrator house or foundry may exploit the security vulnerabilities to pirate original designs or make false claims of IP ownership, underscoring the critical need for robust security mechanisms. From a different perspective, it is crucial for the SoC integrator to select and isolate pirated/counterfeited hardware designs before integration into the final product to uphold the safety and reliability standards for end-users in the face of these hardware threats. Further, implementing embedded secret security marks as watermark emerges as a vital strategy in mitigating ownership conflicts and providing detective countermeasure against IP piracy, thus safeguarding the integrity of the global hardware design supply design process (Sengupta *et al.*, 2023a; Rathor *et al.*, 2023b; Colombier and Bossuet, 2015; Rathor *et al.*, 2024; Anshul and Sengupta, 2023c; Islam *et al.*, 2020; Anshul *et al.*, 2022; Yashin *et al.*, 2016; Sengupta *et al.*, 2023b).

This chapter discusses hardware watermarking-based security methodology using protein molecular biometrics (Sengupta *et al.*, 2023c). The discussed method involves exploiting the IP seller's unique protein sequence (comprising 20 distinct amino acid combinations) to generate a protein molecular signature. Additionally, the chapter outlines the AES encryption of the protein molecular signature using the IP seller's facial biometric signature (encryption key). By incorporating two different classes of biometrics from IP sellers, the discussed encrypted protein molecular-based security (watermarking) approach ensures robust and unique authentication of genuine IPs as well as detection of IP piracy (Sengupta *et al.*, 2023c).

2.2 Discussion on the state-of-the-art watermarking approaches

The literature presents various security methodologies to safeguard hardware IP cores (used in computing and multimedia gadgets) against IP piracy and false IP ownership assertion/claim using hardware watermarking and steganography. These watermarking methods include dynamic watermarking (Koushanfar *et al.*, 2005), steganography (Sengupta and Rathor, 2019), multi-variable watermarking (Sengupta and Bhadauria, 2016), triple-phase watermarking (Sengupta *et al.*, 2018), encrypted signature (Castillo *et al.*, 2008), digital signature (Sengupta *et al.*, 2019), fingerprint biometric (Sengupta and Rathor, 2020), facial biometric (Sengupta and Rathor, 2021), hardware watermarking (Chen and Schafer, 2021), handwritten signature (Rathor and Rathor, 2024), and chromosomal DNA (Sengupta and Chaurasia, 2022). Hardware watermarking, as described in Koushanfar *et al.* (2005), Sengupta and Bhadauria (2016), Chen and Schafer (2021), Sengupta *et al.* (2018), and Sengupta *et al.* (2019), involves the integration of a unique signature chosen by the IP seller. This signature is determined by several variables, for example, signature variable types, their arrangement, signature size, and embedding principles. But, if the variable details and embedding rules are compromised, watermarking protection can also be compromised. IP steganography, outlined in Sengupta and Rathor (2019), generates covert constraints based on hardware design information (register allocation information), covert steganographic keys, and thresholding factors/variables. Sengupta and Rathor (2019) discuss a signature-free security approach, utilizing threshold entropy factor and register allocation information to derive security constraints. These hardware security constraints are then embedded as covert information (digital evidence) into the target hardware design. The sophisticated process of generating covert constraints using covert steganographic keys and thresholding factors enhances the resilience of the steganography approach compared to watermarking-based approaches. However, if the entropy thresholding factors or steganographic key value is compromised by an adversary, then it could undermine the overall security. Similarly, Sengupta *et al.* (2019) discuss a digital signature-based hardware security approach for safeguarding hardware (such as DSP) applications. Sengupta *et al.* (2019) determine hardware security constraints by employing the RSA cryptosystem and SHA-512 hash digest computation. However, vulnerability arises if the RSA key is compromised, limiting the security of Sengupta *et al.* (2019). Additionally, Castillo *et al.* (2008) propose an encrypted signature-based methodology utilizing AES-128 and MD-5 cryptosystems to protect hardware systems. Nevertheless, Castillo *et al.* (2008) fall short of providing robust protection because of limited security constraints production.

Further, the fingerprint biometric security methodology (Sengupta and Rathor, 2020) exploits IP seller's fingerprint minutiae points to generate the fingerprint signature, which is further translated into covert security constraints. The fingerprint biometric security approach (Sengupta and Rathor, 2020) outperforms hardware watermarking (Koushanfar *et al.*, 2005; Sengupta and Bhadauria, 2016) and hardware steganography (Sengupta and Rathor, 2019) in terms of the strength of generated covert security constraints. Next, the chromosomal DNA-based security methodology (Sengupta and Chaurasia, 2022) employs the generation of the IP seller's DNA signature based on the sequence of chromosomal DNA (composed of chemical constituents (four) present in human DNA samples). Additionally, Sengupta and Chaurasia (2022) utilize DES encryption to encrypt the produced DNA template. In contrast, the encrypted protein molecular signature approach (*encrypted protein signature*) (Sengupta *et al.*, 2023c) described in this chapter involves the generation of IP seller's protein molecular biometric signature consisting of a protein sequence comprising 20 different permutations of amino acid. Sengupta *et al.* (2023c) employ AES encryption, which provides more robust encryption than the DES used in Sengupta and Chaurasia (2022). Additionally, the determined security constraints using the security approach (Sengupta *et al.*, 2023c) result in a lower probability of coincidence (indicating stronger digital evidence) and higher tolerance to tampering due to the production of a larger key space. Forgery and precise reproduction of the encrypted protein signature (Sengupta *et al.*, 2023c) are not feasible for an adversary owing to the need for numerous intricate and covert information such as the length of the polypeptide chain (large in size) containing the 20 unique amino acids, amino acid encoding rule, AES key, watermarking/security constraints generation rule, and information of *s-box*. Moreover, the encrypted protein signature approach (Sengupta *et al.*, 2023c) described in this chapter is beneficial in providing robust security for large-size designs (such as JPEG-CODEC, etc.) due to the creation of greater signature space (Sengupta *et al.*, 2023c).

2.3 Discussion on the threat model of IP piracy and false IP ownership claim

The elaborate process of design and distribution of hardware IPs is highly vulnerable to a multitude of security risks, largely due to the intricate network of multiple third-party entities and units participating in the global design supply chain process. One significant area of concern revolves around the piracy of hardware IP design once they are transferred from an IP seller to a buyer (typically an SoC integrator). The risk arises when a malicious actor within the SoC integrator's organization endeavors to illicitly/unlawfully replicate/pirate the original design, subsequently marketing it either under the same or a different brand name. This not only complicates the authentication of genuine products but also leads to financial losses for the original IP seller. Apart from IP piracy, adversaries may falsely assert ownership rights over the IP. Furthermore, within a rogue foundry, an attacker might unlawfully pirate the IP without the designer's knowledge or consent. From the opposite perspective, it is also crucial for the SoC integrator to isolate pirated/counterfeited hardware IP designs before integration into the final product to uphold the safety and reliability standards for end-users (Anshul and Sengupta, 2023a; Rizzo *et al.*, 2019; Anshul and Sengupta, 2023b).

Moreover, the issue of IP counterfeiting becomes evident when individuals within a rogue foundry, in collaboration with new or secondary IP sellers, engage in the unauthorized replication or imitation of the original IP design (Rostami *et al.*, 2014). This illicit activity not only undermines the IP rights of the original IP seller but also poses grave risks to consumers. The compromised quality and performance of counterfeit products, often stemming from the use of inferior materials or outdated technology, have a detrimental impact on the overall functionality and reliability of the systems in which they are incorporated. Moreover, the proliferation of counterfeit components tarnishes the reputation of authentic IP sellers, casting doubt on the integrity of their products and services. This erosion of trust can have far-reaching consequences, particularly in critical sectors such as military systems, aviation, automotive industries, and beyond. These vital applications rely heavily on the authenticity and quality of the components they integrate, making them especially vulnerable to the repercussions of IP counterfeiting. Furthermore, the ease with which intentional hardware Trojans or malicious logic can be incorporated into counterfeit/pirated IPs exacerbates the security risks inherent in the integrated design supply chain system. This presents a significant challenge for ensuring the integrity and safety of the products and systems reliant on these components (Sengupta *et al.*, 2021; Chaurasia *et al.*, 2022; Sengupta *et al.*, 2023c; Anshul and Sengupta, 2022).

2.4 Extracting protein biometric sequence of IP vendor

2.4.1 Process of generating protein signature

Overview of the encrypted protein-based security approach: This chapter discusses a hardware security methodology (encrypted protein signature) using an amalgamation of IP seller's "molecular biometric" (protein molecular signatures) and "physical biometric" (i.e., facial biometrics) for securing hardware IP designs (Sengupta *et al.*, 2023c). This methodology uses a distinct IP seller's molecular/cellular level signature (extracted from the IP seller's body sample) for embedding it as digital evidence in the hardware design. The overview of the encrypted protein signature-based security methodology is highlighted in Figure 2.1. As shown in Figure 2.1, the IP seller's protein sample, facial biometric image, control data flow graph (CDFG) of hardware application (extracted from the transfer function/high-level code), module library, and resource constraints are taken as the primary inputs. A protein molecular signature is generated by performing protein sequencing on the input IP seller's protein sample. The produced protein signature comprises a sequence chain of 20 unique amino acids (a chemical component) and serves as the primary input to the AES module. Next, the input IP seller's facial biometric image is used to generate the facial biometric signature, which is used as an AES key. Subsequently, the protein molecular signature is encrypted using AES based on IP seller's facial encryption key, resulting in an encrypted protein signature. Next, covert security constraints are generated from the produced encrypted signature using the IP seller's mapping/embedding rule. These determined covert security constraints are then embedded into the hardware design using the HLS process (by exploiting the register allocation phase of the HLS). Ultimately, the encrypted protein molecular signature embedded secured register transfer level (RTL) design datapath of hardware application is obtained as the final output. The embedding of the encrypted IP seller's protein molecular signature protects hardware IP designs against piracy and false IP ownership assertion (Sengupta *et al.*, 2023c). Further, the details are discussed in the next section/subsections.

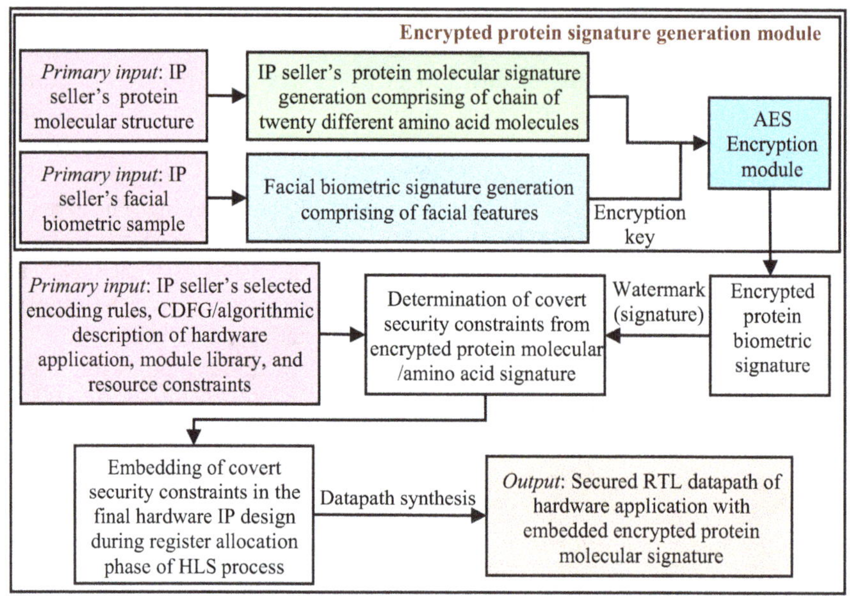

Figure 2.1 Overview of encrypted protein molecular biometric-based HLS watermarking approach

Details of the protein molecular biometric sequence: The human body's proteins are comprised of linear sequences of amino acids. They are bonded together through peptide bonds. This linkage forms a polypeptide chain or protein sequence (Duong *et al.*, 2021). Each protein consists of a distinct amino acid in series. Within a protein series/sequence, there are 20 different amino acids that link together to form the chain. The lengthiest polypeptide chain includes up to 5,000 amino acids (Duong *et al.*, 2021). Every unique amino acid present in the protein series is represented by a specific letter/symbol. Protein sequencing is used to determine the amino acid sequence from collected IP seller's body samples. Additionally, partial sequencing can also provide adequate data for identification purposes. Different human body sources, such as saliva, muscle, bone, hair, etc., can be used to collect protein samples. However, hair, bone, and muscle samples are particularly efficient for identification of individuals (Merkley *et al.*, 2019; Parker *et al.*, 2016; Chen *et al.*, 2000). After collection, samples undergo protein sequencing, where peptides are generated after trypsin digestion. Trypsin, an enzyme, initiates protein digestion by breaking longer amino acid chains into smaller fragments. Mass spectrometry and Edman degradation are commonly used protein sequencing practices (Steendam *et al.*, 2013). The advancements in proteomics, including the improved sequencing methodologies and increased availability of high-resolution mass spectrometers, have been critical in enabling the use of protein molecules for unique identification, particularly in forensic and identification contexts (Steendam *et al.*, 2013). Generally, the protein sequencing is performed as follows: (a) breaking of the disulfide bridges within the protein using a reducing agent, (b) separating the protein complex into the individual chain(s), (c) identifying the terminal and amino acid composition within each chain, (d) segmenting each chain into smaller fragments, with each fragment containing up to 50 amino acids, (f) isolating and purifying each resulting fragment, (g) utilizing the fragments to ascertain the sequence of amino acids, (h) iterating through steps 1–6 to generate the complete protein sequence (Sengupta *et al.*, 2023c).

The utilization of protein sequencing offers several advantages compared to DNA sequencing (Duong *et al.*, 2021; Sengupta and Chaurasia, 2022):

(i) Proteomics, the comprehensive study of an organism's or system's proteins using mass spectrometry, offers a more precise and distinctive analysis of human body samples compared to genomic analysis. Protein analysis serves as an orthogonal and confirmatory technique aiding forensic identification.
(ii) Proteins within a sample exhibit greater resilience and chemical stability compared to DNA, allowing them to persist for longer periods. Conversely, DNA can degrade in various environmental conditions.
(iii) While DNA samples contain genomic information, proteins largely determine a cell's characteristics and functions.
(iv) Protein sequencing demonstrates higher accuracy than DNA sequencing due to proteins being comprised of twenty amino acids, whereas DNA comprises only four chemical components. Consequently, the signal-to-noise ratio in protein sequence chains is significantly superior to that of DNA.

The resulting protein molecular signature is notably distinct from DNA signatures, enabling unique sample identification.

Details of protein molecular signature generation process: Figure 2.2 illustrates the generation process of protein molecular biometric signature. As shown in Figure 2.2 and also mentioned in earlier sections, a protein sequence is generated corresponding to the input IP seller's protein sample after protein sequencing. In this sequencing process, a polypeptide chain consisting of 20 different types of amino acids is generated. The protein sequence shown in Figure 2.2 depicts the polypeptide chain comprised of 380 amino acids, each denoted by a unique alphabet (Sengupta *et al.*, 2023c). It's worth noting that the length of the polypeptide chain is adjustable based on the size of the amino acid series derived from the human protein sample (as selected by the IP seller). Following the selection of the polypeptide chain's amino acid sequence length, their respective encoding takes place. Each amino acid is encoded uniquely according to its alphabetical position. For instance, methionine (M) holds the alphabetical position of 13, represented by the binary bits 1101 in encoding. Similarly, all remaining amino acids, along with their respective encoded bits, are shown in Figure 2.2. Finally, a protein signature is produced after concatenating (as per IP seller's concatenation fashion) the obtained encoded bits corresponding to the amino acids of the polypeptide chain (Sengupta *et al.*, 2023c).

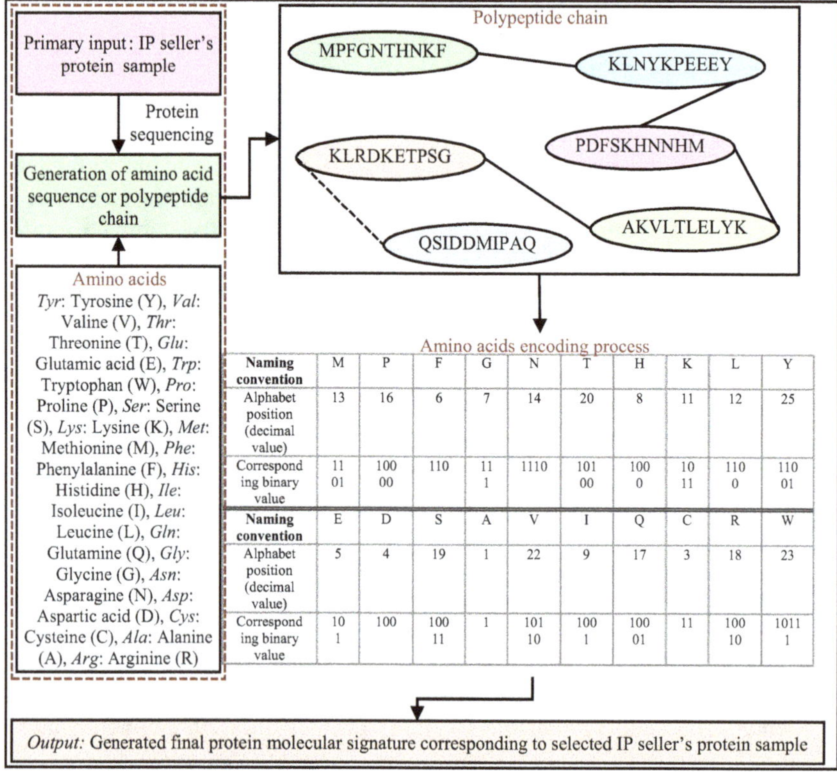

Naming convention	M	P	F	G	N	T	H	K	L	Y
Alphabet position (decimal value)	13	16	6	7	14	20	8	11	12	25
Corresponding binary value	1101	10000	110	111	1110	10100	1000	1011	1100	11001
Naming convention	E	D	S	A	V	I	Q	C	R	W
Alphabet position (decimal value)	5	4	19	1	22	9	17	3	18	23
Corresponding binary value	101	100	10011	1	10110	1001	10001	11	10010	10111

Figure 2.2 Protein molecular signature generation process

2.4.2 Encrypting protein molecular signature using facial biometric of IP vendor

Facial biometric signature (encryption key) generation process: Here, the facial biometric signature is used as the AES encryption key, which is used to encrypt protein molecular biometric signature. Figure 2.3 illustrates the details of the facial biometric signature generation process (Sengupta *et al.*, 2023c). The procedure for creating the facial encryption key from the facial features of the IP seller is outlined below:

(i) A high-resolution facial image of the IP seller is captured using a digital camera.
(ii) The captured image is subjected to the IP seller's specified grid size and spacing. This helps in precisely determining the nodal points on the facial image and ensures accurate facial feature extraction.
(iii) Nodal points on the facial image are generated using chosen feature set (as depicted in Figure 2.3).
(iv) Naming conventions are assigned to nodal points generated on the facial image.

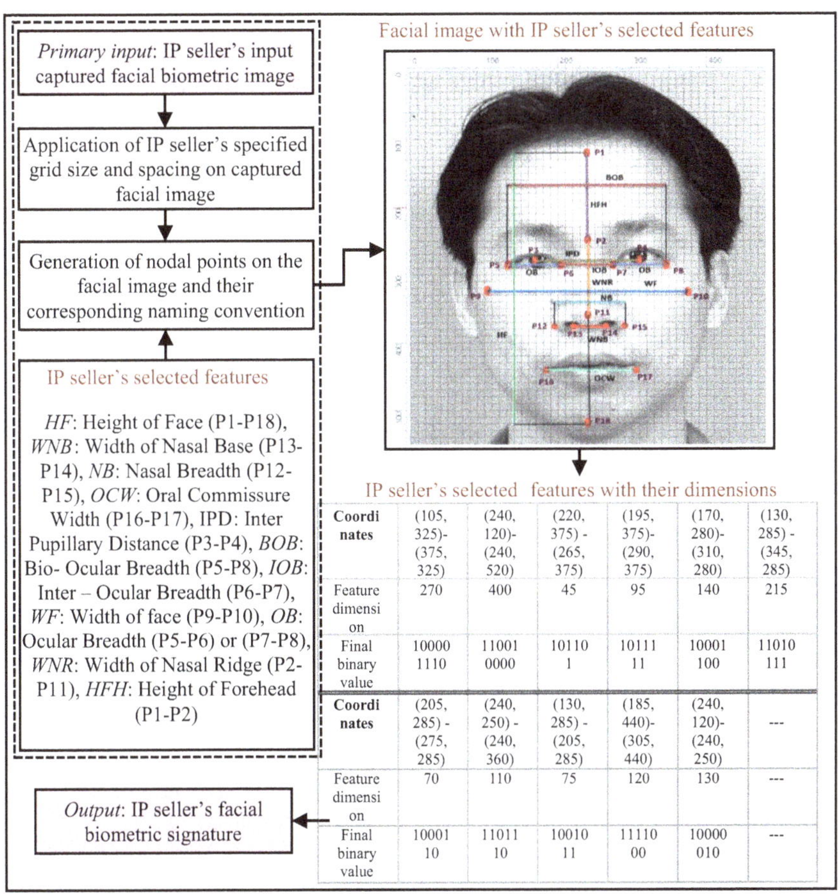

Coordinates	(105, 325)-(375, 325)	(240, 120)-(240, 520)	(220, 375) - (265, 375)	(195, 375)-(290, 375)	(170, 280)-(310, 280)	(130, 285) - (345, 285)
Feature dimension	270	400	45	95	140	215
Final binary value	100001110	110010000	101101	1011111	10001100	11010111
Coordinates	(205, 285) - (275, 285)	(240, 250) - (240, 360)	(130, 285) - (205, 285)	(185, 440)-(305, 440)	(240, 120)-(240, 250)	---
Feature dimension	70	110	75	120	130	---
Final binary value	1000110	1101110	1001011	1111000	10000010	---

Figure 2.3 IP seller's facial signature (encryption key) generation process

(v) Finally, a facial image containing the IP seller's chosen feature set is generated (shown in Figure 2.3).

(vi) The dimensions corresponding to generated features (distance between the nodal points) are computed. This involves determining the coordinates of every nodal point and subsequently using Manhattan distance to compute the feature dimensions.

(vii) Every determined feature dimension is converted into its equivalent binary value. Finally, the facial encryption key is generated based on the IP seller's feature concatenation order. Ultimately, the generated IP seller's facial signature is used as AES key (Sengupta *et al.*, 2023c).

Note: Exact replication of the facial encryption key by an adversary is impossible, as various intricate parameters are decodable only to an authentic IP seller, including the grid size/spacing used for nodal point determination, the chosen feature set, and their concatenation order (Sengupta *et al.*, 2023c).

AES module: The encrypted protein security methodology utilizes the AES technique to produce an encrypted protein molecular biometric signature. As shown in Figure 2.4, the AES module takes the IP seller's protein molecular signature as the primary input and the IP seller's facial signature as the encryption key. The final output is an encrypted protein molecular biometric signature. AES employs substitution boxes as the first layer of security, introducing non-linear bit transformations to add confusion to the output bitstream. The second layer involves row and mix column diffusion, contributing to the diffusion property of the resulting output bitstream. Subsequently, an XOR operation is conducted between the output and the facial biometric-based encryption key. The output of the initial round is fed into subsequent rounds of the AES encryption module, which comprises a total of '*n*' rounds in AES 128 (Sengupta *et al.*, 2023c). Ultimately, the security module produces an encrypted protein molecular biometric signature, which is further translated into covert security constraints based on the IP seller's chosen mapping/embedding rule (Sengupta *et al.*, 2023c).

Creation of initial register allocation table (RAT) and embedding of generated encrypted protein signature (watermark): The generated encrypted protein signature is first translated into covert security constraints and then embedded into the hardware design as a watermark (digital evidence) (Sengupta *et al.*, 2023c). The generation of covert security constraints corresponding to the generated encrypted signature relies on the SDFG of the hardware design and the IP seller's chosen mapping/embedding rule. The scheduled data flow graph (SDFG) of the hardware design helps in the determination of the number of storage variables utilized to construct pairs of hardware security constraints. The SDFG corresponding to the 8-point DCT application is illustrated in Figure 2.5, where storage variables J_0 to J_{22} and their corresponding registers are highlighted. As shown in Figure 2.5, nine control steps (T0–T8) are necessary for scheduling and attaining the output. Subsequently, a RAT for the SDFG is constructed. The IP seller selected encoding rule is as follows (Sengupta *et al.*, 2023c):

- For a signature bit '0', the security constraints are embedded between even-even storage variable pairs $<J_X, J_Y>$ of the SDFG, where '*X*' and '*Y*' denote

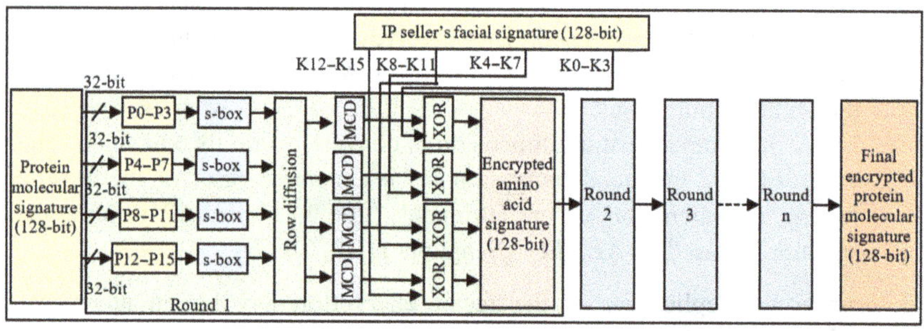

Figure 2.4 Details of AES process. Note: MCD: mix column diffusion, s-box: substitution box.

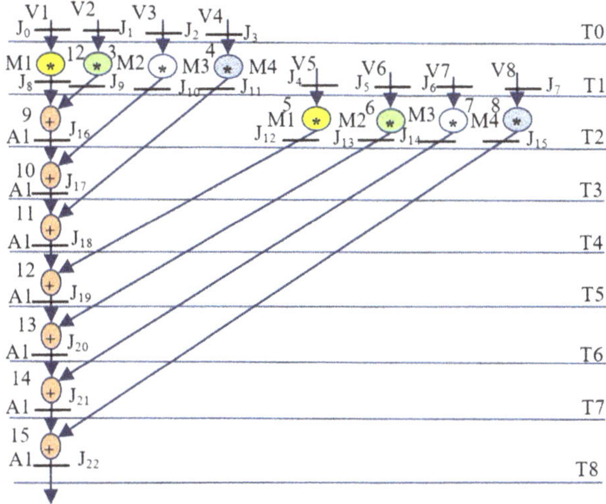

Figure 2.5 Scheduled DFG of 8-point DCT using one adders and four

the storage variable numbers. Thus, the resulting security constraints for 38 ‘0’s in the encrypted protein molecular biometric signature (derived earlier) are $<J_0, J_2>, <J_0, J_4>$,—, $<J_0, J_{22}>, <J_2, J_4>$,—, $<J_2, J_{22}>$,—, $<J_{12}, J_{16}>$.

- For a signature bit ‘1’, the security constraints are embedded between odd-odd storage variable pairs of the scheduled DFG. Consequently, the resulting security constraints for 37 ‘1’s in the encrypted protein molecular biometric signature are $<J_1, J_3>, <J_1, J_5>$,—, $<J_1, J_{21}>, <J_3, J_5>$,—, $<J_3, J_{21}>, <J_{19}, J_{21}>$.

Then, the obtained covert security constraints are integrated into the hardware design during the register allocation phase of the HLS process without incurring any additional design costs. As discussed above, the register allocation table (RAT) for the hardware design (i.e., 8-point DCT) has been generated (containing details of storage variables, control steps, and register allocation information). Subsequently, the register allocation information and covert security constraints are fed into the constraints embedding module, which generates the encrypted protein molecular watermark embedded RAT. **To embed the covert security constraints into the RAT, each storage variable pair is assigned to distinctive registers**. In order to realize this, local adjustments are made among available registers such that each storage variable pair does not share the same register. If any security constraints (storage variable pair) cannot be accommodated distinctively within available registers then a new register is allocated (Sengupta *et al.*, 2023c). However, security constraints that inherently satisfy the distinct register assignment rule do not require such local adjustments (Sengupta *et al.*, 2023c). Table 2.1 shows the register allocation information after embedding the generated encrypted signature (watermark), along with the required number of control steps for obtaining the watermark-embedded schedule hardware design (Sengupta *et al.*, 2023c).

Table 2.1 Register allocation table after embedding encrypted protein biometric-based security constraints corresponding to 8-point DCT

Control steps	T0	T1	T2	T3	T4	T5	T6	T7	T8
V1	J0	J9	—	—	—	—	—	—	—
V2	J1	J8	—	—	—	—	—	—	—
V3	J2	J11	J11	J11	—	—	—	—	—
V4	J3	J10	J10	—	—	—	—	—	—
V5	J4	J4	J13	J13	J13	J13	—	—	—
V6	J5	J5	J12	J12	J12	—	J20	—	J22
V7	J6	J6	J15	J15	J15	J15	J15	J15	—
V8	J7	J7	J14	J14	J14	J14	J14	—	—
V9	—	—	J16	J17	J18	—	—	—	—
V10	—	—	—	—	—	J19	—	—	—
V11	—	—	—	—	—	—	—	J21	—

2.5 Detection of IP piracy and nullifying false IP ownership claim

The security constraints for the encrypted protein molecular biometric approach are regenerated during piracy detection. From the suspected chip, the layout file is reverse engineered to retrieve the RTL information of the IP design. Using the regenerated encrypted protein molecular information, matching of the security constraints (watermark) is performed with the retrieved RTL IP design. In case of a complete matching, IP piracy is detected. Note: The regeneration process does not require the reacquisition of the genuine IP seller's protein and facial biometric information. Instead, the original pre-stored (in an encrypted format in a safe database) protein biometric signature and facial encryption key of the genuine IP seller are utilized for regeneration during the matching process. The same encrypted protein molecular signature (watermark) and its corresponding security constraints can be precisely regenerated from the pre-stored protein molecular biometric signature and facial encryption key during IP piracy detection (Sengupta *et al.*, 2023c).

Further, in case an attacker attempts to forge the original protein molecular watermark (security constraints) to realize evasion of IP piracy detection, he/she would be unable to succeed in regenerating the original watermark for implanting it in his/her fake version. This is because the regeneration of the original encrypted protein molecular watermark depends upon various security factors, such as the IP seller's protein molecular signature, IP seller's facial encryption key, concatenation rule, encoding rule, mapping/embedding rule, etc., discussed in prior sections (that are only decodable by the original IP seller). Therefore, these covert security constraints make it complex for an adversary to evade IP piracy detection (Sengupta *et al.*, 2023c).

Finally, the inclusion of the encrypted protein molecular watermark within the IP design also safeguards against the assertion of false IP ownership. Instances of fraudulent IP ownership claim may arise in the SoC design house and foundry. In the case of ownership conflict resolution, the original watermark (security constraints) is regenerated and then compared against the extracted security constraints from the register allocation of the IP design under test. Ownership right is awarded to the entity that establishes watermark matching (Sengupta *et al.*, 2023c).

2.6 Case study and analysis on benchmarks

2.6.1 Security analysis

The security analysis of the encrypted protein molecular security methodology employs the probability of coincidence and tamper tolerance security metrics (Sengupta *et al.*, 2023c). The probability of coincidence refers to the likelihood of detecting the same covert security constraints (watermark) in an unsecured design (non-watermarked design). In other words, the probability of coincidence is also a measure of the false positive. It is represented as (Sengupta *et al.*, 2023c; Koushanfar *et al.*, 2005; Sengupta and Bhadauria, 2016; Hu *et al.*, 2021; Potkonjak, 2006):

$$\text{Probability of coincidence} = (1 - 1/u)^s \tag{2.1}$$

where "u" represents the number of registers in the schedule design before incorporating covert security constraints, and "s" denotes the total implanted covert security constraints. A lower probability of coincidence value denotes more robust security, thus providing stronger digital evidence and credible authorship proof. The incorporation of a distinct encrypted protein-based watermark into the hardware IP design significantly contributes to the effective IP piracy detection process and legal proof of IP ownership. Figure 2.6 illustrates

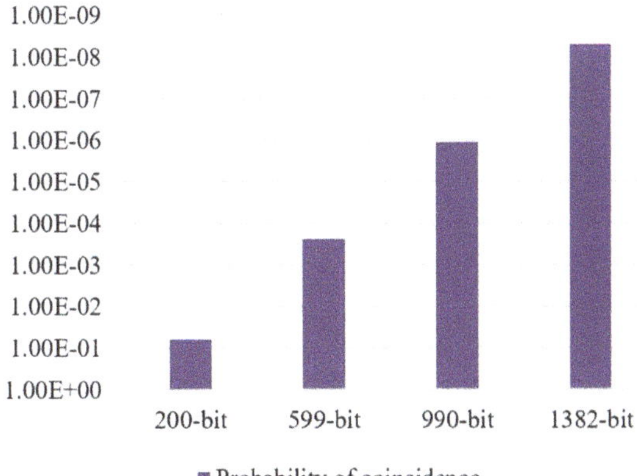

Figure 2.6 Variation in probability of coincidence with varying signature size for JPEG-CODEC application

the variation in the probability of coincidence with varying watermark/signature strength for JPEG-CODEC hardware design. As the number of embedded signature bits increases, the probability of coincidence decreases (Sengupta *et al.*, 2023c). With an increase in the embedded signature bits, the robustness of the security approach increases with a lower probability of coincidence value. Further, Figure 2.7 presents a comparative analysis of the probability of coincidence between the encrypted protein molecular signature approach (Sengupta *et al.*, 2023c) and chromosomal DNA (Sengupta and Chaurasia, 2022). Next, Figure 2.8 presents a comparative analysis of the probability of coincidence between the encrypted protein molecular signature approach (Sengupta *et al.*, 2023c), digital signature (Sengupta *et al.*, 2019), triple-phase watermarking (Sengupta *et al.*, 2018), dynamic watermarking (Koushanfar *et al.*, 2005), and encrypted signature (Castillo *et al.*, 2008). The encrypted protein molecular watermarking approach outperforms all above mentioned contemporary approaches with a lower probability of coincidence value (Sengupta *et al.*, 2023c). This is due to the fact that the protein watermark enables the IP seller to create greater watermarking constraints compared to all above mentioned contemporary approaches.

Further, tamper tolerance refers to the ability of a security mechanism to withstand and resist unauthorized tampering (removal/tinkering) using brute-force attack. It is represented as (Sengupta *et al.*, 2023c; Koushanfar *et al.*, 2005; Sengupta and Bhadauria, 2016; Hu *et al.*, 2021; Potkonjak, 2006):

$$\text{Tamper tolerance} = v^s \tag{2.2}$$

where "s" denotes the total implanted covert security constraints, and "v" represents the IP seller's selected distinct signature mapping variables. A higher tamper

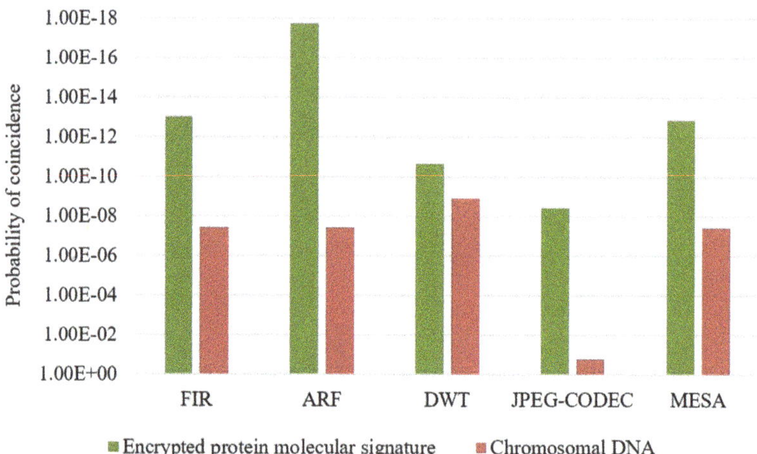

Figure 2.7 Comparison of probability of coincidence between encrypted protein molecular signature (Sengupta et al., 2023c) and chromosomal DNA (Sengupta and Chaurasia, 2022)

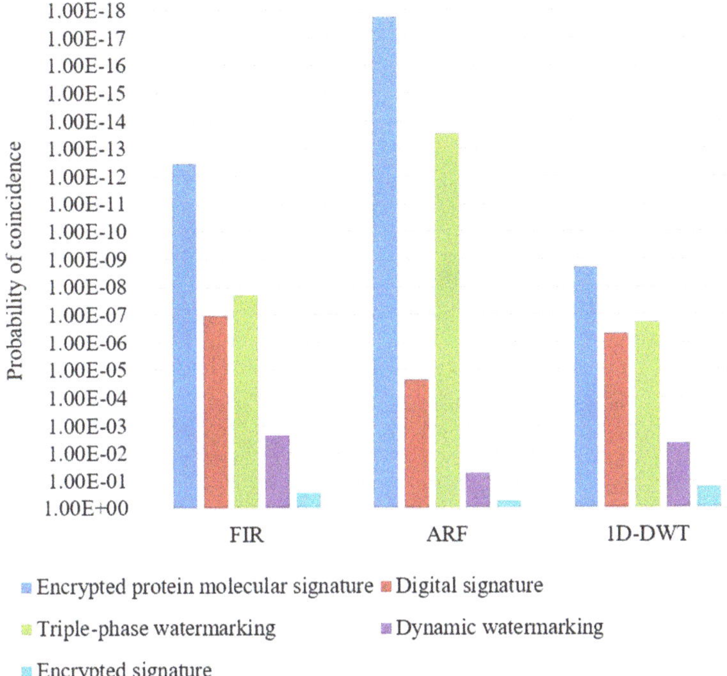

Figure 2.8 Comparison of probability of coincidence between encrypted protein molecular signature (Sengupta et al., 2023c), digital signature (Sengupta et al., 2019), triple-phase watermarking (Sengupta et al., 2018), dynamic watermarking (Koushanfar et al., 2005), and encrypted signature (Castillo et al., 2008)

tolerance signifies a larger signature space, indicating more robust security. With a higher tamper tolerance value, it is possible to generate various signature combinations. With a larger signature combination, it becomes difficult for an attacker to tamper with the exact watermark. An adversary aims to tamper, copy, or remove the embedded watermark. Further, an adversary may attempt to regenerate the exact watermark to insert in fake IPs, such that IP evasion of IP piracy detection can be performed. Thus, a higher tamper tolerance value impedes an adversary from realizing his malicious intent. Figure 2.9 presents a comparative analysis of the tamper tolerance between the encrypted protein molecular signature approach and chromosomal DNA (Sengupta and Chaurasia, 2022). Next, Figure 2.10 presents a comparative analysis of the tamper tolerance between the encrypted protein molecular signature approach (Sengupta *et al.*, 2023c), digital signature (Sengupta *et al.*, 2019), triple-phase watermarking (Sengupta *et al.*, 2018), dynamic watermarking (Koushanfar *et al.*, 2005), and encrypted signature (Castillo *et al.*, 2008). The encrypted protein molecular watermarking approach outperforms all above

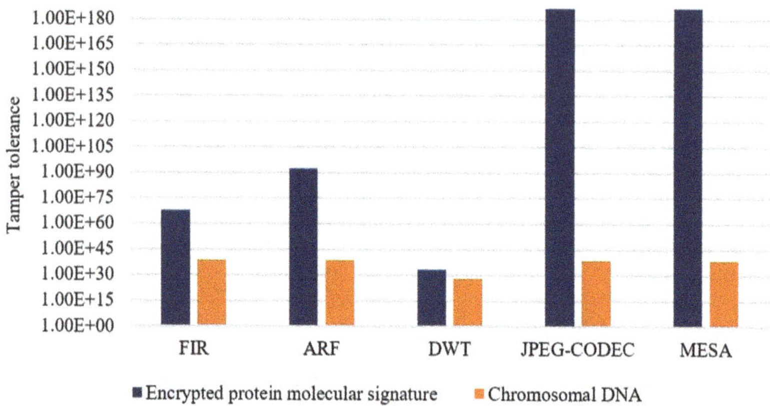

Figure 2.9 Comparison of tamper tolerance between encrypted protein molecular signature (Sengupta et al., 2023c) and chromosomal DNA (Sengupta and Chaurasia, 2022)

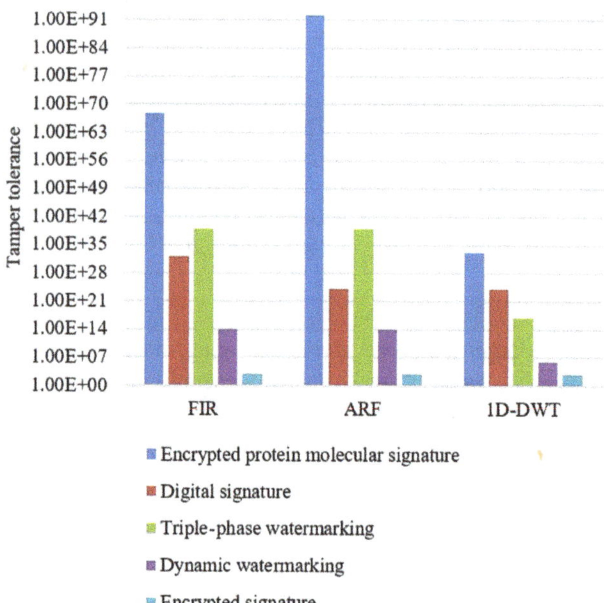

Figure 2.10 Comparison of tamper tolerance between encrypted protein molecular signature (Sengupta et al., 2023c), digital signature (Sengupta et al., 2019), triple-phase watermarking (Sengupta et al., 2018), dynamic watermarking (Koushanfar et al., 2005), and encrypted signature (Castillo et al., 2008)

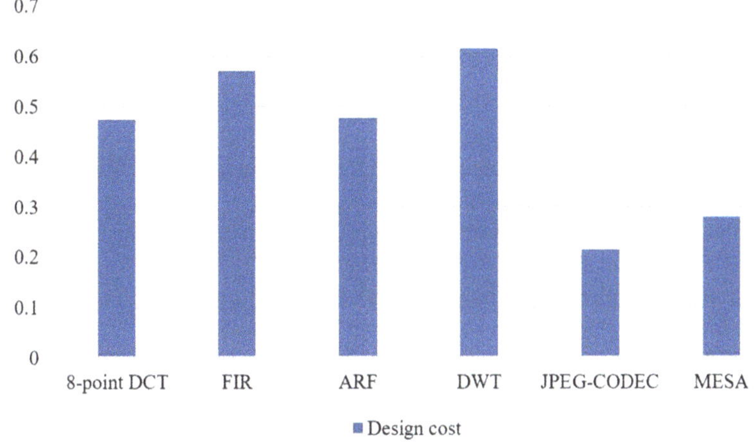

Figure 2.11 Design cost post embedding of the encrypted protein molecular signature-based watermarking approach (Sengupta et al., 2023c) corresponding to various benchmarks

mentioned contemporary approaches with a higher tamper tolerance ability (Sengupta *et al.*, 2023c).

2.6.2 Design cost analysis

The design cost of the encrypted protein watermarking approach is evaluated using (2.3) (Sengupta *et al.*, 2023c; Anshul and Sengupta, 2023b; Anshul and Sengupta, 2023c).

$$\text{Design cost} = f1 * (((Area))/A_{\max}) + f2 * (((Latency))/L_{\max}) \tag{2.3}$$

Here, *f1* and *f2* are set to 0.5, indicating equal importance for both area and latency parameters during watermark embedding process. "*Area*" and "*Latency*" denote the final hardware area and latency (delay) of the watermarked design. Next, "$A_{\max}$" and "$L_{\max}$" represent the maximum design area and latency, respectively. Evaluation of area and latency for the selected hardware benchmarks is conducted on a 15 nm scale using the NanGate library (OCL, 2024). Figure 2.11 presents the design cost for the encrypted protein molecular watermarking approach on different benchmarks after embedding the generated watermark (Sengupta *et al.*, 2023c).

2.7 Conclusion

In this chapter, IP piracy and false IP ownership assertion were discussed as the primary threats in the hardware global design supply chain process. An encrypted protein molecular biometric-based hardware watermarking approach has been discussed in this chapter for offering robust protection against the mentioned threats. A reader learns the following concepts by reading this chapter:

(a) Contemporary approaches for IP watermarking that address threats of IP piracy and false IP ownership assertion.
(b) Generation of protein biometric signature and encryption of generated signature using facial biometric encryption key.
(c) Translation of generated signature (watermark) into covert security constraints and its embedding into the hardware design.
(d) Detection of IP piracy and nullification of false IP ownership assertion using embedded encrypted protein molecular signature (watermark).

Comparative analysis of various IP watermarking techniques against existing attacks of brute force attack, tampering attack, false IP ownership claim using probability of coincidence metric and tamper tolerance metric.

2.8 Questions and exercises

1. Briefly explain the importance of application-specific hardware circuits in the context of multimedia and electronic systems.
2. Briefly explain the different hardware security threats associated with the global design supply chain process of reusable hardware IP cores.
3. Summarize the prior contemporary hardware watermarking methodology available in the literature.
4. Explain the overview of the encrypted protein molecular biometric-based hardware security (watermarking) approach.
5. List primary inputs and outputs of the encrypted protein molecular watermarking approach.
6. Briefly explain the IP seller's protein molecular signature generation process along with the details of the protein molecular biometric sequence.
7. Explain the advantages of protein sequencing over DNA sequencing.
8. Briefly explain the facial signature generation process using IP seller's facial biometric image.
9. Explain the encrypted protein molecular signature generation process using AES.
10. What are some of the advantages/merits of the encrypted protein molecular watermarking approach that helps in achieving robust security?
11. Demonstrate the encrypted protein molecular watermark embedding process with a suitable example.
12. Explain the IP piracy detection and nullification of false IP ownership claim.
13. What are the security metrics used to compare and analyze the security strength of the discussed security approaches?
14. Discuss the advantages of the encrypted protein molecular watermarking approach over contemporary approaches in the literature that helps in achieving robust security.

References

A. Anshul and A. Sengupta, (2022) "IP Core Protection of Image Processing Filters with Multi-Level Encryption and Covert Steganographic Security Constraints," *2022 IEEE International Symposium on Smart Electronic Systems (iSES)*, Warangal, India, pp. 83–88.

A. Anshul and A. Sengupta, (2023a) "A Survey of High Level Synthesis Based Hardware Security Approaches for Reusable IP Cores [Feature]," *IEEE Circuits and Systems Magazine*, vol. 23, no. 4, pp. 44–62.

A. Anshul and A. Sengupta, (2023b) "Exploration of Optimal Crypto-chain Signature Embedded Secure JPEG-CODEC Hardware IP during High Level Synthesis," *Microprocessors and Microsystems*, vol. 102, 104916.

A. Anshul and A. Sengupta, (2023c) "PSO Based Exploration of Multi-phase Encryption Based Secured Image Processing Filter Hardware IP Core Datapath during High Level Synthesis," *Expert Systems with Applications*, vol. 223, 119927.

A. Anshul, K. Bharath and A. Sengupta, (2022) "Designing Low Cost Secured DSP Core Using Steganography and PSO for CE systems," *2022 IEEE International Symposium on Smart Electronic Systems (iSES)*, Warangal, India, 95–100.

E. Castillo, L. Parrilla, A. Garcia, U. Meyer-Baese, G. Botella and A. Lloris, (2008) "Automated Signature Insertion in Combinational Logic Patterns for HDL IP Core Protection," *2008 4th Southern Conference on Programmable Logic*, Bariloche, Argentina, pp. 183–186.

R. Chaurasia, A. Anshul, A. Sengupta and S. Gupta, (2022) "Palmprint Biometric Versus Encrypted Hash Based Digital Signature for Securing DSP Cores Used in CE Systems," *IEEE Consumer Electronics Magazine*, vol. 11, no. 5, pp. 73–80.

J. Chen and B. C. Schafer, (2021) "Watermarking of Behavioral IPs: A Practical Approach," *2021 Design, Automation & Test in Europe Conference & Exhibition (DATE)*, Grenoble, France, pp. 1266–1271.

L. H. Chen, C. B. White, P. C. Babbitt, M. J. McLeish and G. L. Kenyon, (2000) "A Comparative Study of Human Muscle and Brain Creatine Kinases Expressed in *Escherichia coli*," *Journal of Protein Chemistry*, vol. 19, no. 1, pp. 59–66.

B. Colombier and L. Bossuet, (2015) "Survey of Hardware Protection of Design Data for Integrated Circuits and Intellectual Properties," *IET Computers & Digital Techniques*, vol. 8, no. 6, pp. 274–287.

V.-A. Duong, J.-M. Park, H.-J. Lim and H. Lee, (2021) "Proteomics in Forensic Analysis: Applications for Human Samples," *Applied Sciences*, vol. 11, no. 8, pp. 3393.

W. Hu, C. Chang, A. Sengupta, S. Bhunia, R. Kastner and H. Li, (2021) "An Overview of Hardware Oriented Security and Trust: Threats, Countermeasures and Design Tools," *IEEE Transactions on Computer-Aided Design of Integrated Circuits and Systems,* Invited Paper, vol. 40, no. 6, pp. 1010–1038.

S. A. Islam, L. K. Sah and S. Katkoori, (2020) "High-Level Synthesis of Key-Obfuscated RTL IP with Design Lockout and Camouflaging," *ACM Transactions on Design Automation of Electronic Systems*, vol. 26, no. 1, 1–35.

F. Koushanfar, I. Hong and M. Potkonjak, (2005) "Behavioral Synthesis Techniques for Intellectual Property Protection," *ACM Transactions on Design Automation of Electronic Systems*, vol. 10, no. 3, pp. 523–545.

B. Le Gal and L. Bossuet, (2012) "Automatic Low-cost IP Watermarking Technique Based on Output Mark Insertions," *Design Automation for Embedded Systems*, vol. 16, no. 2, pp. 71–92.

E. D. Merkley, D. S. Wunschel, K. L. Wahl and K. H. Jarman, (2019) "Applications and Challenges of Forensic Proteomics," *Forensic Science International*, vol. 297, pp. 350–363.

OCL, (2024) 15 nm open cell library. [Online], Available: https://si2.org/open-cell-library/, last accessed on Jan. 2024.

G. J. Parker, T. Leppert, D. S. Anex, *et al.*, (2016) "Demonstration of Protein-based Human Identification Using the Hair Shaft Proteome," *PLoS ONE*, vol. 11, no. 9, Art. no. e0160653.

M. Potkonjak, (2006) "Methods and Systems for the Identification of Circuits and Circuit Designs," *United States Patent*, US7017043B1.

M. Rathor and G. P. Rathor, (2024) "Hard-Sign: A Hardware Watermarking Scheme Using Dated Handwritten Signature," *IEEE Design & Test*, vol. 41, no. 2, pp. 75–83.

M. Rathor, A. Sengupta, R. Chaurasia and A. Anshul, (2023a) "Exploring Handwritten Signature Image Features for Hardware Security," *IEEE Transactions on Dependable and Secure Computing*, vol. 20, no. 5, pp. 3687–3698.

M. Rathor, A. Anshul, K. Bharath, R. Chaurasia and A. Sengupta, (2023b) "Quadruple Phase Watermarking during High Level Synthesis for Securing Reusable Hardware Intellectual Property Cores," *Computers and Electrical Engineering*, vol. 105, 108476.

M. Rathor, A. Anshul and A. Sengupta, (2024) "Securing Reusable IP Cores using Voice Biometric Based Watermark," *IEEE Transactions on Dependable and Secure Computing*, vol. 21, no. 4, pp. 2735–2749.

S. Rizzo, F. Bertini, and D. Montesi, (2019) "Fine-grain Watermarking for Intellectual Property Protection," *EURASIP Journal on Information Security*, vol. 2019, Article no. 10.

M. Rostami, F. Koushanfar and R. Karri, (2014) "A Primer on Hardware Security: Models, Methods, and Metrics," *Proceedings of the IEEE*, vol. 102, no. 8, pp. 1283–1295.

R. Schneiderman, (2010) "DSPs Evolving in Consumer Electronics Applications," *IEEE Signal Processing Magazine*, vol. 27, no. 3, pp. 6–10.

A. Sengupta and S. Bhadauria, (2016) "Exploring Low Cost Optimal Watermark for Reusable IP Cores during High Level Synthesis," *IEEE Access*, vol. 4, pp. 2198–2215.

A. Sengupta and R. Chaurasia, (2022) "Securing IP Cores for DSP Applications Using Structural Obfuscation and Chromosomal DNA Impression," *IEEE Access*, vol. 10, 50903–50913.

A. Sengupta and M. Rathor, (2019) "IP Core Steganography for Protecting DSP Kernels Used in CE Systems," *IEEE Transactions on Consumer Electronics*, vol. 65, no. 4, pp. 506–515.

A. Sengupta and M. Rathor, (2020) "Securing Hardware Accelerators for CE Systems Using Biometric Fingerprinting," *IEEE Transactions on Very Large Scale Integration (VLSI) Systems*, vol. 28, no. 9, pp. 1979–1992.

A. Sengupta and M. Rathor, (2021) "Facial Biometric for Securing Hardware Accelerators," *IEEE Transactions on Very Large Scale Integration (VLSI) Systems*, vol. 29, no. 1, pp. 112–123.

A. Sengupta, D. Roy and S. P. Mohanty, (2018) "Triple-Phase Watermarking for Reusable IP Core Protection during Architecture Synthesis," *IEEE Transactions on Computer-Aided Design of Integrated Circuits and Systems*, vol. 37, no. 4, pp. 742–755.

A. Sengupta, E. R. Kumar and N. P. Chandra, (2019) "Embedding Digital Signature Using Encrypted-Hashing for Protection of DSP Cores in CE," *IEEE Transactions on Consumer Electronics*, vol. 65, no. 3, pp. 398–407.

A. Sengupta, R. Chaurasia and T. Reddy, (2021) "Contact-Less Palmprint Biometric for Securing DSP Coprocessors Used in CE Systems," *IEEE Transactions on Consumer Electronics*, vol. 67, no. 3, pp. 202–213.

A. Sengupta, R. Chaurasia and A. Anshul, (2023a) "Hardware Security of Digital Image Filter IP Cores against Piracy Using IP Seller's Fingerprint Encrypted Amino Acid Biometric Sample," *2023 Asian Hardware Oriented Security and Trust Symposium (AsianHOST)*, Tianjin, China, pp. 1–6.

A. Sengupta, A. Anshul and R. Chaurasia, (2023b) "Exploration of Optimal Functional Trojan-resistant Hardware Intellectual Property (IP) Core Designs during High Level Synthesis," *Microprocessors and Microsystems*, vol. 103, 104973.

A. Sengupta, R. Chaurasia and A. Anshul, (2023c) "Robust Security of Hardware Accelerators Using Protein Molecular Biometric Signature and Facial Biometric Encryption Key," *IEEE Transactions on Very Large Scale Integration (VLSI) Systems*, vol. 31, no. 6, pp. 826–839.

K. Van Steendam, M. De Ceuleneer, M. Dhaenens, D. Van Hoofstat and D. Deforce, (2013) "Mass Spectrometry-based Proteomics as a Tool to Identify Biological Matrices in Forensic Science," *International Journal of Legal Medicine*, vol. 127, no. 2, pp. 287–298.

M. Yasin, J. J. Rajendran, O. Sinanoglu and R. Karri, (2016) "On Improving the Security of Logic Locking," *IEEE Transactions on Computer-Aided Design of Integrated Circuits and Systems*, vol. 35, no. 9, pp. 1411–1424.

Chapter 3

High-level synthesis-based watermarking using retinal biometrics

Anirban Sengupta[1] *and Rahul Chaurasia*[1]

The chapter describes a high-level synthesis (HLS) driven methodology for designing secure hardware intellectual property (IP) cores using retinal biometrics (Chaurasia and Sengupta, 2023a, 2023b). In this methodology, firstly, the retinal biometrics of the original IP vendor is captured, which is subsequently exploited for extracting unique retinal features and generating the retinal digital template. The generated retinal template (binarized) is then encoded into its corresponding hardware security constraints using IP vendor-specified encoding algorithm. Next, these generated secret constraints are covertly embedded into the design during the register allocation phase of the HLS framework. The embedding process results in the retinal signature implanted hardware IP core design at the register transfer level (RTL). These embedded, retinal biometric-driven encoded security constraints serve as digital evidence, fortifying the IP core design against piracy by providing a detective countermeasure. Thus, HLS-driven hardware security methodology based on retinal biometrics enables robust and seamless detective control against pirated versions of the IP design, thereby enabling integration of authentic designs into the end system (through isolation of pirated/fake versions, not containing authentic IP vendor security mark).

The rest of the chapter has been organized as follows: Section 3.1 provides the introduction of the chapter; Section 3.2 outlines the benefits to both consumer and consumer electronics (CE) system designer; Section 3.3 encapsulates the key highlights of the chapter; Section 3.4 discusses prior similar works on biometrics and non-biometrics based hardware security; Section 3.5 delves into the realm of retinal biometrics-based hardware security and trust; Section 3.6 elucidates the process of embedding security constraints, demonstrated through a case study; Section 3.7 shows detection of pirated IPs using retinal biometric; Section 3.8 explores the security properties of retinal biometrics, emphasizing robust hardware security; Section 3.9 provides a detailed analysis and discussion; and Section 3.10 concludes the chapter by summarizing the chapter's findings and implications.

[1]Department of Computer Science and Engineering, Indian Institute of Technology Indore, India

3.1 Introduction

Integration of hardware IP cores within various consumer electronic (CE) devices like tablets, smartphones, and cameras enables their accelerated performance and efficiency by performing computationally intensive tasks dedicatedly. This is achieved by offloading the computationally intensive tasks to hardware, especially to accelerate that task. The hardware can perform the task more efficiently compared to relying solely on a general-purpose processor. However, the process of designing these IP cores exposes them to potential security vulnerabilities. These threats stem from multiple sources, including untrusted IP vendors, adversaries within offshore design houses, and risks at the foundry level, all of which pose concerns during the design phase (Pilato *et al.*, 2018; Schneiderman, 2010; Mohanty and Meher, 2016; Roy *et al.*, 2008; Mahdiany *et al.*, 2001; Plaza and Markov, 2015; Wang *et al.*, 2015; Koushanfar *et al.*, 2005; Zhang, 2016; Koushanfar *et al.*, 2012; Castillo *et al.*, 2007; Colombier and Bossuet, 2015; Arafin *et al.*, 2017). IP piracy stands out as a significant threat among these concerns, potentially compromising the authenticity/integrity of the hardware. Hence, among the primary objectives in designing hardware IP cores such as optimization, security, and reliability, ensuring security emerges as a pivotal factor in validating the legitimacy of the IP design. A pirated version of the IP core exposes it to various security risks such as counterfeiting and cloning. Therefore, embedding a resilient secret security mark becomes crucial for identifying pirated IP cores during the IP piracy detection process, before integration into the System on Chip (SoC) platform (Sitjongsataporn *et al.*, 2021; Rajendran *et al.*, 2016; Chen *et al.*, 2020; Saeed *et al.*, 2019a, 2019b; Rizo *et al*, 2022; Potluri *et al.*, 2020; Mouris *et al.*, 2022; Hroub and Elrabaa, 2022; Karmakar and Chattopadhyay, 2020; Rathor and Sengupta, 2021; Hu *et al.*, 2021).

Some existing hardware security approaches such as watermarking, steganography, and digital signature insertion are used for securing the hardware IPs against piracy. However, these approaches do not uniquely associate the identity of the original IP vendor and are vulnerable to compromise/adversarial attacks. Therefore, biometric-based approaches are being increasingly used for hardware security and trust in the state of the art. This is because it offers the following advantages:

- **Uniqueness**: Biometric characteristics like retinal features, fingerprints, facial and palm features are unique to individuals, making them highly reliable for generating unique IP vendor signature and can be used for hardware IP authentication as well as robust proof of ownership. This uniqueness adds a strong layer of security.
- **Difficult to Replicate**: Unlike passwords or PINs, which can be forged, shared, or stolen, it is much more challenging to replicate biometric features by an adversary for evading IP piracy detection.
- **Multi-Factor Authentication**: Biometrics can complement other authentication methods, creating a multi-factor authentication system that significantly

enhances security. For example, combining a fingerprint scan with other forms of IP vendor signatures or biometric features of the same IP vendor can provide a higher level of robustness (IP vendor digital evidence) and authorship proof.

Further, the desirable properties of biometric-based techniques are the followings:

- **Revocability**: A compromised template can be replaced with a completely new and different one if needed.
- **Unlinkability**: Multiple templates that are different from each other can be computed from the same biometric trait.
- **Security**: Exact regeneration of original hardware security constraint information from authentic IP vendor biometric features is infeasible for an adversary, for evading IP piracy detection process.

3.1.1 Motivation on retinal biometrics as watermark

Retina biometrics not only offers enhanced security but also boasts numerous advantages owing to its innate strength and capability compared to other biometric and non-biometric techniques. Now, we discuss the advantages of exploiting retinal biometrics for hardware security over others such as fingerprint (Sengupta and Rathor, 2020), facial (Sengupta and Chaurasia, 2022; Chaurasia and Sengupta, 2022), and palmprint biometrics (Sengupta *et al.*, 2021; Chaurasia *et al.*, 2022):

(a) Retinal biometrics during security mark generation does not get influenced by external factors like dirt and grease since the retina, situated at the back of the eye, remains shielded from the external environment. This ensures capturing clear and distinct IP vendor retinal biometrics for generating signature. Conversely, fingerprint and facial biometrics are susceptible to interference from grease, dust, and other external elements, potentially affecting the accuracy of biometric feature extraction. This may lead to an IP vendor biometric that does not provide a clear and distinct mapping into fingerprint/facial signature.

(b) The heightened signature strength of retinal biometrics leads to a lower false positive (probability of coincidental matches with an unsecured version) and increased resistance to tampering compared to hardware security-based techniques using facial (Sengupta and Chaurasia, 2022; Chaurasia and Sengupta, 2022) and fingerprint biometrics (Sengupta and Rathor, 2020).

(c) Without an IP vendor's consent, his/her retinal biometric cannot be captured for generating retinal signature. This makes it harder for an adversary to regenerate the same retinal signature or create a replica of the original retinal signature.

(d) Retinal biometric characteristics exhibit higher distinctiveness, even among twins, compared to characteristics of facial and finger biometrics.

(e) Retina images offer higher accuracy than fingerprints and do not necessitate image enhancement techniques like Fast Fourier transform (FFT).

(f) Additionally, unlike palmprint biometrics, retinal biometrics provides stronger security against IP piracy for the following reasons:

Retinal biometrics consist of numerous highly distinctive feature points, resulting in stronger security mark. This provides the flexibility for embedding a secret mark of higher strength into the target hardware design, thus ensuring robustness against piracy and brute force attacks with lower chances of coincidence (Pc) and higher tamper tolerance (TT) respectively. Therefore, due to the inherent security of retinal biometrics in terms of highly distinctive nature/uniqueness and extensive biometric template, it offers a more robust and seamless means of detecting and controlling against IP piracy compared to palmprint biometrics.

Next, we discuss the merits of retinal biometrics compared to hardware security-based techniques using digital signatures (Sengupta *et al.*, 2019; Chaurasia *et al.*, 2022). Retinal biometrics offers the following merits:

(a) Retinal biometrics ensure the uniqueness of the generated signature by utilizing naturally distinct retinal features. Conversely, in digital signature-based approaches, ensuring uniqueness isn't always guaranteed, despite employing a more complex algorithm compared to the retinal biometric method.
(b) Creating a retinal signature involves leveraging distinctive retinal features chosen by the IP vendor, which cannot be replicated. In contrast, the generation of digital signatures relies on multiple factors like encoding rules, hashing algorithms, and private keys for RSA encryption, which can be potentially compromised by skilled adversaries with diligent efforts.
(c) Replicating a retinal signature by an adversary is highly implausible. Conversely, with digital signature-based approaches, there might be a possibility of compromising the encoding rule through brute force attacks or leaking the encryption key using side-channel attack.
(d) The retinal biometric signature remains non-vulnerable to side-channel attacks. Conversely, digital signature-based approaches relying on key-based security methods are susceptible to theft and attacks targeting the key itself.

3.1.2 Threat model

Retinal biometric-based hardware security methodology (Chaurasia and Sengupta, 2023a, 2023b) utilizes unique retinal features to ensure detective control against pirated versions and false IP ownership claim. In the context of providing detective control against pirated IPs, an SoC integrator gains the ability to distinguish between fake and authentic IPs using authentic IP vendor retinal signature. Original IPs shall contain the authentic retinal signature while fake IPs will not. The covertly implanted security constraints generated through retinal biometrics serve as robust digital evidence to detect pirated IPs. Therefore, retinal biometric-based secret mark is effective in identifying/isolating pirated IP designs with malicious logic (in the scenario where a rogue IP supplier/broker has already implanted malicious logic in his fake IPs and sold it to the SoC integrator). The method helps identify such fraudulent IPs lacking the genuine vendor's authentic security signatures.

The primary emphasis of this chapter revolves around protecting hardware IP cores from external hardware threats of IP piracy. It explores a security methodology based on retinal biometrics to counteract and protect hardware designs against IP piracy (Chaurasia and Sengupta, 2023a, 2023b). In order to do so, it exploits the HLS framework for generating the secure RTL datapath by transforming the behavioral description of the design into control data flow graph and covertly embedding the retinal biometric-driven hardware security constraints during the register allocation phase. The embedding is performed in such a way that it yields negligible design cost overhead (through local alteration of design storage variables among registers uniquely). Hence, the embedded retinal biometric signature serves as a detective countermeasure against pirated IP cores before their integration into the end systems (Chaurasia and Sengupta, 2023a).

3.2 Benefits to consumer and CE system designer

IP cores stand as crucial components within consumer electronics systems. Ensuring the security of these cores becomes imperative in safeguarding CE systems and, consequently, protecting end consumers. The suggested security strategy, centered on retinal biometric features, guarantees this protection through its integrated security measures. By leveraging retinal biometric signatures, it establishes a robust means to detect counterfeit IP cores before their integration into system-on-chips within CE systems. This process ensures the incorporation of secure and authentic CE systems, effectively shielding end consumers from the usage of counterfeit or fake designs. Such counterfeit designs might harbor malicious elements that compromise their reliability and could pose safety risks to end users. On the flip side, IP cores infused with genuine retinal biometric signatures stand out as truly authentic, distinguishing between originals and counterfeits. By employing retinal biometrics, counterfeit IP cores can be easily identified, thwarting their integration into consumer electronics systems and guaranteeing the exclusive use of legitimate designs. As a result, this security measure not only safeguards the integrity of IP cores but also puts the safety of end users in consumer electronics systems first. Hardware security approach using retinal biometrics offers several benefits to consumers such as:

Trust and reliability: Knowing that their devices are protected by robust hardware security instills confidence in consumers regarding the reliability and integrity of the products they purchase.

Anti-counterfeiting: Hardware security can aid in identifying counterfeit products, ensuring consumers receive genuine, trusted devices.

Moreover, hardware security through retinal biometric offers several benefits to CE system designers such as:

Enhanced brand reputation: Prioritizing security measures reflects positively on the brand, establishing trust and credibility among consumers, leading to increased brand loyalty.

Reduced revenue loss: By handling piracy, hardware security measures can reduce revenue losses resulting from counterfeit products or unauthorized use of intellectual property.

Protection of intellectual property: Implementing hardware security measures enables detective control against pirated design, thereby hindering an adversary from piracy evasion. It further protects the original IP vendor from false claim of IP ownership. By using the retinal biometric-based embedded secret mark, an authentic IP vendor can seamlessly nullify fraud IP ownership claim in IP court, in case of legal conflict.

Thus, by incorporating hardware security measures, both consumers and CE system designers can benefit from increased protection against IP piracy, thereby leading to a more secure and trustworthy ecosystem.

3.3 Salient features of the chapter

The salient features of the chapter, comprising of the discussion and analysis on the significance of hardware security using retinal biometrics and HLS, are the following:

- This chapter discusses the HLS-based design methodology for designing secure hardware IPs for computationally intensive application(s) (Sengupta and Sedaghat, 2011).
- This chapter explains the register allocation phase of HLS for enabling security in terms of ensuring seamless detective control against piracy and nullifying fraudulent IP ownership claim.
- Explanation of how retinal biometrics of the IP vendor is exploited for extracting unique retinal features and generating retinal signature is included in this chapter. Retinal features such as branching and bifurcation are considered for generating unique retinal signatures (Aleem *et al.*, 2019; Chaurasia and Sengupta, 2023a, 2023b).
- This chapter provides a case study on "Joint Photographic Experts Group-compression decompression" (JPEG-codec) application and generates its secure IP design using retinal biometrics of IP vendor (Chaurasia *et al.*, 2023).
- The discussed secure hardware IP design methodology attains higher robustness in the authentication/verification process by creating numerous secret security constraints and leveraging the highly distinctive nature of the retinal structure, surpassing other similar security techniques.
- The discussed methodology using retinal biometrics is capable of achieving lower false positive (probability that an unsecured IP design coincidentally contains the same embedded security constraints) and increased resistance to tampering attacks.
- The discussed methodology incurs zero design cost overhead leveraging HLS methodology and retinal biometrics for hardware security (Chaurasia and Sengupta, 2023a, 2023b).

3.4 Similar works on biometrics and non-biometrics-based hardware watermarking

In order to secure hardware coprocessors or IP cores, numerous methods have been introduced in the literature. Among these, some of the security approaches employ non-biometric and some biometric-based hardware security. Non-biometric methods, used to secure hardware IP cores include watermarking (Sengupta and Bhadauria, 2016; Le Gal and Bossuet, 2012; Rai *et al.*, 2019; Karmakar and Chattopadhyay, 2020), crypto-digital signatures using encrypted hash (Sengupta *et al.*, 2019), and steganography (Rathor and Sengupta, 2020). Additionally, biometric-based hardware security methods introduced in the literature include fingerprint biometrics (Sengupta and Rathor, 2020), facial biometrics (Sengupta and Chaurasia, 2022; Chaurasia and Sengupta, 2022), and retinal biometrics (Chaurasia and Sengupta, 2023a, 2023b) for hardware security of IP cores. Now, we discuss each of these approaches and their security strengths. Following that, we discuss how the retinal biometric-based method enhances the security of hardware coprocessors in comparison to other hardware security approaches.

Sengupta and Bhadauria (2016) presented a hardware watermarking approach to protect reusable IP cores during the register allocation phase of HLS (Behavioral Level). In this approach, watermark is formed using four signature variables chosen from IP vendor. However, the generated signature hinges on various factors, including the quantity of signature variables, their combinations, signature length, and encoding regulations. Relying on these intermediate elements makes the watermarking approach susceptible to vulnerabilities, as they could be readily compromised. The hardware security method outlined in Le Gal and Bossuet (2012) aims to tackle security concerns associated with IP reuse, like unauthorized duplication and counterfeiting, by introducing an IP watermarking technique incorporated within HLS. This involves encoding the IP watermark through mathematical sub-marks placed on design output ports, automatically integrated during synthesis. However, this method comes with a drawback that it adds design overhead to the process. Additionally, Rai *et al.* (2019) presented two watermarking techniques aimed at combating integrated circuit counterfeiting and IP piracy. These approaches employ encoding via polymorphic inverter configurations, capitalizing on adaptable nanowire technologies. Through an exclusive fabrication process reliant on nanowires, specific nodes in the logical framework can be permanently designated as either 0 or 1. This deliberate manipulation of nodes directs polymorphic inverters in predetermined manners, contributing to the watermarking process. In essence, this research delves into leveraging the unique attributes of reconfigurable emerging nanotechnologies to bolster hardware security by incorporating an encoding mechanism to embed the designer's signature within an IC. However, it results in an increased spatial requirement, even for a 64-bit watermark signature. Furthermore, (Karmakar and Chattopadhyay, 2020) introduced a watermarking strategy based on cellular automata-driven finite state machines (FSM) to identify potential theft of designers' intellectual properties by an adversary. Additionally, Sengupta *et al.* (2019) presented a crypto-digital signature method for hardware security. In this method, digital signature created undergoes encoding, utilizing the secure hashing algorithm (SHA-512), and undergoes RSA encryption employing the vendor's 128-bit key (private). This process involves intricate computations during signature generation to impede an adversary from replicating the digital signature. However, its reliance solely on the standard SHA-512 and private key introduces vulnerability to potential compromise. Further, hardware steganography method (Rathor and Sengupta, 2020) offers a signature-free technique for securing hardware IP cores. Steganography provides robust security with minimal overhead compared to watermarking methods (Sengupta and Bhadauria, 2016; Le Gal and Bossuet, 2012; Rai *et al.*, 2019). However, there exists a vulnerability wherein an attacker could compromise steganography method by exploiting secret stego-keys, stego-encoder, and mapping rules. Moreover, both the crypto-digital signatures (Sengupta *et al.*, 2019) and hardware steganography method (Rathor and Sengupta, 2020) involve encryption keys susceptible to key-based threats like side-channel attacks.

The primary drawback of the non-biometric methods mentioned earlier lies in the potential replication and regeneration of the signature by compromising limited security variables such as private keys, encoding algorithms, and signature combinations. Consequently, these methods fail to ensure effective security of hardware IP cores against piracy. In contrast, the retinal biometric approach is not reliant on external secret information and remains impervious to key-based attacks. It relies on numerous security variables, including the diversity of retinal features, and vessel structure, types and numbers of retinal features, size of kernel matrix and pattern, grid size specifications, encoding methods, selected region of interest (ROI) on the retina, signature strength and their combinations, etc. These parameters are absent in the discussed non-biometric approaches and cannot be decoded from a potential adversary's perspective. Consequently, it becomes highly improbable for an adversary to precisely replicate the secret mark to evade detection of IP piracy. Moreover, hardware security approach using the retinal biometric provides stronger digital evidence by reducing false positives (the probability of coincidence) and enhancing tamper tolerance. Additionally, hardware using retinal biometrics offers stronger entropy than similar non-biometric approaches. Further, in the contemporary non-biometric-based approaches, the generation of unique secret constraints is not always assured, unlike retinal biometrics.

The biometric-based approach using fingerprint (Sengupta and Rathor, 2020) enables the security of hardware IP cores by embedding specific security constraints generated from corresponding fingerprint templates, harnessing the naturally distinct features of IP vendor fingerprint. The process initiates with capturing fingerprint impression using an optical scanning device. Subsequently, to accurately extract minutiae points from the fingerprint biometric image, a pre-processing module is applied. Minutiae points considered in Sengupta and Rathor (2020) are classified as ridge endings where ridge lines abruptly end and ridge bifurcations where lines split into branches. The pre-processing involves three key sub-processes: (a) Image enhancement: utilizing Fast Fourier Transform (FFT), the captured image undergoes enhancement. This process operates on pixel sets, magnifying and reconnecting broken ridges within the fingerprint. (b) Binarization: the enhanced image is represented with only two intensity levels (low and high) by comparing pixel intensities against a threshold value. (c) Thinning: reducing the ridge line thickness to a single pixel width, thinning the image occurs as the final step of pre-processing. After this pre-processing stage, the thinned fingerprint image is analyzed to extract minutiae points, leveraging the unique features of the IP designer's fingerprint. The generation of the target fingerprint signature involves representing each minutia point in its corresponding binary form, representing the signature associated with each one. Each minutiae point's signature includes its coordinates/location, the crossing number (CN) value indicating the minutiae type, and the ridge angle. Subsequently, a final digital template is constructed by concatenating the signatures of each of the minutiae points. This fingerprint template is then employed for enabling the security of the design against piracy. Yet, creating an accurate fingerprint signature involves a complex image enhancement phase using Fast Fourier Transform, adding complexity to the process, and necessitates the use of an optical scanner. Furthermore, it is susceptible to injuries and external factors that might hinder precise fingerprint capturing and generation. Further, facial biometric approaches (Sengupta and Chaurasia, 2022; Chaurasia and Sengupta, 2022) leverage the naturally unique features of IP vendor facial image for signature generation. In this approach, firstly a high-quality imaging device is utilized for capturing the facial biometrics of IP vendor. Subsequently, this image undergoes a precise grid-sizing process to extract biometric information with accuracy. Next, the nodal points are generated, based on facial features selected by the IP vendor (for final signature generation), capturing pivotal and distinctive facial information. Then, a facial image incorporating the IP vendor's chosen feature set is produced. Next, in order to create facial biometric-driven digital template, the dimensions of each feature (as selected by the IP vendor) are determined. This involves computing the Manhattan distance between nodal points corresponding to each facial feature. Subsequently, the dimensions of each feature are transformed into its binarized format. Finally, the facial signature is constructed by concatenating the binarized information related to each facial feature.

Both these biometric methods (Sengupta and Rathor, 2020; Sengupta and Chaurasia, 2022; Chaurasia and Sengupta, 2022) expose their features to the external environment which may lead to an IP vendor biometric that does not provide a clear and distinct mapping into fingerprint/facial signature (especially in case of external damage and injury). On the contrary, retinal biometric-based hardware security approach not only achieves stronger digital evidence by reducing the probability of coincidence but also exhibits stronger tamper tolerance compared to fingerprint (Sengupta and Rathor, 2020) and facial biometric (Sengupta and Chaurasia, 2022; Chaurasia and Sengupta, 2022) hardware security approaches. This is because retinal biometrics yield a larger number of security constraints which in turn enables robust security by enabling higher tamper tolerance and ensuring a lesser probability of coincidently detecting authentic security constraints in a pirated design version (unsecured). Moreover, retinal biometrics are non-injury prone and not affected by external factors like grease and dirt, during capturing and security constraints generation. They are securely positioned at the back end of the human eye, inaccessible directly. Furthermore, the distinctive and rich features within the retinal vascular structure make it highly suitable for generating unique retinal signature that enables robust security against piracy in terms of detective control. It therefore enables sturdy isolation of pirated IPs from its authentic versions as well as is capable of nullifying false IP ownership claim during conflict resolution in IP court. Furthermore, for ensuring definitive protection of hardware IP core, retinal biometric stands out as the most secure, stable, distinctive (easily distinguishable even among twins), and accurate (unaffected by external factors) approach.

3.5 Retinal biometrics-based hardware security and trust

So far, we have discussed and analyzed the effectiveness of both the biometric and non-biometric-based hardware security techniques for securing an IP core design. As apparent, retinal biometrics offer more robust security than other techniques. Now, in this section, we discuss the process of generating retinal digital template (by extracting unique biometric features) from retinal image of IP vendor and its corresponding constraints for enabling security of IP core design.

3.5.1 Summary

The utilization of retinal biometric for hardware security offers a robust defense mechanism for securing IP cores against piracy and false IP ownership claim. The approach (Chaurasia and Sengupta, 2023b) ensures the resilient isolation of pirated IPs during piracy detection process. To accomplish this, it covertly integrates the genuine IP vendor's retinal biometric information into the design, producing a secure hardware IP core (soft IP core or RTL datapath) as depicted in Figure 3.1. The process of creating a secured IP core using the retinal biometric approach involves two primary modules (Chaurasia and Sengupta, 2023a):

(a) Module-1: Digital template generation and
(b) Module-2: RTL datapath generation.

The initial module is tasked with generating retinal digital template by performing the process of feature extraction. In this module, the captured retinal biometric image of a genuine IP vendor is utilized as input and pre-processed for extracting unique retinal features using the feature extraction block. Then, the output undergoes the signature generation block, which generates the digital template based on retinal biometrics. Subsequently, covert security constraints are derived from the digital template using a specific encoding method employed by the IP vendor.

The subsequent module focuses on embedding the generated covert security parameters during the phase of register allocation (of the HLS framework),

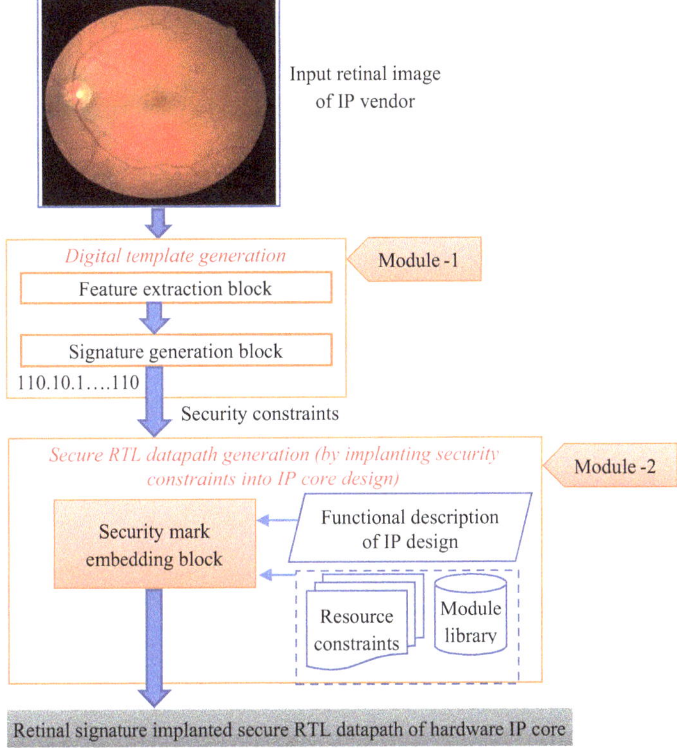

Figure 3.1 Overview of retinal biometric methodology for hardware security

resulting in the creation of the signature-embedded secured hardware IP core. In order to do so, this module accepts the following inputs:

(a) Library module (comprising parametric information corresponding to hardware components)
(b) Resource constraints (useful for scheduling the DFG of the design and are obtained from the design space exploration process)
(c) Functional description (behavioral description represented as data flow graph) of IP design
(d) The covert security constraints generated from the preceding module.

The resulting output from this block is a secured RTL datapath design of hardware IP core implanted with the retinal signature of IP vendor. Thus, enabling enhanced security against IP piracy and false IP ownership claim form an adversary that may present in untrustworthy design houses.

3.5.2 Major steps in the framework

Now, let us discuss the major steps of retinal biometric-based hardware IP security, including their input(s), functionality, and corresponding output. As shown in

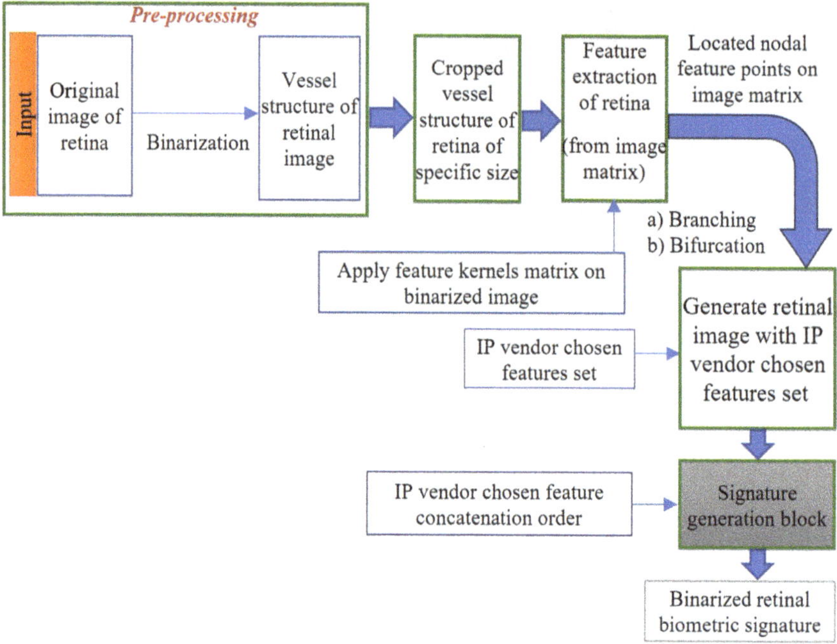

Figure 3.2 Retinal biometric signature generation for hardware security

Figure 3.2, the retinal signature generation framework comprises the following major steps: "***pre-processing***" which results in binarized vessel structure of retinal image of IP vendor, "***ROI cropping***" which results in retinal image of specific size, "***feature extraction***" which results in designated feature points on retinal image matrix, "***featured retinal image generation***" which results in retinal image with IP vendor chosen features set, "***signature generation***" which results in binarized retinal biometric generation. Binarized retinal biometric generation comprises two subprocesses including "*feature dimension computation*" and "*concatenation*" of generated binarized retinal features based on IP vendor-chosen ordering. Further, a detailed description of each of the steps of the framework for generating retinal biometric signature is provided in Figure 3.3.

3.5.3 Explanation of the hardware security framework

Extracting features and producing nodal feature points: In this section, we discuss the process of extracting features from the captured retinal image of the original IP vendor and generating nodal feature points in detail. The retina, located at the back of the eye, is digitally represented by the retinal image, which includes the retina itself, the optic nerve, and blood vessels. Initially, the retinal image of the original IP vendor is captured using a fundus camera, adhering to specific parameters: a field of view of approximately 45 degrees and a resolution of 565 × 584 pixels for capturing the retinal biometric image of the IP vendor (Aleem *et al.*, 2019). *Note: sample retinal*

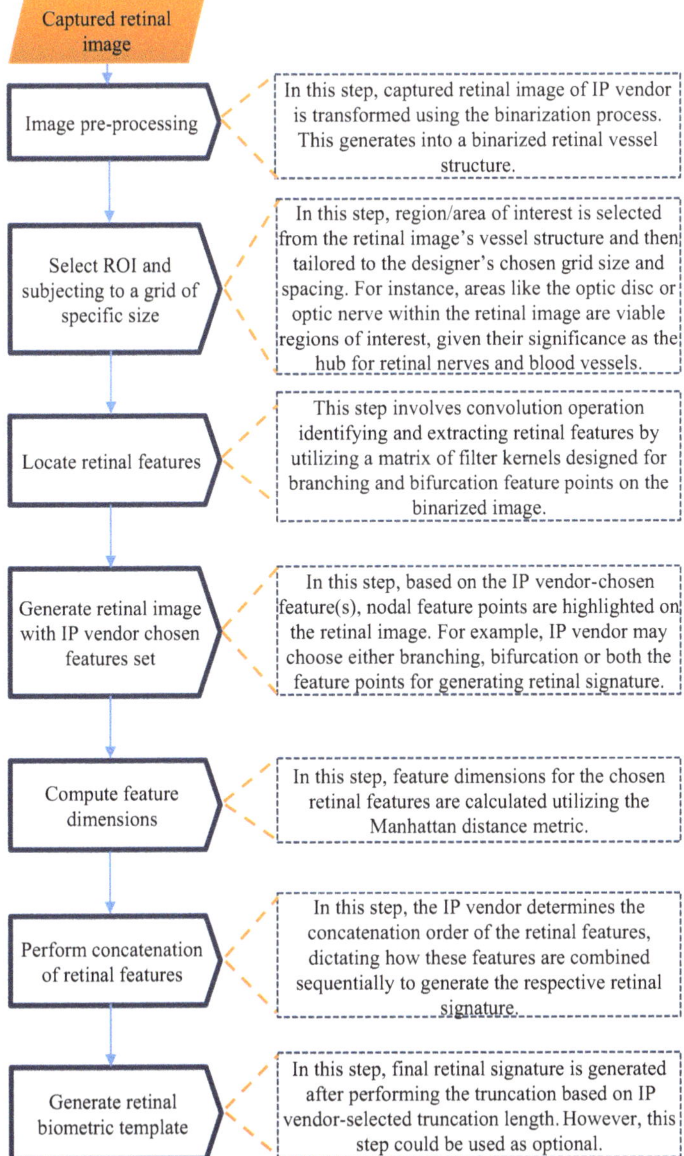

Figure 3.3 Details of the major steps of the framework for retinal biometric signature generation for hardware security

images available at Multimedia Laboratory datasets. However, it is noteworthy that advancements in technology offer user-friendly devices for retinal scanning, including options like capturing the retina using a 20D lens and a standard-quality camera. This stands in contrast to the discomfort associated with conventional retinal capturing devices that expose the eye to infrared light (Aleem *et al.*, 2019). Subsequently, the captured image undergoes a transformation into a binarized format to facilitate accurate feature extraction. The size of the captured retinal image, denoted as $p \times q$ = '565 $\times$ 584', signifies the pixel dimensions, from which the Region of Interest (ROI) is selected. This ROI comprises the optic disc/optic nerve and forms the root of the retinal blood vessels. While '565 $\times$ 584' serves as a sample dimension for the sake of demonstration, variations in the size of retinal images (either smaller or larger) can also be chosen. Subsequently, the resultant binarized retinal image is then processed with the vendor-specified grid size and spacing. This enhances the accuracy in terms of locating retinal features precisely. This process is pivotal in generating accurate feature nodal points such as bifurcation and branching, along with their corresponding coordinates. Moreover, this aids in the creation of retinal signatures for hardware security, allowing for easy regeneration of feature coordinates and dimensions from the pre-stored original retinal image, which holds specific grid size and spacing. The generation of the cropped retinal image, employing the specified grid size and spacing, from the binarized retinal image is illustrated in Figure 3.4. Here, Figure 3.4(a) and Figure 3.4(b) depict the captured retinal image and its binarized version, respectively. Meanwhile, Figure 3.4(c) and Figure 3.4(d) showcase the cropped binarized image and its depiction with the IP vendor's specified grid size and spacing.

Next, the retinal image in Figure 3.4(d) is used for extracting features and generating nodal feature points. The retinal security framework in (Chaurasia and Sengupta, 2023a) has considered branching and bifurcation points as the main feature points of the retinal image for generating binarized digital template. Vascular bifurcation and branching within the optical disk of the retina are pivotal geometric features that encapsulate essential data about vessel centerlines and widths. These points hold significance as specific junctures in the retinal blood vessel network. Now, we discuss on what basis a retinal feature point is considered as branching or bifurcation (Chaurasia and Sengupta, 2023a).

Bifurcation retinal feature: it refers to the instance where a vessel divides into two approximately equal-width vessels.

Branching retinal feature: it refers to the instance where a new vessel emerges or grows from a major vessel (with wider pixel width), resulting in a minor vessel (with smaller pixel width).

The feature extraction block identifies critical nodal points like branching and bifurcation points within the retinal image matrix. This feature extraction module requires two inputs: *generated matrix of the cropped binarized retinal image* and *the feature kernel matrix*. Sample kernel matrices representing branching and bifurcation nodal feature points are illustrated in Figure 3.5, each with a size of $m \times n = 11 \times 5$. These matrices define the dimensions for the convolution operation, enabling the accurate localization of retinal features. They consist of binary values: '0' denotes low-intensity pixels, while '255' signifies high-intensity pixels. It is essential to note that smaller kernel matrices might not effectively detect retinal features due to the wider pixel length of retinal blood vessels. The nodal feature points are established by convolving the cropped binarized retinal image matrix (as depicted in Figure 3.5(a))

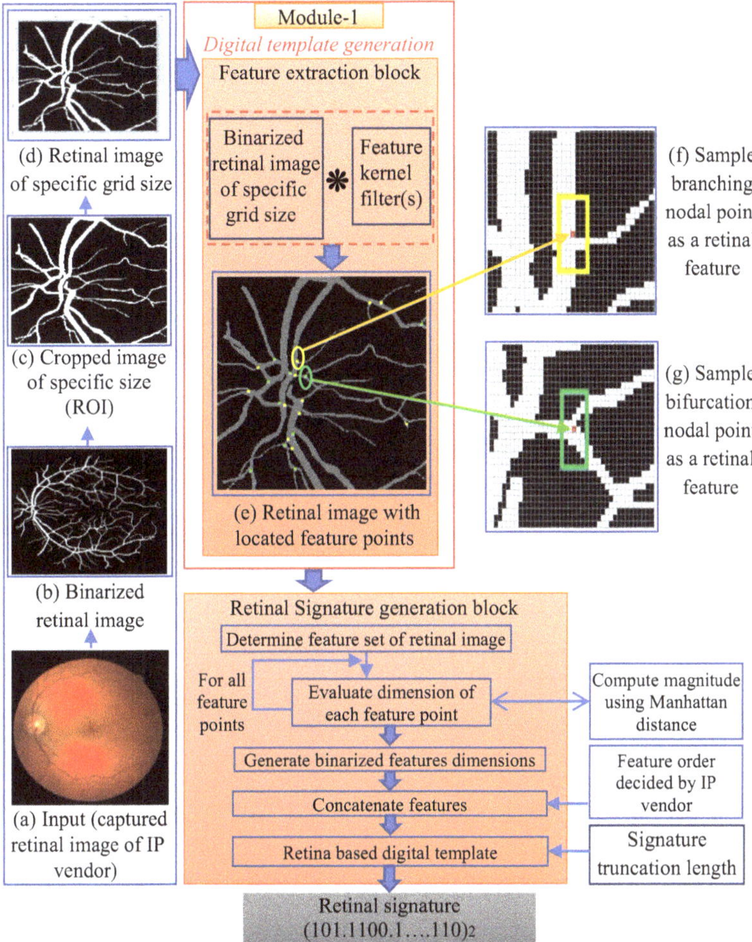

Figure 3.4 Detailed process flow of retinal biometric signature generation

with the sample kernel matrix. This convolution process automatically identifies the retinal image's feature points corresponding to the feature kernel matrix of branching and bifurcation, scanning the image from top to bottom and left to right. In Figure 3.5(b), the highlighted yellow points denote branching points, while the green points represent bifurcation points. Thus, the obtained output retinal image (image matrix) with located nodal feature points, serves as the foundation for generating a security signature.

However, in case an IP vendor intends to broaden the feature set, he/she could include a third feature type, "crossover" into the set. A crossover feature point refers to the instance where two different vessels cross each other. Further, as observed in Figure 3.5(a), the retinal image shows some discontinuities in vessel structure. This may arise due to illumination variations or loss of intensity. It necessitates a procedure involving the assembly of segments to construct complete

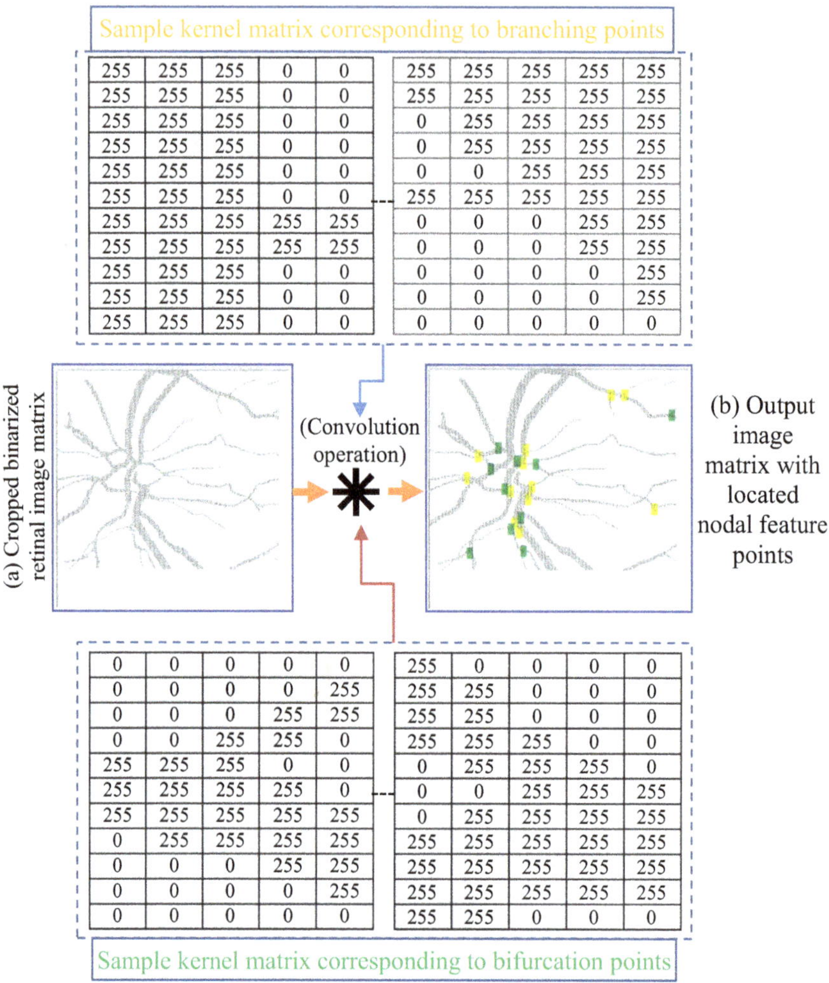

Figure 3.5 Detection of key nodal points (such as bifurcation and branching) through automated processes. Note: Convolution operation is performed either between cropped binarized retinal image and branching kernel matrix (blue arrow) or between cropped binarized retinal image and bifurcation kernel matrix (red arrow).

vessels before conducting the feature extraction and analysis (Ortega *et al.*, 2009). In the discussed retinal biometric-based hardware security approach (Chaurasia and Sengupta, 2023a), "branching" and "bifurcation" features are mainly considered, as these alone yield sufficient nodal points, resulting in a robust retinal signature. Moreover, sometimes a single crossover may be classified as two crossover points or in other words capturing the crossover may not be accurate (Bhuiyan *et al.*, 2012). This is because in crossover, one vessel overlaps the other, therefore making it challenging to locate crossover accurately and illumination variations or loss of intensity also cause inconsistency. Therefore, to avoid such inconsistencies in unique and distinctive retinal signature generation, only branching and bifurcation features are preferred to be extracted/located.

Computing feature dimensions and its corresponding signature: Once the convolution process is completed, retinal features are located. Consequently, a retinal image with located nodal feature points is generated, as presented in Figure 3.5(b). The IP vendor can choose any of the possible options for selecting the retinal features for generating retinal signature: (i) selecting bifurcation points only, (ii) choosing branching points only, or (iii) opting for both. Additionally, the IP vendor can decide the number of features, influencing the signature's size and strength. The more the nodal features selected, the larger/stronger the generated signature.

The next step involves determining the dimensions of retinal features in order to generate their corresponding binarized template. Next, the final retinal signature can be formed using any of the several possible ways of concatenating the binarized signature of features (branching/bifurcation). The concatenation process results in generating the biometric template based on IP vendor chosen retinal features based on their type. Further, depending on the size (small/big) of the target hardware IP design (to be secured), an IP vendor can truncate the length of the retinal signature. Thus, based on IP vendor chosen features set, concatenation order, and truncation length, the retinal biometric signature is generated.

For instance, if the IP vendor opts for both branching and bifurcation nodal feature points for signature generation, only the nodal points corresponding to these features are generated on the retinal image. Figure 3.5(b) illustrates these nodal feature points, with branching points shown in yellow and bifurcation points in green, across the captured retina's ROI. The convolution process results in 22 nodal feature points, comprising 10 bifurcation points and 12 branching points, as shown in Figure 3.5(a) for the captured retinal binarized image. Subsequently, the dimensions of these features are computed. Initially, the coordinates related to all these selected features, chosen by the IP vendor for signature generation, are determined. However, the determination of feature dimensions focuses on the coordinates of the center pixel within this feature matrix. For instance, as previously illustrated in Figure 3.4(f) and (g), the center coordinates of the feature matrix associated with branching and bifurcation features respectively are utilized to compute the feature dimensions by measuring Manhattan distance. The computed feature dimensions (magnitude) for all the feature points corresponding to bifurcation and branching are tabulated as per the IP vendor's selection in Table 3.1 and Table 3.2, respectively. Further, binarized feature dimensions corresponding to bifurcation and branching points are presented in Table 3.3. The retinal signature based on IP vendor chosen features set and their concatenation order is generated as follows:

(a) *Chosen number of features*: considering that the IP vendor has chosen all the extracted/located retinal features for strengthening the signature.
(b) *Concatenation Order*: The retinal features are concatenated in a specific order, such as F1 ⧺ F1 ⧺ F2 ⧺ F2 ⧺ F3... ⧺ F12. Here, 'F' denotes feature points, '⧺' signifies the concatenation operation, and bifurcation and branching feature

Table 3.1 *Determining dimensions corresponding to bifurcation feature points*

Feature #	Bifurcation feature points	Dimension (magnitude)
F1	(45,196)	201.09
F2	(78,56)	96.02
F3	(91,71)	115.42
F4	(94,86)	127.40
F5	(99,50)	110.90
F6	(119,61)	133.72
F7	(148,74)	165.46
F8	(159,66)	172.15
F9	(181,76)	196.30
F10	(183,33)	185.95

Table 3.2 *Determining dimensions corresponding to branching feature points*

Feature #	Branching feature points	Dimension (magnitude)
F1	(25,147)	149.11
F2	(25,158)	159.96
F3	(80,78)	111.73
F4	(86,40)	94.84
F5	(93,76)	120.10
F6	(108,31)	112.36
F7	(117,83)	143.45
F8	(121,66)	137.82
F9	(129,80)	151.79
F10	(139,182)	229.00
F11	(153,69)	167.83
F12	(162,73)	177.68

Table 3.3 *Binarized feature dimensions corresponding to bifurcation and branching points*

Feature #	Binarized bifurcation retinal feature	Feature #	Binarized branching retinal feature
F1	11001001.00010111000010100l	F1	10010101.0001110000101000 1111
F2	1100000.00000101000111 1011	F2	10011111.1111010111000010 1001
F3	1110011.01101011100001 01001	F3	1101111.101110101110000101
F4	1111111.01100110011001 10011	F4	1011110.110101110000101001
F5	1101110.11100110011001 10011	F5	1111000.0001100110011001101
F6	10000101.1011100001010001 1111	F6	1110000.0101110000101000 1111
F7	10100101.0111010111000010 1001	F7	10001111.0111001100110011 0011
F8	10101100.00100110011001 10011	F8	10001001.110100011110 10111
F9	11000100.0100110011001100 1101	F9	10010111.1100101000111101 0111
F10	10111001.1111001100110011 0011	F10	11100101
		F11	10100111.110101000111101 0111
		F12	10110001.101011100001 01001

points are represented using green and yellow colors, respectively. The IP vendor has the flexibility to choose among various possible concatenation orders.

(c) *Signature generation*: Post concatenating the binarized retinal features, the generated final signature corresponding to the IP vendor's retinal image (shown in Figure 3.4(a)) is as follows:

11001001.00010111000010100110010101.0001110000101000111111100000.00000
10100011110111001111.11110101110000101001111001l.0110101110000101001
1101111.10111010111000010l 1111111.0110011001100110011101111 0.11010111
0000101001 1101110.1110011001100110011—10110001.10101110000101001

(589-bit size).

However, the strength of the retinal signature to be generated can be further customized/tailored by the IP vendor through the truncation process. Thus, the variation in retinal signature can be achieved by varying cropped image sizes, chosen number of retinal features, concatenation order, and truncation length. However, once the retinal signature is established and fixed, it remains static for the embedding process into the design and cannot be altered. It is then securely stored in an encrypted format for subsequent piracy detection processes. Further, variations in cropped image sizes, minor camera tilts, or changes in resolution do not impact into process of piracy detection because the retinal biometric data is not captured again for piracy detection or IP ownership resolution.

3.6 Embedding process of the security constraints and its demonstration in a case study

So far, we have discussed the process for generating binarized retinal biometric signature from the captured retina image of the IP vendor. Now, let us discuss the embedding process of the generated retinal signature (security mark) into the design. Here, we are taking JPEG-codec IP core application as a case study for demonstrating the hardware security methodology using retinal biometric (Chaurasia and Sengupta, 2023a).

JPEG-codec hardware IPs are extensively utilized in digital imaging devices like cameras, smartphones, and several other electronic gadgets for conducting tasks related to image and video processing, including compression and decompression. Further, JPEG codec finds its utility beyond the realm of consumer electronics, including its applications in medical imaging fields such as magnetic resonance imaging (MRI) and computed tomography (CT) scans, among others. Therefore, its security becomes crucial for ensuring authentic/reliable functionality of the end system/application. The details of JPEG-codec algorithm are discussed in Sengupta *et al.* (2018).

Now, we discuss behavioral/functional description of JPEG-codec application framework, followed by scheduling the design and subsequently generating its corresponding register allocation information. *Note: The retinal biometric-based hardware security methodology exploits the register allocation phase of HLS for covertly embedding the security constraints in order to enable security against external threats (as discussed earlier in Section 3.1.2).*

Behavioral description of JPEF-codec: We first explain the transfer function that governs the relationship between input and output, followed by deducing its data flow graph from its transfer function. Transfer function governs the computation of the first pixel of the compressed image using the JPEG compressor through the following steps (Chaurasia and Sengupta, 2023a):

Step-1: Visualize the input image into a matrix comprising pixel intensity values and perform slicing to generate sub-matrices of size 8×8, for operation with discrete cosine transform (DCT) function of JPEG compressor.

Step-2: Transform each of the generated sub-matrices using the following 2D-DCT function:

$$T = (M_c \times M_s) \times M_c{}' \tag{3.1}$$

Here, in this context, 'M_c' represents the 2D-DCT coefficient matrix (e4 represents the first-row elements of the matrix), 'M_s' denotes a submatrix of the input image (represented using p11 to p81 as image matrix elements), '$M_c{}'$' signifies the transpose matrix corresponding to matrix 'M_c', and 'T' denotes the resulting transformed matrix.

The details of the input image matrix and the DCT coefficient matrix are explained further in Sengupta *et al.* (2018). Elements of '$M_c \times M_s$' matrix multiplication ($t11$, $t12$, …, $t88$) indicates column wise transformed elements of input block. Further, elements of 'T' matrix ($T11$, $T12$, …, $T88$) indicate both row and column-wise transformed elements of the input block.

Step-3: Compute the first pixel of the compressed image using the following function:

$$\begin{aligned} T11 = {} & (e4 * t11) + (e4 * t12) + (e4 * t13) + (e4 * t14) \\ & + (e4 * t15) + (e4 * t16) + (e4 * t17) + (e4 * t18) \end{aligned} \tag{3.2}$$

Where the first operand represents the coefficient value of the first column of matrix $M_c{}'$, and the second operands represent elements from the first row of matrix $(M_c \times M_s)$ and $t11$ represents the output of the first micro-unit (IP-1) and is computed as follows:

$$\begin{aligned} t11 = {} & (e4 * p11) + (e4 * p21) + (e4 * p31) + (e4 * p41) \\ & + (e4 * p51) + (e4 * p61) + (e4 * p71) + (e4 * p81) \end{aligned} \tag{3.3}$$

where the first operand represents the coefficient value of the first row of the coefficient matrix M_c, while the second operand represents elements from the first column of the matrix M_s.

Step-4: Generate a compressed output pixel ($T11'$) by employing a quantization matrix on every submatrix of size 8 × 8 that has undergone DCT transformation.

This involves multiplying the first pixel of the DCT-transformed matrix ($T11$) with the quantization coefficient 'Q_x'.

Next, the transfer function in (3.2) is now transformed into DFG. The DFG corresponding to JPEG-codec is shown in Figure 3.6. This includes the computation of eight micro-IPs. The data flow graph corresponding to one micro-IP is shown in Figure 3.7. The JPEG hardware IP design computing compressed image shown in Figure 3.6 consists of eight micro-IPs (IP-1 to IP-8). Each micro-IP

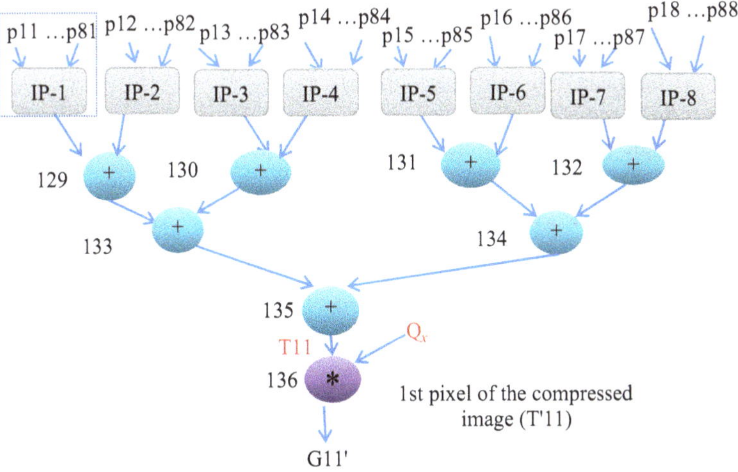

Figure 3.6 Data flow graph of JPEG-codec framework comprising 8 micro-IPs

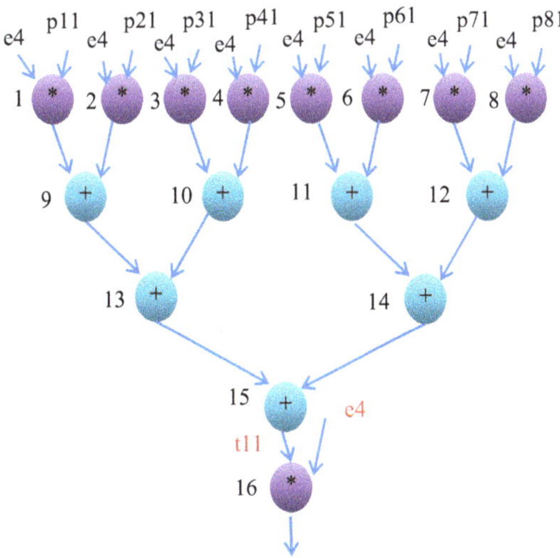

Figure 3.7 Sub-graph for output computation of first micro-IP module (IP-1)

encompasses 16 operations (for instance, IP-11 contains 9 (*) and 7 (+)). Hence, the design of the JPEG-codec IP involves a total of 136 operations to determine the output pixel for the compressed image.

Scheduling the DFG of the design: Next, the design methodology involves the process of scheduling the operations, allocating hardware resources, and establishing their bindings. Scheduling transforms the DFG into a control flow graph (comprising required control steps detail). The scheduling requires the DFG of the design, a scheduling algorithm, and resource constraints as its input. The DFG depicted in Figure 3.6, a LIST scheduling algorithm, and '*3 adders and 3 multipliers*' as IP vendor specified resources are used as inputs in the scheduling process. The scheduled design necessitates 30 control steps to schedule 136 operations of the design (Chaurasia and Sengupta, 2023a).

Generating register allocation information: Subsequently, following the scheduling of the DFG corresponding to JPEG-codec design, a register allocation information is extracted as shown in Figure 3.8. It encompasses the following details about the design such as:

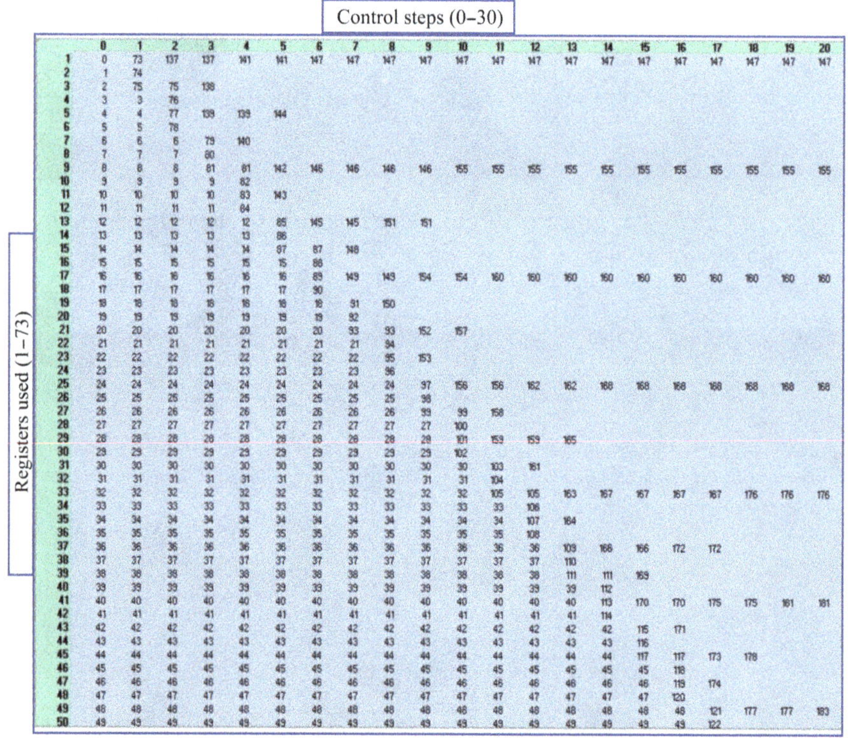

Control steps (0–30)

Registers used (1–73)

	0	1	2	3	4	5	6	7	8	9	10	11	12	13	14	15	16	17	18	19	20
1	0	73	137	137	141	141	147	147	147	147	147	147	147	147	147	147	147	147	147	147	147
2	1	74																			
3	2	75	75	138																	
4	3	3	76																		
5	4	4	77	139	139	144															
6	5	5	78																		
7	6	6	6	79	140																
8	7	7	7	80																	
9	8	8	8	81	81	142	146	146	146	146	155	155	155	155	155	155	155	155	155	155	155
10	9	9	9	9	82																
11	10	10	10	10	83	143															
12	11	11	11	11	84																
13	12	12	12	12	12	85	145	145	151	151											
14	13	13	13	13	13	86															
15	14	14	14	14	14	87	87	148													
16	15	15	15	15	15	15	88														
17	16	16	16	16	16	16	89	149	143	154	154	160	160	160	160	160	160	160	160	160	160
18	17	17	17	17	17	17	90														
19	18	18	18	18	18	18	18	91	150												
20	19	19	19	19	19	19	19	92													
21	20	20	20	20	20	20	20	93	93	152	157										
22	21	21	21	21	21	21	21	21	94												
23	22	22	22	22	22	22	22	22	95	153											
24	23	23	23	23	23	23	23	23	96												
25	24	24	24	24	24	24	24	24	24	97	156	156	162	162	168	168	168	168	168	168	168
26	25	25	25	25	25	25	25	25	25	98											
27	26	26	26	26	26	26	26	26	26	99	99	158									
28	27	27	27	27	27	27	27	27	27	27	100										
29	28	28	28	28	28	28	28	28	28	28	101	159	159	165							
30	29	29	29	29	29	29	29	29	29	29	102										
31	30	30	30	30	30	30	30	30	30	30	30	103	161								
32	31	31	31	31	31	31	31	31	31	31	31	104									
33	32	32	32	32	32	32	32	32	32	32	32	105	105	163	167	167	167	167	176	176	176
34	33	33	33	33	33	33	33	33	33	33	33	33	106								
35	34	34	34	34	34	34	34	34	34	34	34	34	107	164							
36	35	35	35	35	35	35	35	35	35	35	35	35	108								
37	36	36	36	36	36	36	36	36	36	36	36	36	36	109	166	166	172	172			
38	37	37	37	37	37	37	37	37	37	37	37	37	37	110							
39	38	38	38	38	38	38	38	38	38	38	38	38	38	111	111	169					
40	39	39	39	39	39	39	39	39	39	39	39	39	39	39	112						
41	40	40	40	40	40	40	40	40	40	40	40	40	40	40	113	170	170	175	175	181	181
42	41	41	41	41	41	41	41	41	41	41	41	41	41	41	114						
43	42	42	42	42	42	42	42	42	42	42	42	42	42	42	42	115	171				
44	43	43	43	43	43	43	43	43	43	43	43	43	43	43	43	116					
45	44	44	44	44	44	44	44	44	44	44	44	44	44	44	44	117	117	173	178		
46	45	45	45	45	45	45	45	45	45	45	45	45	45	45	45	45	118				
47	46	46	46	46	46	46	46	46	46	46	46	46	46	46	46	46	119	174			
48	47	47	47	47	47	47	47	47	47	47	47	47	47	47	47	47	120				
49	48	48	48	48	48	48	48	48	48	48	48	48	48	48	48	48	48	121	177	177	183
50	49	49	49	49	49	49	49	49	49	49	49	49	49	49	49	49	49	122			

Figure 3.8 Register allocation information of JPEG IP. Note: For the sake of brevity, only information for 50 registers out of the total 73 and 21 control steps out of the total 31 have been provided.

(a) Required storage variables corresponding to primary and intermediate input/output values.
(b) Required registers for accommodating/storing the storage variables.
(c) Required control steps (necessary to generate the output pixel value of the compressed image through the scheduled design using the hardware resources ‘*3 adders and 3 multipliers*’).

As depicted in Figure 3.8, the register allocation indicates the utilization of 73 distinct registers, aligning with 209 storage variables (0–208), and the requirement of 31 control steps (0–30) required for computing the first pixel output of the compressed image.

Generating secret hardware security constraints of retinal signature: So far, we have discussed the process of deriving register allocation information corresponding to JPEG-codec IP, followed by scheduling its DFG, allocation, and binding of hardware resources. Now, we discuss how the security constraints for hardware security are derived corresponding to the generated signature from IP vendor retinal biometric. These hardware security constraints are used for covertly implanting into the design by performing local alteration of storage variables among the design registers uniquely.

The process of generating the security constraints for hardware security involves accepting the following specific inputs:

a) Retinal biometric digital signature to be embedded into the design (of IP vendor chosen strength/size).
b) Scheduled DFG, utilized for extracting storage variables information.
c) IP vendor-specified encoding rules for generating the security constraints corresponding to the signature bits.

The generated retinal signature (in Section 3.5.3) is of size 589-bit magnitude (comprises 306 1s, 262 0s, and 21 binary points ‘.’), derived directly from the 22 nodal points, which symbolize bifurcation and branching point from the retinal image, shown in Figure 3.5(b).

Assuming the following encoding rules specified by the IP vendor for generating the security constraints:

Corresponding to encoding of signature bit ‘0’: formulate a security constraint by pairing the ‘<even-even>’ storage variable in sorted ascending order (starting from 0) and repeat the process for all the remaining signature bits of 0s.

Corresponding to encoding of signature bit ‘1’: formulate a security constraint by pairing the ‘<odd-odd>’ storage variable in sorted ascending order (starting from 1) and repeat the process for all the remaining signature bits of 1s

Corresponding to encoding of signature bit ‘.’: formulate a constraint by pairing the ‘<zero-integer>’ storage variable in sorted ascending order and repeat the process for all the remaining signature bits of binary points.

Thus, the security constraints corresponding to signature bits ‘0’, ‘‘1’, and binary points are generated using the above encoding rules as presented in Table 3.4, Table 3.5, and Table 3.6, respectively. As evident, 262 security

Table 3.4 Retinal signature-based secret hardware security constrains corresponding to signature bit '0'

<0,2>	<–,–>	<–,–>	<–,–>
<0,4>	<–,–>	<–,–>	<–,–>
<0,6>	<–,–>	<–,–>	<–,–>
<0,8>	<0,208>	<2,208>	<4,98>
<0,10>	<2,4>	<4,6>	<4,100>
<0,12>	<2,6>	<4,8>	<4,102>
<0,14>	<2,8>	<4,10>	<4,104>
<0,16>	<2,10>	<4,12>	<4,106>
<0,18>	<2,12>	<4,14>	<4,108>
<0,20>	<2,14>	<4,16>	<4,110>
<0,22>	<2,16>	<4,18>	<4,112>
<0,24>	<2,18>	<4,20>	<4,114>

Table 3.5 Retinal signature-based secret hardware security constraints corresponding to signature bit '1'

<1,3>	<–,–>	<–,–>	<–,–>
<1,5>	<–,–>	<–,–>	<–,–>
<1,7>	<–,–>	<–,–>	<–,–>
<1,9>	<1,207>	<3,207>	<5,191>
<1,11>	<3,5>	<5,7>	<5,193>
<1,13>	<3,7>	<5,9>	<5,195>
<1,15>	<3,9>	<5,11>	<5,197>
<1,17>	<3,11>	<5,13>	<5,199>
<1,19>	<3,13>	<5,15>	<5,201>
<1,21>	<3,15>	<5,17>	<5,203>
<1,23>	<3,17>	<5,19>	<5,205>
<1,25>	<3,19>	<5,21>	<5,207>

Table 3.6 Encoded hardware security constraints corresponding to binary points of signature bit corresponding to retinal signature

<0,1>	<0,13>	<0,25>	<0,37>
<0,3>	<0,15>	<0,27>	<0,39>
<0,5>	<0,17>	<0,29>	<0,41>
<0,7>	<0,19>	<0,31>	<0,19>
<0,9>	<0,21>	<0,33>	<0,21>
<0,11>	<0,23>	<0,35>	<0,23>

constraints correspond to signature bit '0', 306 security constraints correspond to signature bit '1', and 21 security constraints correspond to binary points generated.

Generating retinal security constraints embedded secured design: Next, these security constraints are covertly embedded into the design. In order to do so, the following constraints embedding rules are followed (Chaurasia and Sengupta, 2023a, 2023b):

If two storage variables are alive or are executing in the same control step, they cannot share the same register. However, if they operate in different control steps, they can use the same registers. The resulting register allocation framework, showcased in Table 3.7, presents the locally altered position of storage variables due to embedding the signature constraints. Here for example (in Table 3.7), consider the storage variable pair (0–2). Since they are already assigned to different registers, register R1 and register R3, there is no conflict. However, conflict arises for the storage variable pairs (0–196), (0–202), and (0–208). Consequently, storage variables 196, 202, and 208 cannot share the same register R1 with variable 0. Therefore, their positions are locally altered and placed in register R2. *Note*: *In case if placing them in register R2 again causes conflict, then they are checked against other available registers. This process continues until all the conflicts are satisfied within the available design registers. However, in case it is not possible to satisfy all the storage variable conflicts among the available registers, then a new register is to be allocated to accommodate them.* Likewise, a conflict arises for the pair (2–138), preventing them from being accommodated with storage variables 2 in register R3. Therefore, it is also altered to register R4. In this table, storage variables highlighted in red signify updated positions, while those marked in blue indicate the previous positions of the storage variables. The register allocation framework reveals that embedding all secret security constraints for the retinal signature (image_1) into the JPEC-codec design does not necessitate any additional registers. Subsequently, a secured JPEG-codec datapath incorporating retinal biometric implants can be designed post synthesis using classical HLS design flow

Table 3.7 Register allocation information of retinal biometric-driven encoded hardware security constraints embedded design

CS	Pre-embedding							Post-embedding						
	R1	**R2**	**R3**	**R4**	**R5**	**R6**	**R7**	**R1**	**R2**	**R3**	**R4**	**R5**	**R6**	**R7**
1	0	1	2	3	4	5	6	0	1	2	3	4	5	6
2	73	74	75	3	4	5	6	73	74	75	3	4	5	6
3	137	–	75	76	77	78	6	137	–	75	76	77	78	6
4	141	–	138	–	139	–	79	141	–	–	138	139	–	79
5	141	–	–	–	139	–	140	141	–	–	–	139	–	140
23	196	–	–	–	144	–	–	–	196	–	–	144	–	–
24	196	–	–	–	–	–	–	–	196	–	–	–	–	–
25	202	–	–	–	–	–	–	–	202	–	–	–	–	–
26	202	–	–	–	–	–	–	–	202	–	–	–	–	–
27	202	–	–	–	–	–	–	–	202	–	–	–	–	–
28	202	–	–	–	–	–	–	–	202	–	–	–	–	–
29	207	–	–	–	–	–	–	207	–	–	–	–	–	–
30	208	–	–	–	–	–	–	–	208	–	–	–	–	–

(Sengupta and Sedaghat, 2011; Chaurasia *et al.*, 2023; Rathor *et al.*, 2023; Sengupta and Bhadauria, 2016; Rajendran *et al.*, 2016; Koushanfar *et al.*, 2005; Chaurasia and Sengupta, 2023a, 2023b; Chaurasia *et al.*, 2022; Sengupta and Chaurasia, 2023).

3.7 Detection of pirated IPs and ownership validation using retinal biometrics

The retinal biometric-based hardware design methodology (Chaurasia and Sengupta, 2023a) involves capturing the retina biometric of an IP vendor just once. The corresponding image (retinal biometric), along with grid specifications, is securely stored for later use in detecting IP piracy by the system integrator. This methodology does not necessitate recapturing the retina biometric of IP vendor for the piracy detection process. Instead, the pre-stored retinal biometric image (safely stored in encrypted format) is employed to recreate the retinal signature and its associated hardware security parameters, effectively identifying/isolating pirated IP designs. All characteristics and measurements of the retina biometric can be precisely identified and calculated from the pre-stored retinal image. Since the biometric image is only captured once, factors like eye vascular damage, fatigue, minor camera angle changes, resolution variations, or differences in cropping size do not affect the IP piracy detection process. The pre-stored retinal image, along with its grid specifications, proves adequate for IP piracy detection. Furthermore, in case if an adversary manages to compromise/forge the stored retinal image/retinal signature (for evading IP piracy detection), the methodology still shows resiliency against adversarial attacks to exactly regenerate the authentic hardware security constraints due to encryption involved in the retinal signature. Further, from the perspective of forging the pre-stored retinal image for regenerating the security constraints, it becomes highly challenging for an adversary as it involves several additional security layers/parameters specified by the original IP vendor. This significantly complicates the adversarial attempt to decode security parameters in order to evade piracy detection.

During the piracy detection process, an SoC integrator needs to perform the following steps:

(a) Generate authentic hardware security constraints corresponding to retinal biometrics of original IP vendor (pre-stored retinal image with IP vendor-specified grid size and spacing).
(b) Extract the security constraints information from the authentic/pirated IP (design under test).
(c) Perform bit-by-bit matching of the authentic hardware security constraints generated corresponding to retinal signature of genuine IP vendor against the extracted security constraints from the authentic/pirated IP (by analyzing register allocation information from backtracking the design up to the level of register allocation phase).
(d) If the matching results in a true value, then the IP is considered authentic else a pirated version.

Furthermore, the covertly embedded encoded security constraints of the original IP vendor generated using retinal biometric methodology serve as a highly authentic, naturally unique secret mark for the nullification of false ownership during litigation in IP court. This is because the employed security parameters are non-decodable easily from an adversarial perspective in order to exactly regenerate the retinal security constraints successfully. Further, due to the inherit uniqueness of an individual's retina, an adversary cannot successfully claim/prove the retinal biometric information of the original IP vendor as their biometric. Thus, the retinal biometric-based hardware security methodology offers robust security against false IP ownership claim. It is important to note that only the genuine IP vendor's retinal signature matches the embedded digital signature due to the distinctiveness of each individual's retinal vessel structure. As a result, it becomes virtually impossible for an adversary to replicate the identical retinal characteristics of the genuine IP vendor.

3.8 Analysis and discussion of case studies

So far, we have discussed HLS-based methodology for designing hardware IPs (demonstrated using JPEG-codec application framework as a case study) and how IP vendor's retinal biometric can be exploited for covertly integrating into the design in the form of encoded hardware security constraints during register allocation phase. This therefore enables hardware IP security in terms of detective control against piracy and protecting IP ownership rights of genuine IP owner by employing multi-layer security in order to hinder exact signature regeneration from an adversarial perspective.

Now, firstly we discuss the security properties of the retinal biometric-based hardware IP security methodology (Chaurasia and Sengupta, 2023a). Subsequently, we discuss the security offered by retinal biometrics (Chaurasia and Sengupta, 2023a, 2023b) and analyze its robustness against other biometric approaches such as fingerprint (Sengupta and Rathor, 2020), facial biometrics (Sengupta and Chaurasia, 2022; Chaurasia and Sengupta, 2022), and non-biometric approach based on digital signature (Sengupta *et al.*, 2019).

3.8.1 Security properties/attributes of the retinal biometric-based secure IP design methodology

The HLS-based secure IP design methodology using retinal biometric offers several security properties (involving multi-layer security attributes specified by the genuine IP vendor) as follows:

(a) An adversary cannot replicate the exact retinal signature due to numerous critical security parameters essential for signature generation. These parameters, completely non-decodable to any potential attacker, include:
    1. Selection of the ROI in the captured image of retina by the IP vendor.
    2. Technical specifications set by the IP designer regarding grid size and spacing of the retina biometric image.

3. The size of the kernel matrix for capturing retinal features.
4. Orientation specifics of kernel matrices based on the retinal features decided by the IP vendor.
5. Type and quantity of nodal feature points employed for generating the signature.
6. Name designation protocol and sequencing of retinal attributes (potentially aligned with the convolution process or specific to the IP vendor).
7. Locations/co-ordinates of retinal nodal feature points.
8. Determination of truncation length for the generated signature, a decision made by the IP vendor before encoding it into covert constraints for embedding into the design.

(b) The efficacy of hardware security using retina biometric has been assessed using False Accept Rate (FAR) and False Reject Rate (FRR) criteria, in the literature. In the retinal biometric-based methodology, the FAR is 0% for an adversary, signifying that unauthorized authentication is entirely prevented. Additionally, the FRR is 0% for the authentic IP vendor. This outcome is attributed to the fact that even if an adversary gains access to a pre-stored retinal image/retinal signature, they are incapable of reproducing the exact security constraints. This inability stems from the fact that there exist several non-decodable security parameters enumerated earlier in (a)1 to 9.

(c) Even in the unlikely event that an adversary manages to forge the exact signature, the generation of secret constraints becomes unattainable due to several critical details that remain non-decodable from any potential attacker:
1. Due to storage of retinal signature in an encrypted format, an attacker needs to decrypt the signature to derive the security constraints.
2. The truncation length of the utilized retinal signature, specific to the original IP vendor, is crucial for deriving the final retinal signature, which is essential for extracting covert hardware security constraints.
3. The specific encoding method specified by the genuine IP vendor is utilized for generating constraints for hardware security corresponding to the strength of the retinal signature.
4. The ordering protocol of the storage variables within the design is non-decodable to an adversary. This protocol, whether sorted in ascending or descending order, organized according to control steps in scheduling, alternate arrangements based on functional units, among other possibilities, contributes to creating storage variable pairs essential for the security constraints.
5. The methodology governing retinal signature generation encompasses security facets including the type of retinal features utilized, the number of nodal points generated, the order of feature concatenation, feature dimensions, among other factors, all exclusive knowledge retained solely by the original IP vendor. As a result, potential adversary face significant challenges in circumventing detection of IP piracy. This makes the retina biometric method highly resilient for safeguarding hardware IP cores, even if an adversary manages to compromise retinal image.

(d) The adversary faces an impassable hurdle in finding an exact match between the covert security constraints of the regenerated signature with the extracted register allocation information of the target design under test, during false IP ownership proof.
(e) The retinal biometric method for fortifying the hardware IP core operates autonomously for signature generation, devoid of reliance on any external keys. Consequently, it remains impervious to attacks exploiting keys.
(f) The retinal biometric methodology boasts a more resilient generation of covert security constraints owing to the inclusion of a greater number of encoding digits within the generated retinal signature in comparison to facial and fingerprint biometrics, which typically comprise only two-digit encoding.
(g) Furthermore, leveraging the retinal structures from both eyes pertaining to the genuine IP vendor can significantly enhance the generation of a more robust retinal signature template. This utilization allows for an even stronger and more intricate security measure when safeguarding large size hardware IP cores such as JPEG-codec.
(h) The retinal biometric signature indeed outperforms embedding a random secret key within the design in terms of security. If a random signature were compromised, an adversary could manipulate counterfeit IPs by leveraging this leaked information, potentially evading IP piracy detection. However, the retinal biometric signature operates uniquely, abstaining from storing the key, retinal signature, or hardware security constraints. Therefore, even if the retinal image were compromised, it would not breach the security concerning IP piracy evasion. This heightened security emanates from several intricate security layers discussed earlier.

3.8.2 Analyzing the strength of achieved security robustness

The strength of security achieved by the retinal biometric is assessed through metrics like probability of coincidence (X_c) and tamper tolerance (Chaurasia and Sengupta, 2023a; Sengupta and Rathor, 2020). 'X_c' quantifies the likelihood of coincidentally detecting the genuine retinal security constraints within an unsecured hardware design. Thus, a lower X_c value is preferable, signifying higher security strength and proof of authorship (or lower probability of false positive) in terms of providing definitive distinction between authentic and pirated designs. Additionally, a lower X_c value ensures robust security, enhancing the strength of digital evidence, which is measured by the following metric (Chaurasia and Sengupta, 2023a; Sengupta and Rathor, 2020; Koushanfar *et al.*, 2005):

$$X_c = \left(1 - \frac{1}{Rn}\right)^{S_c} \tag{3.4}$$

Here, 'Rn' represents the number of registers needed to accommodate all storage variables in the baseline IP design, while 'S_c' denotes the signature strength of the retinal biometric. Figure 3.9 demonstrates the X_c value for various retinal signature sizes corresponding to different numbers of retinal features for a particular retina (image_1). Additionally, Figure 3.10 presents the X_c values for different

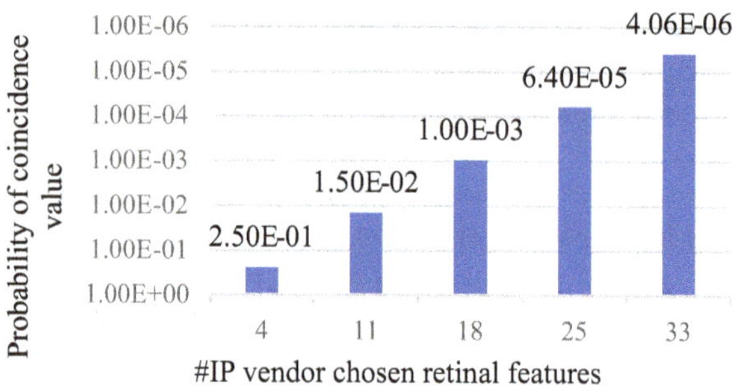

Figure 3.9 Exhibiting variation X_c concerning different size of retinal signature corresponding to a particular retina sample image

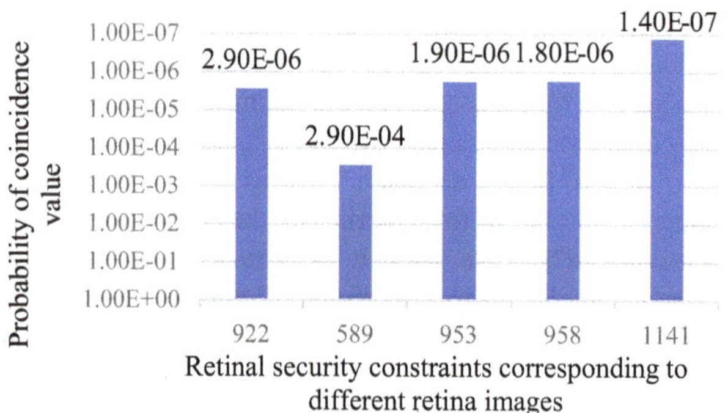

Figure 3.10 Exhibiting variation X_c concerning retinal signature corresponding to different retina sample image

retina sample images. It is apparent from Figures 3.9 and 3.10 that a larger retinal signature size (enabling the IP vendor to create numerous security constraints) leads to a lower X_c value, and conversely, a smaller signature size results in a higher X_c value. Therefore, the methodology is capable of generating a greater number of security constraints and is more suitable for enhancing the security strength against IP piracy and false IP ownership. Further, Figure 3.11 presents the comparison of X_c among the different hardware security approaches. As apparent, the retinal biometric-based hardware security methodology is capable of offering much lower value of X_c, which is desirable.

Next, we discuss the security of a methodology in terms of tamper tolerance. Tamper tolerance serves as a measure of a design's security strength against brute force attack (tampering), reflecting its robustness. Greater tamper tolerance signifies the inability of an adversary to guess the precise retinal signature through brute force attack with the intension of tampering/removal of the embedded signature constraints. This impedes an adversary from evading the piracy detection

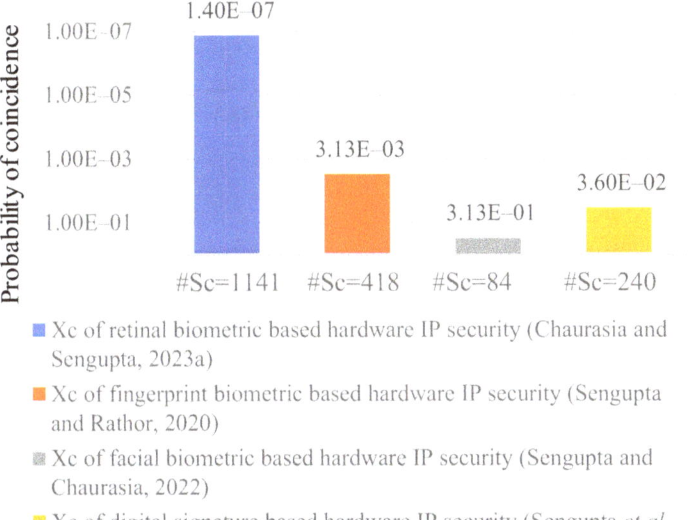

Figure 3.11 Comparison of X_c among different hardware security methodologies

Table 3.8 Variations in T_R corresponding to different sizes of retina signature

# Retinal features	# Constraints (S_c)	T_R
33	900	$2.56e^{429}$
25	700	$9.65e^{333}$
18	500	$3.63e^{238}$
11	300	$1.36e^{143}$
4	100	$5.15e^{47}$

process by inserting the replicated signature into its pirated version. The evaluation of tamper tolerance (T_R) or resiliency against tampering attack relies on a specific metric (Chaurasia and Sengupta, 2023a; Sengupta and Rathor, 2020):

$$T_R = (S_b)^{S_c} \tag{3.5}$$

Here, 'S_b' denotes the encoded signature variable types (0s, 1s, and the binary point), while 'S_c' indicates the strength of the signature for embedding. Table 3.8 displays the tamper tolerance provided by the retinal biometric for varying signature size. Additionally, Table 3.9 presents tamper tolerance for different retina sample images. It is apparent from both tables that a larger retinal signature size (enabling the IP vendor to create numerous security constraints) leads to a higher

Table 3.9 Variations in T_R concerning retinal signature corresponding to different retina sample image

# Retinal images	# Constraints (S_c)	T_R
Image_1	922	$8.05e^{439}$
Image_2	589	$1.05e^{281}$
Image_3	953	$4.97e^{454}$
Image_4	958	$1.20e^{457}$
Image_5	1141	$2.48e^{544}$

Table 3.10 Comparison of T_R among different hardware security approaches

Retinal biometric-based hardware IP security (Chaurasia and Sengupta, 2023a)		**Fingerprint biometric-based hardware IP security (Sengupta and Rathor, 2020)**		**Facial biometric-based hardware IP security (Sengupta and Chaurasia, 2022)**		**Digital signature-based hardware IP security (Sengupta *et al.*, 2019)**	
S_c	T_R	S_c	T_R	S_c	T_R	S_c	T_R
922	$8.05e^{439}$	526	$9.24e^{250}$	75	$6.08e^{35}$	15	$1.43e^{7}$
589	$1.05e^{281}$	350	$9.8e^{166}$	80	$1.47e^{38}$	30	$2.05e^{14}$
953	$4.97e^{454}$	538	$4.91e^{256}$	81	$4.43e^{38}$	60	$4.23e^{28}$
958	$1.20e^{457}$	555	$6.34e^{264}$	83	$3.99e^{39}$	120	$1.79e^{57}$
1141	$2.48e^{544}$	418	$2.73e^{199}$	84	$1.19e^{40}$	240	$3.22e^{114}$

T_R value, and conversely, a smaller signature size results in a lower T_R value. Therefore, the methodology capable of generating and embedding a greater number of security constraints is more suitable for enhancing the security strength in terms of robustness against tampering attacks using brute-force attacks. Further, Table 3.10 presents the comparison of T_R among the different hardware security approaches such as retinal biometrics, fingerprint biometrics, facial biometrics, and digital signature. As apparent retinal biometric-based methodology (Chaurasia and Sengupta, 2023a) surpasses the related approaches (Sengupta and Rathor, 2020; Sengupta and Chaurasia, 2022; Chaurasia and Sengupta, 2022; Sengupta *et al.*, 2019) in terms of robustness against tampering attack.

3.8.3 Analysis of the design cost

The impact of the design cost for enabling the robust security of hardware IP (JPEG-codec) is analyzed using the following normalized function (Chaurasia and Sengupta, 2023a; Sengupta and Rathor, 2020):

$$C_f(R_{\text{conf}}) = y_1 \frac{L_d}{L_{\max}} + y_2 \frac{F_d}{F_{\max}} \tag{3.6}$$

where 'R_{conf}' signifies the IP vendor selected resource configuration, 'F_d' and 'L_d' stand for area and latency of the design (infused with security mark), respectively. Additionally, 'F_{max}' and 'L_{max}' denote the maximum area and latency of the design. 'y_1' and 'y_2' serve as weighing factors used to normalize the impact of both the parameters in the cost function. The evaluation utilizes the 15-nm NanGate library. The design cost analysis corresponding to generating secure JPEG-codec hardware IP using retinal biometrics is presented in Table 3.11. It demonstrates the design cost of the JPEG-codec hardware IP core before and after embedding the signature for various strengths corresponding to a particular retina sample image (Image_1). Notably, no design overhead is observed across different retinal signature strengths. This absence of overhead is due to the non-requirement of additional registers during the embedding process of all hardware security constraints into the JPEG-codec design. Furthermore, Table 3.12 showcases the design cost for

Table 3.11 Design cost analysis corresponding to embedding the security constraints of varying size into the design (corresponding to a particular retina image)

S_c	# of registers in retinal signature implanted design (pre embedded, post embedded)	Design cost of baseline	Design cost of retinal signature implanted design	% Cost overhead
100 bits	73, 73	0.214	0.214	0.0%
300 bits	73, 73	0.214	0.214	0.0%
500 bits	73, 73	0.214	0.214	0.0%
700 bits	73, 73	0.214	0.214	0.0%
900 bits	73, 73	0.214	0.214	0.0%

Table 3.12 Design cost analysis corresponding to embedding the security constraints of varying size into the design (corresponding to different retina sample images)

Retinal images	# of registers in baseline design and retinal signature implanted design (pre embedded, post embedded)	Design cost of baseline	Design cost of retinal signature implanted design	% Cost overhead
Image_1	73, 73	0.214	0.214	0.0%
Image_2	73, 73	0.214	0.214	0.0%
Image_3	73, 73	0.214	0.214	0.0%
Image_4	73, 73	0.214	0.214	0.0%
Image_5	73, 73	0.214	0.214	0.0%

different retina sample images (Image_1 to Image_5) post-embedding security constraints. In this case also, no design overhead is incurred following the inclusion of security constraints for various retinal biometric images. Consequently, the approach using retinal biometric demonstrates enhanced security against IP piracy and fraudulent IP ownership claim compared to other hardware security approaches, all without incurring any design cost overhead.

3.9 Conclusion

This chapter discussed an HLS-based secure hardware IP design methodology using retinal biometric-based security (Chaurasia and Sengupta, 2023a, 2023b). Among the other biometric and non-biometric approaches, retinal biometric approach offers robust security of data intensive applications against the threat of IP piracy and fraudulent IP ownership claim. Specifically, it presented a secured JPEG-codec hardware IP core embedding a distinctive retinal signature from the authentic IP vendor, achieving heightened security without incurring any design overhead. This retinal biometric approach guarantees the smooth and strong identification of pirated design versions before their integration into System-on-Chips (SoCs) for Consumer Electronics (CE) systems. Ultimately, this methodology protects the IP rights of the genuine IP vendor against false IP ownership claim and enables seamless detective control against pirated IP versions during SoC integration level. Thus, it ensures reliable functionality of the end systems comprising only the authentic reusable IPs (pirated versions are isolated by piracy detective control mechanism), thereby ensuring the root of trust from the perspective of end consumers as well.

At the end of this chapter, a reader gains the knowledge about the following:

- End-to-end description and demonstration of HLS-based hardware IP design methodology and its different phases.
- Importance of HLS for integrating the security into the design covertly.
- Significance of retinal biometrics compared to other hardware security techniques like fingerprint and facial biometrics.
- Basic ideology and consequences of hardware threats of IP piracy and false IP ownership claim.
- How a robust detective control against IP piracy is established using retinal biometric? Further, how a genuine IP vendor can seamlessly prove his/her ownership in case of conflict resolution in the IP court?
- A retinal biometric-based secured design flow of JPEG-codec hardware.

3.10 Questions and exercise

1. Discuss the need to design secure hardware IPs.
2. Discuss the advantages of using biometric-based approaches for hardware security as compared to non-biometric approaches.

3. Discuss the motivation of retinal biometrics over facial and fingerprint biometrics.
4. Discuss the potential hardware security threats that an adversary can induce during IC design chain.
5. Discuss the motivation of retinal biometrics over non-biometric-based approaches.
6. Discuss the benefits of designing secure IPs from the perspective of consumer and CE system designer.
7. Discuss the benefits offered by retinal biometrics-based hardware security to consumers.
8. Discuss the comparative analysis between different hardware security approaches.
9. Explain the overview of retinal biometric-based hardware security
10. Explain the major steps involved in the retinal biometric-based hardware IP security framework.
11. Discuss different types of retinal features and their significance for hardware authentication.
12. Discuss the retinal feature extraction mechanism.
13. Discuss the retinal biometric signature generation process.
14. Discuss the behavioral description of JPEG-codec application framework.
15. Discuss the generation of encoded constraints corresponding to a sample retinal signature for hardware security.
16. Discuss the process of detecting pirated IP versions and ownership validation.
17. Examine the potential constraints associated with utilizing retinal biometrics for hardware security, if applicable.
18. Explain the security attributes of retinal biometric methodology for hardware security.
19. Discuss the factors impacting the tamper tolerance of the design.
20. Discuss the factors impacting the probability of coincidence metric.
21. Discuss the desired properties of security metrics such as tamper tolerance and probability of coincidence.
22. Explain the design flow for generating a secured version of JPEG-codec IP using a sample retinal signature.
23. Discuss the significance of HLS for hardware IP security.

References

S. Aleem, B. Sheng, P. Li, P. Yang and D. D. Feng, (2019) "Fast and Accurate Retinal Identification System: Using Retinal Blood Vasculature Landmarks," *IEEE Transactions on Industrial Informatics*, vol. 15, no. 7, pp. 4099–4110, doi:10.1109/TII.2018.2881343.

M. T. Arafin, A. Stanley and P. Sharma, (2017) "Hardware-Based Anti-counterfeiting Techniques for Safeguarding Supply Chain Integrity," *IEEE International Symposium on Circuits and Systems (ISCAS)*, Baltimore, MD, USA, pp. 1–4, doi:10.1109/ISCAS.2017.8050605.

A. Bhuiyan, B. Nath, and K. Ramamohanarao, (2012) "Detection and Classification of Bifurcation and Branch Points on Retinal Vascular Network," *2012 International Conference on Digital Image Computing Techniques and Applications, DICTA*, pp. 1–8, 10.1109/DICTA.2012.6411742.

E. Castillo, U. Meyer-Baese, A. Garcia, L. Parrilla and A. Lloris, (2007) "IPP@HDL: Efficient Intellectual Property Protection Scheme for IP Cores," *IEEE Transactions on Very Large Scale Integration (VLSI) Systems*, vol. 15, no. 5, pp. 578–591, doi: 10.1109/TVLSI.2007.896914.

R. Chaurasia and A. Sengupta, (2022) "Protecting Trojan Secured DSP Cores against IP Piracy Using Facial Biometrics," *2022 IEEE 19th India Council International Conference (INDICON)*, India, pp. 1–6, doi:10.1109/INDICON56171.2022.10039864.

R. Chaurasia and A. Sengupta, (2023a) "Retinal Biometric for Securing JPEG-Codec Hardware IP Core for CE Systems," *IEEE Transactions on Consumer Electronics*, vol. 69, no. 3, pp. 441–457, doi:10.1109/TCE.2023.3264669.

R. Chaurasia and A. Sengupta, (2023b) "Designing Optimized and Secured Reusable Convolutional Hardware Accelerator against IP Piracy Using Retina Biometrics," *Proceedings of 9th IEEE International Symposium on Smart Electronic Systems (IEEE – iSES)*, India, Dec. 2023.

R. Chaurasia, A. Anshul, A. Sengupta and S. Gupta, (2022) "Palmprint Biometric Versus Encrypted Hash Based Digital Signature for Securing DSP Cores Used in CE Systems," *IEEE Consumer Electronics Magazine*, vol. 11, no. 5, pp. 73–80, doi:10.1109/MCE.2022.3153276.

R. Chaurasia, A. Sengupta and P. Pradeeprao, (2022) "Secured Integrated Circuit (IC/IP) Design Flow," *Nanoelectronics for Next-generation Integrated Circuits,* Boca Raton, FL: CRC Book.

R. Chaurasia, A. Reddy Asireddy and A. Sengupta, (2023) "Fault Secured JPEG-Codec Hardware Accelerator with Piracy Detective Control Using Secure Fingerprint Template," *2023 IEEE International Symposium on Defect and Fault Tolerance in VLSI and Nanotechnology Systems (DFT)*, Juan-Les-Pins, France, pp. 1–6, doi: 10.1109/DFT59622.2023.10313536.

S. Chen, J. Jung, P. Song, K. Chakrabarty and G.-J. Nam, (2020) "BISTLock: Efficient IP Piracy Protection Using BIST," *2020 IEEE International Test Conference (ITC)*, Washington, DC, USA, doi:10.1109/ITC44778.2020.9325210.

B. Colombier and L. Bossuet, (2015) "Survey of Hardware Protection of Design Data for Integrated Circuits and Intellectual Properties," *IET Computers and Digital Techniques*, vol. 8, no. 6, pp. 274–287, https://doi.org/10.1049/iet-cdt.2014.0028.

A. Hroub and M. E. S. Elrabaa, (2022) "SecSoC: A Secure System on Chip Architecture for IoT Devices," *2022 IEEE International Symposium on Hardware Oriented Security and Trust (HOST)*, McLean, VA, USA, pp. 41–44, doi: 10.1109/HOST54066.2022.9839995.

W. Hu, C.-H. Chang, A. Sengupta, S. Bhunia, R. Kastner and H. Li, (2021) "An Overview of Hardware Security and Trust: Threats, Countermeasures, and

Design Tools," *IEEE Transactions on Computer-Aided Design of Integrated Circuits and Systems*, vol. 40, no. 6, pp. 1010–1038, doi:10.1109/TCAD.2020.3047976.

R. Karmakar and S. Chattopadhyay, (2020) "Hardware IP Protection Using Logic Encryption and Watermarking," *2020 IEEE International Test Conference (ITC)*, Washington, DC, USA, pp. 1–10, doi:10.1109/ITC44778.2020.9325223.

F. Koushanfar, S. Fazzari, C. McCants, *et al.*, (2012) "Can EDA Combat the Rise of Electronic Counterfeiting?," *DAC Design Automation Conference 2012*, San Francisco, CA, USA, pp. 133–138, doi:10.1145/2228360.2228386.

F. Koushanfar, I. Hong and M. Potkonjak, (2005) "Behavioral Synthesis Techniques for Intellectual Property Protection," *ACM Transactions on Design Automation of Electronic Systems*, vol. 10, no. 3, pp. 523–545, doi:10.1145/1080334.1080338.

B. Le Gal and L. Bossuet, (2012) "Automatic Low-cost IP Watermarking Technique Based on Output Mark Insertions," *Design Automation for Embedded Systems*, vol. 16, no. 2, pp. 71–92, https://doi.org/10.1007/s10617-012-9085-y.

H. R. Mahdiany, A. Hormati and S. M. Fakhraie, (2001) "A Hardware Accelerator for DSP System Design: University of Tehran DSP Hardware Emulator (UTDHE)," *ICM 2001 Proceedings. The 13th International Conference on Microelectronics*, Rabat, Morocco, pp. 141–144, doi:10.1109/ICM.2001.997507.

B. K. Mohanty and P. K. Meher, (2016) "A High-Performance FIR Filter Architecture for Fixed and Reconfigurable Applications," *IEEE Transactions on Very Large Scale Integration (VLSI) Systems*, vol. 24, no. 2, pp. 444–452, doi: 10.s1109/TVLSI.2015.2412556.

D. Mouris, C. Gouert and N. G. Tsoutsos, (2022) "Privacy-Preserving IP Verification," *IEEE Transactions on Computer-Aided Design of Integrated Circuits and Systems*, vol. 41, no. 7, pp. 2010–2023, doi:10.1109/TCAD.2021.3107251.

Multimedia Laboratory Datasets, (2022) Available: https://www.medicmind.tech/retinal-image-databases, accessed in 2022.

M. Ortega, M. G. Penedo, J. Rouco, *et al.*, (2009) "Retinal Verification Using a Feature Points-Based Biometric Pattern," *EURASIP Journal on Advances in Signal Processing*, vol. 2009, p. 235746, https://doi.org/10.1155/2009/235746.

C. Pilato, S. Garg, K. Wu, R. Karri and F. Regazzoni, (2018) "Securing Hardware Accelerators: A New Challenge for High-level Synthesis," *IEEE Embedded Systems Letters*, vol. 10, no. 3, pp. 77–80, doi:10.1109/LES.2017.2774800.

S. M. Plaza and I. L. Markov, (2015) "Solving the Third-Shift Problem in IC Piracy with Test-Aware Logic Locking," *IEEE Transactions on Computer-Aided Design of Integrated Circuits and Systems*, vol. 34, no. 6, pp. 961–971, doi:10.1109/TCAD.2015.2404876.

S. Potluri, A. Aysu and A. Kumar, (2020) "SeqL: Secure Scan-Locking for IP Protection," *2020 21st International Symposium on Quality Electronic Design (ISQED)*, Santa Clara, CA, USA, doi: 10.1109/ISQED48828.2020.9136991.

S. Rai, A. Rupani, P. Nath and A. Kumar, (2019) "Hardware Watermarking Using Polymorphic Inverter Designs Based on Reconfigurable Nanotechnologies,"

2019 IEEE Computer Society Annual Symposium on VLSI (ISVLSI), Miami, FL, USA, pp. 663–669, doi: 10.1109/ISVLSI.2019.00123.

J. J. Rajendran, O. Sinanoglu and R. Karri, (2016) "Building Trustworthy Systems Using Untrusted Components: A High-Level Synthesis Approach," *IEEE Transactions on Very Large Scale Integration (VLSI) Systems*, vol. 24, no. 9, pp. 2946–2959.

M. Rathor and A. Sengupta, (2020) "IP Core Steganography Using Switch Based Key-Driven Hash-Chaining and Encoding for Securing DSP Kernels Used in CE Systems," *IEEE Transactions on Consumer Electronics*, vol. 66, no. 3, pp. 251–260, doi:10.1109/TCE.2020.3006050.

M. Rathor and A. Sengupta, (2021) "Signature Biometric Based Authentication of IP Cores for Secure Electronic Systems," *2021 IEEE International Symposium on Smart Electronic Systems (iSES)*, Jaipur, India, pp. 384–388, doi: 10.1109/iSES52644.2021.00094.

M. Rathor, A. Anshul, K Bharath, R. Chaurasia and A. Sengupta, (2023) "Quadruple Phase Watermarking during High Level Synthesis for Securing Reusable Hardware Intellectual Property Cores," *Computers and Electrical Engineering*, vol. 105, p. 108476, doi.org/10.1016/j.compeleceng.2022.108476.

A. R. D. Rizo, J. Leonhard, H. Aboushady and H.-G. Stratigopoulos, (2022) "RF Transceiver Security Against Piracy Attacks," *IEEE Transactions on Circuits and Systems–II: Express Briefs*, vol. 69, no. 7, pp. 3169—3173, doi:10.1109/TCSII.2022.3165709.

J. A. Roy, F. Koushanfar and I. L. Markov, (2008) "EPIC: Ending Piracy of Integrated Circuits," *2008 Design, Automation and Test in Europe (DATE)*, Munich, Germany, pp. 1069–1074, doi: 10.1109/DATE.2008.4484823.

S. M. Saeed, A. Zulehner, R. Wille, R. Drechsler and R. Karri, (2019) "Reversible Circuits: IC/IP Piracy Attacks and Countermeasures," *IEEE Transactions on Very Large Scale Integration (VLSI) Systems*, vol. 27, no. 11, pp. 2523–2535, doi:10.1109/TVLSI.2019.2934465.

S. M. Saeed, N. Mahendran, A. Zulehner, R. Wille and R. Karri, (2019) "Identification of Synthesis Approaches for IP/IC Piracy of Reversible Circuits," *Journal on Emerging Technologies in Computing Systems*, vol. 15, no. 3, pp. 23:1–23:17, https://doi.org/10.1145/3289392.

R. Schneiderman, (2010) "DSPs Evolving in Consumer Electronics Applications," *IEEE Signal Processing Magazine*, vol. 27, no. 3, pp. 6–10, doi: 10.1109/MSP.2010.936031.

A. Sengupta and S. Bhadauria, (2016) "Exploring Low Cost Optimal Watermark for Reusable IP Cores during High Level Synthesis," *IEEE Access*, vol. 4, pp. 2198–2215, doi:10.1109/ACCESS.2016.2552058.

A. Sengupta and R. Chaurasia, (2022) "Secured Convolutional Layer IP Core in Convolutional Neural Network Using Facial Biometric," *IEEE Transactions on Consumer Electronics*, vol. 68, no. 3, pp. 291–306, doi:10.1109/TCE.2022.3190069.

A. Sengupta and R. Chaurasia, (2023) "Methodology for Exploration of Security-Design Cost Tradeoff for Signature-Based Security Algorithms," in *Physical Biometrics for Hardware Security of DSP and Machine Learning Coprocessors*. Stevenage: The Institution of Engineering and Technology, Chap. 8, pp. 259–297, doi: 10.1049/PBCS080E_ch8.

A. Sengupta and M. Rathor, (2020) "Securing Hardware Accelerators for CE Systems Using Biometric Fingerprinting," *IEEE Transactions on Very Large Scale Integration (VLSI) Systems*, vol. 28, no. 9, pp. 1979–1992, doi:10.1109/TVLSI.2020.2999514.

A. Sengupta and R. Sedaghat, (2011) "A High Level Synthesis Design Flow from ESL to RTL with Multi-parametric Optimization Objective," *IETE Journal of Research*, vol. 57, no. 2, pp. 169–186.

A. Sengupta, D. Roy, S. P. Mohanty and P. Corcoran, (2018) "Low-Cost Obfuscated JPEG CODEC IP Core for Secure CE Hardware," *IEEE Transactions on Consumer Electronics*, vol. 64, no. 3, pp. 365–374, doi:10.1109/TCE.2018.2852265.

A. Sengupta, D. Roy and S. P. Mohanty, (2018) "Triple-Phase Watermarking for Reusable IP Core Protection during Architecture Synthesis," *IEEE Transactions on Computer-Aided Design of Integrated Circuits and Systems*, vol. 37, no. 4, pp. 742–755, doi:10.1109/TCAD.2017.2729341.

A. Sengupta, E. R. Kumar and N. P. Chandra, (2019) "Embedding Digital Signature Using Encrypted-Hashing for Protection of DSP Cores in CE," *IEEE Transactions on Consumer Electronics*, vol. 65, no. 3, pp. 398–407, doi:10.1109/TCE.2019.2924049.

A. Sengupta, R. Chaurasia and T. Reddy, (2021) "Contact-Less Palmprint Biometric for Securing DSP Coprocessors Used in CE Systems," *IEEE Transactions on Consumer Electronics*, vol. 67, no. 3, pp. 202–213, doi:10.1109/TCE.2021.3105113.

S. Sitjongsataporn, A. Thitinaruemit and S. Prongnuch, (2021) "Implementation of High Level Synthesis for Adaptive FIR Filtering on Embedded System," *2021 7th International Conference on Engineering, Applied Sciences and Technology (ICEAST)*, Pattaya, Thailand, pp. 257–260, doi:10.1109/ICEAST52143.2021.9426296.

X. Wang, Y. Zheng, A. Basak and S. Bhunia, (2015) "IIPS: Infrastructure IP for Secure SoC Design," *IEEE Transactions on Computers*, vol. 64, no. 8, pp. 2226–2238, doi:10.1109/TC.2014.2360535.

J. Zhang, (2016) "A Practical Logic Obfuscation Technique for Hardware Security," *IEEE Transactions on Very Large Scale Integration (VLSI) Systems*, vol. 24, no. 3, pp. 1193–1197, doi:10.1109/TVLSI.2015.2437996.

Chapter 4

HLS-based mathematical watermarks for hardware security and trust

Anirban Sengupta[1] *and Aditya Anshul*[1]

Ensuring security and trust within the global hardware design supply chain process is paramount. Intellectual property (IP) piracy and false IP ownership assertion are two primary threats that exist within the global design supply chain process. This chapter discusses a high-level synthesis (HLS)-based mathematical watermarking approach that exploits the characteristics of the design parameters and design space, *viz.* dispersion matrix and eigen decomposition framework, respectively. In this chapter, the IP seller is considered the defender and the system-on-chip (SoC) integration house is the attacker. The discussed mathematical watermarking methodology demonstrates a security framework that leverages the characteristics of the design space parameters chosen by the IP seller, as well as the design space's size in terms of resource configurations selected by the IP seller, to embed them as unique features for safeguarding the IP design. The discussed approach also employs AES encryption to generate encrypted watermarking constraints, which are further integrated/embedded into the hardware design (serving as digital evidence) and provide detective security countermeasures against potential intellectual property piracy and false IP ownership claim.

4.1 Introduction

In today's digital landscape, the importance of application-specific computing is increasing continuously, as it plays a vital role in addressing the growing need for efficient and specialized processing capabilities. Some critical factors that lead to the increased usage of application-specific computing include the need for enhanced performance, real-time processing capabilities, improved power efficiency, cost-effectiveness, and scalability. The application-specific computing systems serve as the backbone for executing numerous crucial and computation-intensive functions, including complex functions such as digital data filtering, data compression-decompression, etc. Therefore, these application-specific computing systems are carefully designed as dedicated hardware intellectual property (IP)

[1]Department of Computer Science and Engineering, Indian Institute of Technology Indore, India

cores to cater to the unique demands of specific applications, driving innovation and advancement in various domains (Anshul and Sengupta, 2023a). Examples of hardware IPs include the discrete wavelength transform (DWT), JPEG compression-decompression (JPEG-CODEC), finite impulse response (FIR), discrete cosine transform, etc. High-level synthesis (HLS) process is used to design these application-specific computing systems as dedicated reusable hardware IP core (Schneiderman, 2010). HLS serves as a bridge between the high-level behavioral description of hardware (typically expressed in programming languages like C/C++ or transfer function) and the lower-level register-transfer level (RTL) implementation (synthesized into hardware components). This seamless integration between high-level design descriptions and RTL implementation enables efficient and streamlined development processes of reusable hardware IP cores (Rathor *et al.*, 2023a; Le Gal and Bossuet, 2012; Anshul *et al.*, 2022).

Furthermore, within the contemporary global design supply chain process, securing the hardware IP designs has emerged as a critical concern. In this era, where integrated circuits are intricately designed by various entities within the supply chain, the potential for malicious activities targeting hardware IPs has escalated significantly (Rostami *et al.*, 2014). Consider a scenario where an attacker infiltrates a system-on-chip (SoC) integrator's design house within the supply chain. This infiltrator could illicitly pirate original hardware IP design, which were sourced from legitimate IP seller, thereby tarnishing the trustworthiness of the design house. Motivations for such illicit actions can vary, ranging from personal gains, such as financial profit, to malicious intent aimed at defaming the legitimate IP provider. Furthermore, within the SoC integrator's design and fabrication houses, an adversary might fraudulently claim the ownership of a genuine hardware IP design (Rizzo *et al.*, 2019), (Anshul and Sengupta, 2023b). Additionally, IP counterfeiting is another threat, where the adversary tries to counterfeit the original design, jeopardizing the safety of end consumers also (Rostami *et al.*, 2014; Anshul and Sengupta, 2023a). Therefore, it is imperative for the original IP seller to secure the design against the looming threat of IP piracy and false IP ownership claim (Sengupta *et al.*, 2023a; Rathor *et al.*, 2023b; Colombier and Bossuet, 2015; Rathor *et al.*, 2024; Anshul and Sengupta, 2023c; Islam *et al.*, 2020; Anshul *et al.*, 2022; Yasin *et al.*, 2016; Sengupta *et al.*, 2023b; Anshul and Sengupta, 2023d; Sengupta *et al.*, 2023c).

This chapter discusses a mathematical watermark-based security methodology using the design parameter-driven encrypted dispersion matrix coupled with an eigen decomposition-based security framework (Sengupta and Anshul, 2024). The integration/embedding of a robust mathematical watermark (digital evidence) in the form of an encrypted signature provides a detective security countermeasure against both intellectual property piracy and fraudulent claim of IP ownership (Sengupta and Anshul, 2024).

4.2 Salient features of the chapter

- The chapter discusses a mathematical hardware watermarking framework using HLS that incorporates watermarking constraints derived from the characteristics of hardware design parameters and design space. The characteristics of design parameters and the design space as watermarking constraints can be estimated using the statistical model proposed in Sengupta and Anshul (2024), *viz. dispersion matrix* and *eigen decomposition* framework. The discussed framework leverages the characteristics of the hardware's design parameters (such as area, delay, etc.) and the IP seller's resource configuration-based design space characteristics (such as eigen roots) to generate a robust watermark (i.e., signature).

- The unique attributes (eigen roots obtained through eigen decomposition block and elements of dispersion matrix) serve as digital evidence, which are embedded as covert watermarking constraints to secure hardware IP design against piracy and false IP ownership claim.
- The discussed approach outlines the generation and embedding of encrypted signature (generated through dispersion matrix, eigen decomposition, and AES encryption block) during the register allocation phase of the HLS process. Additionally, it showcases the complete end-to-end mathematical watermarking process/algorithm and its implementation on an 8-point DCT hardware design.
- The discussed approach in this chapter encompasses several phases contributing to the generation of a robust mathematical watermark (Sengupta and Anshul, 2024):
  (a) Strengthening the signature post encryption.
  (b) Utilizing statistical models such as dispersion matrix and eigen decomposition to generate binary signatures and corresponding watermarking constraints.
  (c) Employing different z-bit key values to determine the number of input resource configurations.
  (d) Selecting appropriate design parameters for the dispersion matrix and eigen decomposition.
  (e) IP seller's chosen AES encryption key.

4.2.1 Discussion on threat model

A SoC integrator design house and fabrication house may become potential sources of security/trustworthiness breach due to the possibility of an adversary who might engage in IP piracy or false IP ownership claim, particularly pirating the original hardware IP (obtained from a legitimate IP seller) (Rostami *et al.*, 2014). As discussed in the introduction section, this illicit activity could involve the replication of designs for personal gain, such as financial profit or causing harm to the reputation of the authentic IP seller by including malicious logic within the pirated copies. Additionally, there is a risk of fraudulent ownership claim over the hardware IP by an adversary within the SoC integrator design and fabrication houses (Anshul and Sengupta, 2023a; Rizzo *et al.*, 2019; Anshul and Sengupta, 2023b). To mitigate these threats, it's imperative for the authentic IP seller to implement robust security countermeasures to safeguard their hardware IP before distributing it to the SoC integrator. In this chapter, the IP seller is viewed as the defender, while the SoC integrator/foundry house is seen as the potential attacker. This chapter discusses a mathematical watermarking-based security methodology as a detective countermeasure against IP piracy and false IP ownership claim (Chaurasia *et al.*, 2022; Sengupta *et al.*, 2023c; Anshul and Sengupta, 2022).

4.2.2 Motivation for employing mathematical watermark as detective security countermeasure (Sengupta and Anshul, 2024)

- The utilization of statistical techniques (tools) like eigen decomposition framework and dispersion matrix facilitates the creation of distinctive, inimitable, and irreproducible watermark by IP sellers for embedding into the hardware IP designs.
- The process of generating a dispersion matrix includes the evaluation of the variance and covariance of the IP seller's selected design space parameters (such as area, delay, etc.). The utilization of covariance in creating mathematical watermarks serves as a crucial design information, capturing the synergy between area and delay parameters associated with an IP design.
- Similarly, incorporating variance and eigen roots into mathematical watermarks captures the extent of variation in the design search space and the respective parameters, respectively. By computing the eigen roots of selected resource configurations, variance and covariance of selected design space parameters (e.g., area, latency, etc.) for the hardware application, IP seller can generate mathematical watermark based on the unique attributes of the design itself (exclusive of external signatures or biometrics of the IP seller).
- It's worth noting that no other statistical model captures the characteristics of hardware design search space as accurately and lucidly as the discussed approach. The discussed mathematical watermarking methodology extracts the features of the IP seller's selected design space parameters and the characteristics of the design space itself, thereby utilizing them as distinct features for providing digital evidence as watermark.

4.3 HLS-based mathematical hardware watermarking framework

4.3.1 Summary

Figure 4.1 depicts the overview of the HLS-based mathematical watermarking for generating secure RTL IP (Sengupta and Anshul, 2024). The primary inputs consist of (a) control data flow graph (CDFG)/transfer function of the hardware application, (b) IP seller chosen z-bit key, (c) concatenation rule, (d) AES key, (e) embedding/mapping rule, and (f) module library. As depicted in Figure 4.1, the mathematical watermark generation block is responsible for the generation of the watermark signature based on the output produced by the HLS scheduling and allocation block. Subsequently, the obtained watermark is encrypted using AES, and the generated encrypted watermark is fed into the embedding block within the HLS framework. Post-embedding of the encrypted watermark in the hardware design, datapath synthesis (during HLS) is performed to generate the final secure RTL data path of hardware as output. Next, Figure 4.2 illustrates the details of a mathematical watermarking framework (Sengupta and Anshul, 2024). The major components of the mathematical watermarking framework are as follows:

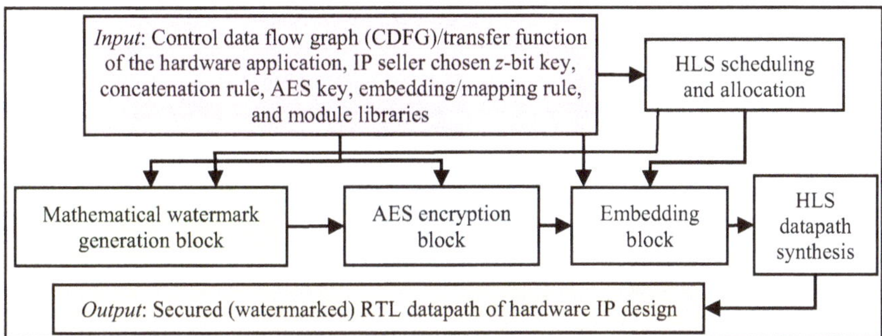

Figure 4.1 Overview of HLS-based mathematical watermarking for secure RTL IP

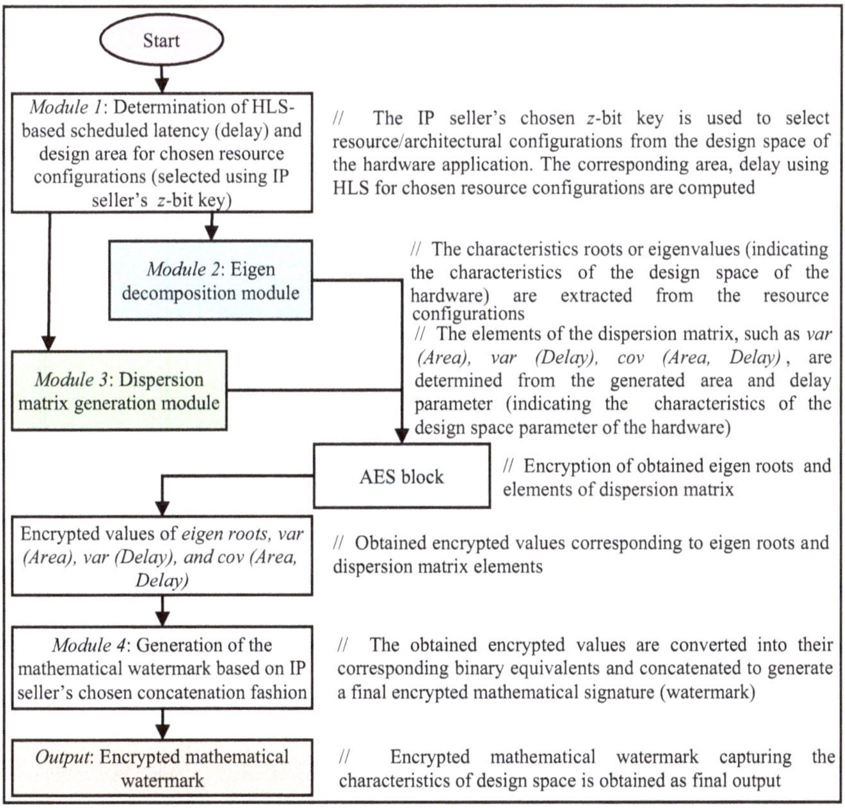

Start

Module 1: Determination of HLS-based scheduled latency (delay) and design area for chosen resource configurations (selected using IP seller's z-bit key)

// The IP seller's chosen z-bit key is used to select resource/architectural configurations from the design space of the hardware application. The corresponding area, delay using HLS for chosen resource configurations are computed

Module 2: Eigen decomposition module

// The characteristics roots or eigenvalues (indicating the characteristics of the design space of the hardware) are extracted from the resource configurations

Module 3: Dispersion matrix generation module

// The elements of the dispersion matrix, such as *var (Area), var (Delay), cov (Area, Delay)*, are determined from the generated area and delay parameter (indicating the characteristics of the design space parameter of the hardware)

AES block

// Encryption of obtained eigen roots and elements of dispersion matrix

Encrypted values of *eigen roots, var (Area), var (Delay), and cov (Area, Delay)*

// Obtained encrypted values corresponding to eigen roots and dispersion matrix elements

Module 4: Generation of the mathematical watermark based on IP seller's chosen concatenation fashion

// The obtained encrypted values are converted into their corresponding binary equivalents and concatenated to generate a final encrypted mathematical signature (watermark)

Output: Encrypted mathematical watermark

// Encrypted mathematical watermark capturing the characteristics of design space is obtained as final output

Figure 4.2 Details of mathematical watermarking framework

(a) *Module 1*: determination of HLS-based scheduled latency (delay) and design area based on key-controlled resource configurations, (b) *Module 2*: eigen decomposition module, (c) *Module 3*: dispersion matrix generation module, (d) AES block, and (e) *Module 4*: generation of the mathematical watermark based on IP seller's chosen concatenation fashion (Sengupta and Anshul, 2024). The details of each component are explained in Figures 4.3–4.5, respectively.

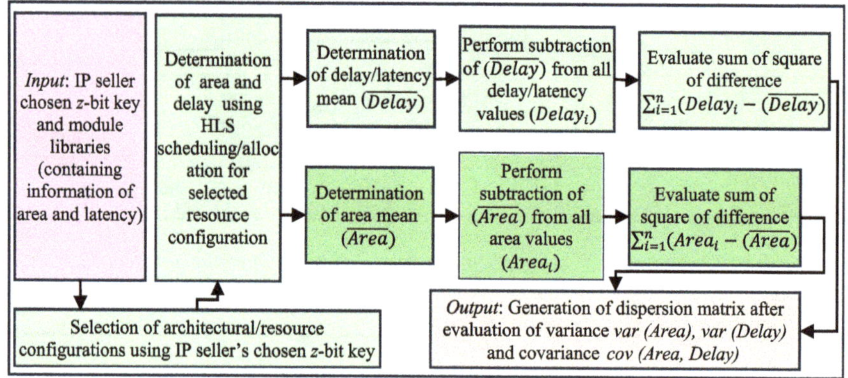

Figure 4.3 Dispersion matrix generation module

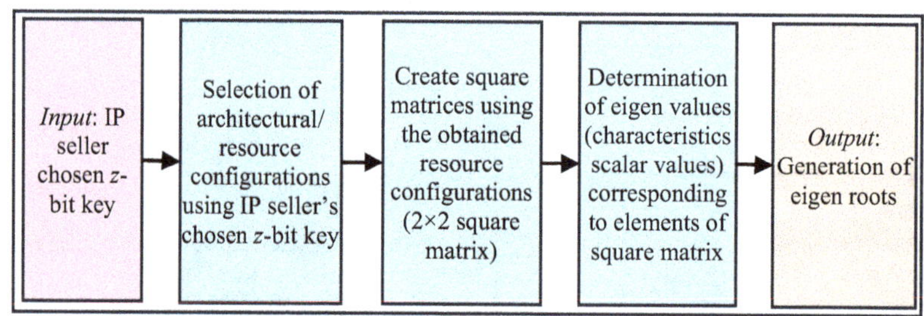

Figure 4.4 Eigen roots computation module

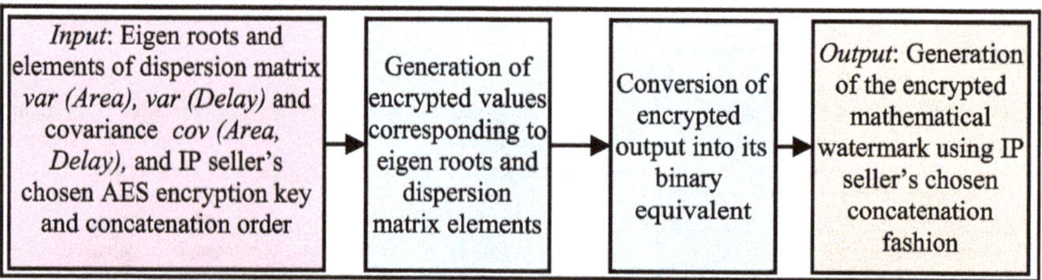

Figure 4.5 AES encryption block and generation of final mathematical watermark

4.3.2 Details of the HLS-based mathematical watermarking framework

This subsection discusses the generation of encrypted mathematical watermark in detail. Figure 4.3 shows the details of the dispersion matrix generation process (module 3). The dispersion matrix generation module accepts the module library and a z-bit key chosen by the IP seller as its primary input. As discussed above, the z-bit input key is responsible for generating initial resource configurations from the hardware's design search space (module 1). The maximum size of this initial key (z-bit) is determined by the size of the design search space corresponding to the hardware. For example, for 8-point DCT, the maximum number of adders and multipliers required for parallel execution is one and eight, respectively. Therefore, the complete design search space size can include $2^3 = 8$ variances ($1*8 = 8$). This design space size can be easily represented by a 3-bit IP seller selected key. As discussed above, the initial resource configurations are generated based on the input z-bit key. Next, the scheduled latency (*Delay*) and area (*Area*) corresponding to the IP seller's chosen resource configurations are determined using the respective module library (module 1). For realizing this, the control data flow graph (derived from the respective mathematical/transfer function) is scheduled based on selected resource configurations using the HLS process, and then the corresponding area and delay are computed (Sengupta and Anshul, 2024). The initial step involves utilizing the determined latency and area parameters as input for the dispersion matrix generation module (module 3). This module aims to extract the characteristics of the IP seller selected design space parameters, focusing on the variance of area (*var (Area)*), the variance of latency (*var (Delay)*), and the covariance (*cov (Area, Latency)*). After the determination of area and delay parameters corresponding to generated resource configurations, the mean values of the latency and area are evaluated. Next, the obtained mean values are subtracted from the corresponding latency and area parameters, respectively. Then, the sum of the squares of the differences corresponding to the latency and design area are calculated. Following this computation, the characteristics of the IP seller's chosen design space parameters (i.e., *var (Area), var (Delay)*, and *cov (Area, Delay)*) are determined. Ultimately, a dispersion matrix corresponding to hardware design is generated as output (Sengupta and Anshul, 2024). The complete process is demonstrated on 8-point DCT in the next section.

Next, Figure 4.4 shows the process employed in the decomposition process (module 2). As shown in Figure 4.4, the eigen decomposition process accepts the z-bit key chosen by the IP seller as its primary input (module 1). As discussed above, the z-bit input key is utilized to generate initial resource configurations. Next, square matrices are created using the generated resource configurations. These matrices are then utilized to determine the eigen roots (i.e., characteristic roots) corresponding to the chosen resource configurations within the design space. The objective of this module is to extract the characteristics of the design space corresponding to the hardware, expressed in terms of eigenvalues (β_n). The final output of this module is the eigen roots corresponding to square matrices of resource configurations (Sengupta and Anshul, 2024).

Figure 4.5 highlights the AES encryption block along with the generation of the final encrypted mathematical watermark signature as output. The distinctive features extracted (i.e., elements of the dispersion matrix and eigen roots) are individually encrypted using the AES. The resulting encrypted values are then produced as output. Following this, each encrypted output is converted into its binary representation. Finally, the obtained binary equivalents are concatenated as per the IP seller's chosen concatenation fashion (module 4) to generate an encrypted mathematical watermark (Sengupta and Anshul, 2024).

4.3.3 Embedding of generated mathematical watermark

The generated mathematical watermark is first converted into watermarking constraints using IP seller's selected mapping/embedding rule, and further embedded into the hardware design during the register allocation phase of the HLS process (Sengupta and Anshul, 2024). The IP seller's chosen mapping/embedding rule is as follows:

- For a watermark (mathematical signature) bit '0', the watermarking constraint comprises even-even storage variable pairs (U_R, U_S) of the scheduled data flow graph (SDFG), where *'R'* and *'S'* denote the storage variable numbers. Thus, the resulting watermarking constraints corresponding to '0's in mathematical watermark are *<U0, U2>, <U0, U4>–<U0, U30>, <U2, U4>–<U2, U30>, <U4, U6>–<U4, U30>, <U6, U8>–<U6, U30>, <U8, U10>–<U20, U28>.*
- For a watermark (mathematical signature) bit '1', the watermarking constraints comprise odd-odd storage variable pairs of the scheduled DFG. Consequently, the resulting watermarking constraints corresponding to '1's in the mathematical watermark are *<U1, U3>, <U1, U5>–<U1, U29>,<U3, U5>–<U3, U29>,<U5, U7>–<U27, U29>.*

The embedding of the determined watermarking constraints on 8-point DCT is demonstrated in the next subsection.

4.3.4 Demonstration of mathematical watermarking framework along an 8-point DCT

This subsection demonstrates the generation and embedding process of encrypted watermarking constraints on 8-point DCT. The watermark signature of module 3 is explained in Figure 4.6 with the help of an IP seller's key-controlled resource configuration values. Figure 4.6 illustrates the watermark signature computation process for the variance of area, variance of delay, and covariance (area, delay). The obtained values (i.e., *var (Area), var (Delay)*, and *cov (Area, Delay)*) are now used to generate dispersion matrix (shown in Figure 4.7). Next, Figure 4.8 illustrates the eigen decomposition framework (the watermark signature of module 2). Subsequently, Figure 4.9 depicts the generation of encrypted mathematical watermarking signature post-AES encryption and concatenation. The determination process of watermarking constraints from generated encrypted mathematical watermark has already been discussed in section 4.3.3.

Step 1: Area and delay/latency evaluation for IP seller's chosen resource configuration (using *z*-bit key)

R_C (Key driven resources selected)	*Area* (IP seller computed)	*Delay* (IP seller computed)
[1, 4]	327 um^2	927 *ps*
[1, 1]	101 um^2	2186 *ps*
[1, 8]	629 um^2	729 *ps*
[1, 2]	177 um^2	1325 *ps*

Step 2

Mean Computation of *Area:*

$$\overline{Area} = \sum_{i=1}^{n} Area_i$$

$$\overline{Area} = \frac{327 + 101 + 629 + 177}{4} = 308.5 = {\sim}309$$

Mean Computation of *Delay:*

$$\overline{Delay} = \sum_{i=1}^{n} Delay_i$$

$$\overline{Delay} = \frac{927 + 2186 + 729 + 1325}{4} = 1291.75 = {\sim}1292$$

Step 3

Subtract the mean (*Area*) from all area parameter values:

$$\Rightarrow \quad (Area_1 - \overline{Area}), (Area_2 - \overline{Area}), \ldots\ldots (Area_n - \overline{Area})$$

$\Rightarrow$ (327–309), (101–309), (629–309), (177–309)

$\Rightarrow$ (18), (–208), (320), (–132)

Subtract the mean (*Delay*) from all delay/latency parameter values:

$$\Rightarrow \quad (Delay_1 - \overline{Delay}), (Delay_2 - \overline{Delay}), \ldots\ldots (Delay_n - \overline{Delay})$$

$\Rightarrow$ (927–1292), (2186–1292), (729–1292), (1325–1292)

$\Rightarrow$ (–365), (894), (–563), (33)

Step 4

Compute sum of the square of the differences:

$$S_A = \sum_{i=1}^{n} (Area_i - \overline{Area})^2$$

$$\Rightarrow \quad S_A = (18)^2 + (-208)^2 + (320)^2 + (-132)^2$$

$$\Rightarrow \quad (324 + 43264 + 102400 + 17424) = 163412$$

Compute sum of the square of the differences:

$$S_D = \sum_{i=1}^{n} (Delay_i - \overline{Delay})^2$$

$$\Rightarrow \quad S_D = (-365)^2 + (894)^2 + (-563)^2 + (33)^2$$

$$\Rightarrow \quad (133225 + 799236 + 316969 + 1089) = 1{,}25{,}0519$$

Step 5

Computation of variance (*Area*):

$$Var(Area) = \frac{\sum_{i=1}^{n} (Area_i - \overline{Area})^2}{n-1}$$

$$\Rightarrow \quad Var(Area) = \left(\frac{163412}{3}\right) = 54470.66 = {\sim}55000$$

Computation of variance (*Delay*):

$$Var(Delay) = \frac{\sum_{i=1}^{n} (Delay_i - \overline{Delay})^2}{n-1}$$

$$\Rightarrow \quad Var(Delay) = \left(\frac{1250519}{3}\right) = 416839.667 = {\sim}417000$$

Covariance computation:

$$Cov\,(Area, Delay) = \sum_{i=1}^{n} \frac{(Area_i - \overline{Area}) \times (Delay_i - \overline{Delay})}{n-1}$$

$$\Rightarrow \quad \frac{-377038}{4-1} = -125679.33 = {\sim}-125679$$

Figure 4.6 Computation process of variance of area, variance of delay, and covariance (area, delay)

Dispersion matrix (*H*):

$$cov(H) = \begin{bmatrix} \frac{\sum_{i=1}^{n}(Area_i - \overline{Area})^2}{n-1} & \sum_{i=1}^{n}\frac{(Area_i - \overline{Area}) \times (Delay_i - \overline{Delay})}{n-1} \\ \sum_{i=1}^{n}\frac{(Area_i - \overline{Area}) \times (Delay_i - \overline{Delay})}{n-1} & \frac{\sum_{i=1}^{n}(Delay_i - \overline{Delay})^2}{n-1} \end{bmatrix}$$

$$cov(H) = \begin{bmatrix} Var(Area) & Cov\,(Area, Delay) \\ Cov\,(Area, Delay) & Var(Delay) \end{bmatrix}$$

$$cov(H) = \begin{bmatrix} 55000 & -125679 \\ -125679 & 417000 \end{bmatrix}$$

Figure 4.7 Generation of dispersion matrix

Square matrices	$X = \begin{bmatrix} 1 & 4 \\ 1 & 2 \end{bmatrix}$	$Y = \begin{bmatrix} 1 & 8 \\ 1 & 1 \end{bmatrix}$
Step 1	$det\,(\beta I - X) = 0$	$det\,(\beta I - Y) = 0$
Step 2	$det\left(\beta \begin{bmatrix} 1 & 0 \\ 0 & 1 \end{bmatrix} - \begin{bmatrix} 1 & 4 \\ 1 & 2 \end{bmatrix}\right) = 0$	$det\left(\beta \begin{bmatrix} 1 & 0 \\ 0 & 1 \end{bmatrix} - \begin{bmatrix} 1 & 8 \\ 1 & 1 \end{bmatrix}\right) = 0$
Step 3	$det\left(\begin{bmatrix} \beta - 1 & -4 \\ -1 & \beta - 2 \end{bmatrix}\right) = 0$	$det\left(\begin{bmatrix} \beta - 1 & -8 \\ -1 & \beta - 1 \end{bmatrix}\right) = 0$
Step 4	$\beta^2 - 3\beta - 2 = 0$	$\beta^2 - 2\beta - 7 = 0$
Step 5	$\beta_1 = -3.56\ and\ \beta_2 = 0.56$	$\beta_3 = 3.82\ and\ \beta_4 = -1.82$

Figure 4.8 Eigen decomposition block: determination of eigen roots

Embedding process: The determined watermarking constraints are embedded in the hardware design during the register allocation phase of the HLS process. The complete HLS-based embedding process is highlighted in Figure 4.10. Initially, an SDFG is generated from the input CDFG, IP seller's chosen resource constraints, and LIST scheduling algorithm. Figures 4.11 and 4.12 depict CDFG of 8-point DCT and SDFG of 8-point DCT using one adder and four multipliers, respectively. The register allocation table (RAT) for the scheduled design (i.e., SDFG) is generated (providing details of storage variables, control steps, and register allocation information). Subsequently, the register allocation information and obtained watermarking constraints are fed into the embedding module, which generates the watermark-embedded register allocation information. Each storage variable pair linked to the generated watermarking constraints must be allocated to a unique register (Sengupta and Anshul, 2024). To embed the watermarking constraints into the scheduled design (i.e., the corresponding RAT), local alterations in register allocations are made to accommodate the watermarking constraints. In instances where there is a conflict in the allocation of storage variable pairs during the embedding of watermarking constraints, adjustments are made either by local swapping of registers or by assigning a new register (Sengupta and Anshul, 2024). Following the embedding of the determined watermarking constraints, the resulting final register allocation information containing secret watermark is obtained as

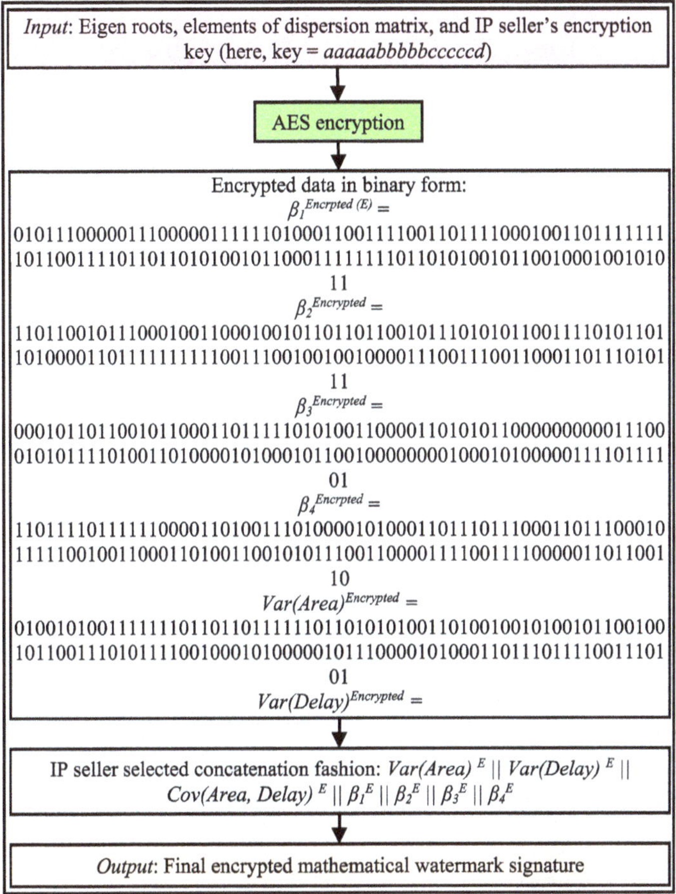

Input: Eigen roots, elements of dispersion matrix, and IP seller's encryption key (here, key = aaaaabbbbbcccccd)
AES encryption
Encrypted data in binary form:
$\beta_1^{Encrpted\ (E)}$ =
0101110000011100000111111010001100111100110111100010011011111111
1011001111011011010100101100011111111011010100101100100010010101
11
$\beta_2^{Encrypted}$ =
1101100101110001001100010010110110110010111010101100111101011011
1010000110111111111100111001001001000011100111001100011011101011
11
$\beta_3^{Encrypted}$ =
0001011011001011000110111111010100110000110101011000000000011100
0101011110100110100001010001011001000000000100010100000111101111
01
$\beta_4^{Encrpted}$ =
1101111011111100001101001110100001010001101110111000110111000010
1111100100110001101001100101011100110000111100111100000110110011
10
$Var(Area)^{Encrypted}$ =
0100101001111111011011011111110110101010011010010010100101100100
1011001110101111001000101000001011100001010001101110111100111010
01
$Var(Delay)^{Encrypted}$ =
IP seller selected concatenation fashion: $Var(Area)^E \,||\, Var(Delay)^E \,||\, Cov(Area, Delay)^E \,||\, \beta_1^E \,||\, \beta_2^E \,||\, \beta_3^E \,||\, \beta_4^E$
Output: Final encrypted mathematical watermark signature

Figure 4.9 Generation of encrypted mathematical watermark signature post-encryption

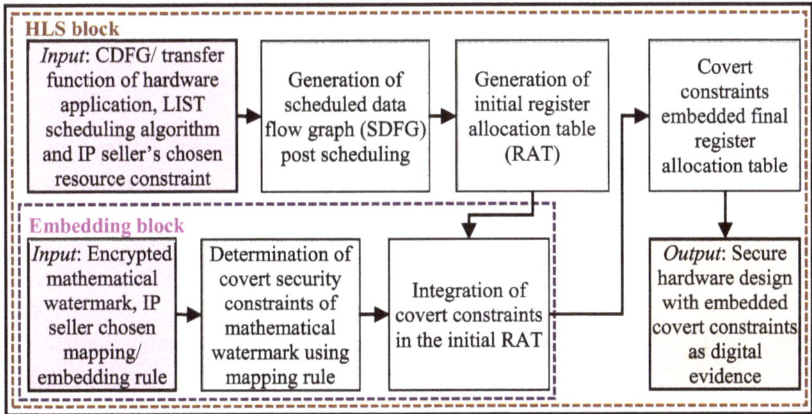

Figure 4.10 Security constraints embedding and HLS block

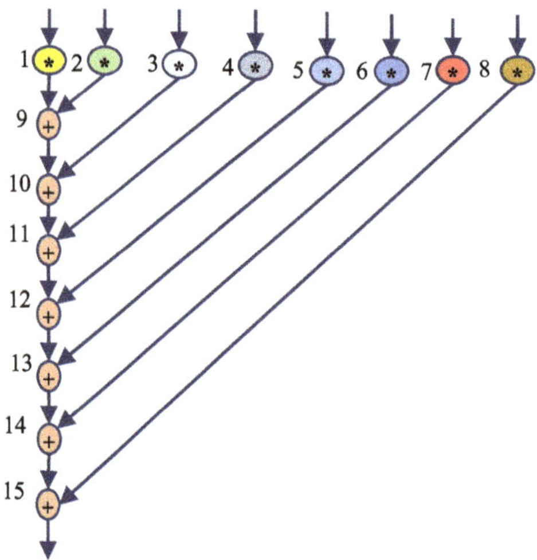

Figure 4.11 Control data flow graph (CDFG) of 8-point DCT

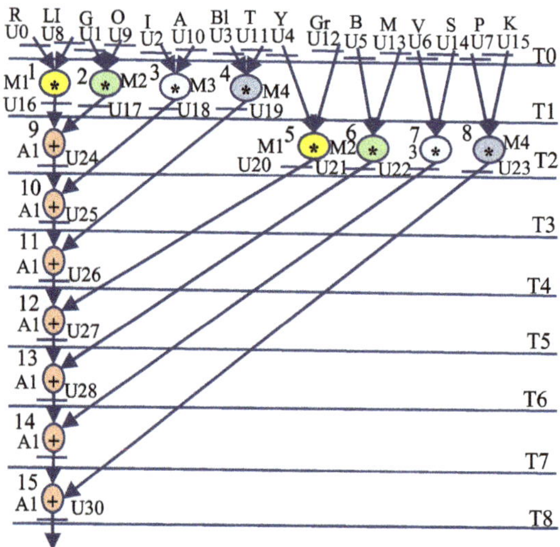

Figure 4.12 Scheduled data flow graph (CDFG) of 8-point DCT based on IP seller's chosen resource constraint (one adder and four multipliers)

output. The obtained final RAT is then subjected to datapath synthesis to generate the corresponding secure register transfer level (RTL) datapath. Table 4.1 shows the register allocation table before and after embedding the generated mathematical watermark, along with the required number of control steps for obtaining the watermark-embedded hardware design.

Table 4.1 Register allocation table pre- and post-embedding mathematical watermarking constraints on 8-point DCT

CS	Red (R)	Green (G)	Indigo (I)	Blue (Bl)	Yellow (Y)	Black (B)	Violet (V)	Pink (P)	Lime (Ll)	Olive (O)	Aqua (A)	Teal (T)	Grey (Gr)	Magenta (M)	Silver (S)	Khaki (K)
0	U0	U1	U2	U3	U4	U5	U6	U7	U8	U9	U10	U11	U12	U13	U14	U15
1	U16/ U17	U17/ U16	U18/ U19	U19/ U18	U4	U5	U6	U7	–	–	–	–	–	–	–	–
2	U24	–	U18/ U19	U19/ U18	U20/ U21	U21/ U20	U22/ U23	U23/ U22	–	U24	–	–	–	–	–	–
3	U25	–	U19	U19	U20/ U21	U21/ U20	U22/ U23	U23/ U22	U25	–	–	–	–	–	–	–
4	U26	–	–	–	U20/ U21	U21/ U20	U22/ U23	U23/ U22	–	U26	–	–	–	–	–	–
5	U27	–	–	–	U21	U21	U22/ U23	U23/ U22	–	–	U27	–	–	–	–	–
6	U28	–	–	–	–	U28	U22/ U23	U23/ U22	–	–	–	–	–	U29	–	–
7	U29	–	–	–	–	–	–	U23	–	–	–	–	–	–	–	–
8	U30	–	–	–	–	U30	–	–	–	–	–	–	–	–	–	–

4.4 Detection of IP piracy and resolving false IP ownership claim

An attacker can attempt to perform IP piracy and fraudulently claim IP ownership. While doing this, the following threats can emerge: (a) watermark collision/false positive/ghost insertion search attack, (b) brute-force attack (tampering/removal). The embedded watermark should be capable to handle the aforementioned threats as well as provide detective security countermeasure against IP piracy and fraudulently claim IP ownership.

In scenarios involving disputes over intellectual property ownership, it is assumed that both the defender (the original IP seller) and the adversary (situated within the SoC/foundry house) possess access to the disputed IP design. The mathematical watermarking methodology aims to resolve IP ownership conflict by utilizing digital evidence (i.e., watermarking constraints) embedded within the design. To achieve this, the authenticity of the watermark is verified by extracting hidden constraints from the RTL file of the IP under test and comparing them with the original regenerated mathematical watermarking constraints. Only the legitimate IP seller would be capable of successfully matching their watermarking constraints with those extracted from the IP design. Conversely, an adversary would be unable to regenerate the watermarking constraints successfully and match them with the extracted constraints, failing to establish ownership in an IP court. The inclusion of multiple parameters in the generation of the mathematical watermark, such as parameters selected by the IP seller and several mathematical computations for the determination of watermarking constraints, significantly increases the complexity for an adversary attempting to regenerate the authentic watermark (Sengupta and Anshul, 2024).

Further, to identify instances of IP piracy, the original watermarking constraints can be compared to those extracted from a suspected chip. If a match is found, it indicates the detection of IP piracy. Moreover, in case an adversary attempts to forge the original mathematical watermarking constraints to realize evasion of IP piracy detection, he/she would be unable to succeed in regenerating the original watermarking constraints for implanting it in his/her fake version. This is because the regeneration of the original watermarking constraints depends upon various security factors, such as the IP seller's input z-bit key, AES key, concatenation rule, mathematical framework, mapping/embedding rule, etc., discussed in prior sections (that are only decodable by the original IP seller). Therefore, these watermarking constraints make it complex for an adversary to evade IP piracy detection (Sengupta and Anshul, 2024).

4.5 Analysis of case studies

4.5.1 Security analysis

The security analysis of the mathematical watermarking methodology employs the probability of coincidence and tamper tolerance metrics (Sengupta and Anshul,

2024). The probability of coincidence refers to the likelihood of detecting identical covert security information (constraints) in an unsecured design. In other words, the probability of coincidence is also a measure of false positive. It is formulated as (Sengupta *et al.*, 2023c; Koushanfar *et al.*, 2005; Anshul and Sengupta, 2023a; Hu *et al.*, 2021; Potkonjak, 2006):

$$\text{Probability of coincidence} = (1 - 1/b)^k \tag{4.1}$$

where 'b' represents the number of registers in the SDFG before implanting mathematical watermarking constraints, and 'k' denotes the total integrated/implanted watermarking constraints. A lower probability of coincidence value denotes more robust security, thus providing credible authorship proof and stronger digital evidence. Figure 4.13 presents a comparative analysis of the probability of coincidence among the mathematical watermarking (Sengupta and Anshul, 2024), facial biometric (Sengupta and Rathor, 2021), encrypted signature (Castillo *et al.*, 2008), dynamic watermarking (Koushanfar *et al.*, 2005), and fingerprint biometric (Sengupta and Rathor, 2020). The mathematical watermarking-based security approach outperforms all similar prior approaches with a lower probability of coincidence value. The discussed approach facilitates the determination and incorporation of higher watermarking constraints, thus making the occurrence of

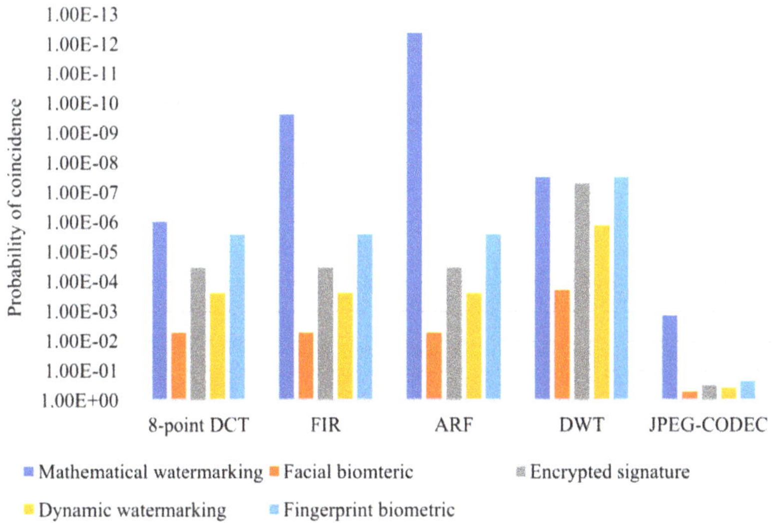

Figure 4.13 Comparison of probability of coincidence among mathematical watermarking (Sengupta and Anshul, 2024), facial biometric (Sengupta and Rathor, 2021), encrypted signature (Castillo et al., 2008), dynamic watermarking (Koushanfar et al., 2005), and fingerprint biometric (Sengupta and Rathor, 2020)

identical security information in an unsecured design highly improbable for an attacker.

Further, tamper tolerance refers to the ability of a security mechanism to withstand and resist unauthorized tampering/removal using brute-force attack. It is formulated as (Sengupta *et al.*, 2023c; Koushanfar *et al.*, 2005; Anshul and Sengupta, 2023a; Hu *et al.*, 2021; Potkonjak, 2006):

$$\text{Tamper tolerance} = v^k \tag{4.2}$$

where 'k' denotes the total implanted watermarking constraints, and 'v' represents the IP seller's selected distinct embedding/mapping variables. A higher tamper tolerance signifies a larger signature space, indicating more robust security. With a higher tamper tolerance value, it is possible to generate various signature combinations. With a larger signature combination, it becomes difficult for attackers to extract the exact watermark (watermarking constraints). Figure 4.14 presents a comparative analysis of the tamper tolerance among the mathematical watermarking (Sengupta and Anshul, 2024), facial biometric (Sengupta and Rathor, 2021), encrypted signature (Castillo *et al.*, 2008), dynamic watermarking (Koushanfar *et al.*, 2005), and fingerprint biometric (Sengupta and Rathor, 2020). The mathematical watermarking approach outperforms all contemporary approaches with a higher tamper tolerance value.

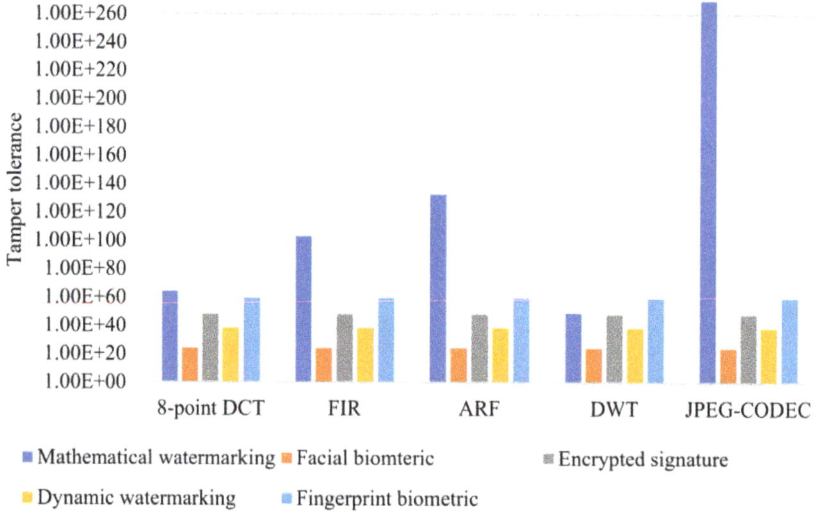

Figure 4.14 *Comparison of tamper tolerance among mathematical watermarking (Sengupta and Anshul, 2024), facial biometric (Sengupta and Rathor, 2021), encrypted signature (Castillo et al., 2008), dynamic watermarking (Koushanfar et al., 2005), and fingerprint biometric (Sengupta and Rathor, 2020)*

Furthermore, Figures 4.15 and 4.16 illustrate the variation in the probability of coincidence and tamper tolerance values with varying mathematical watermark signature size. As the number of embedded watermark signature bits increases, the probability of coincidence decreases, whereas the value of tamper tolerance increases.

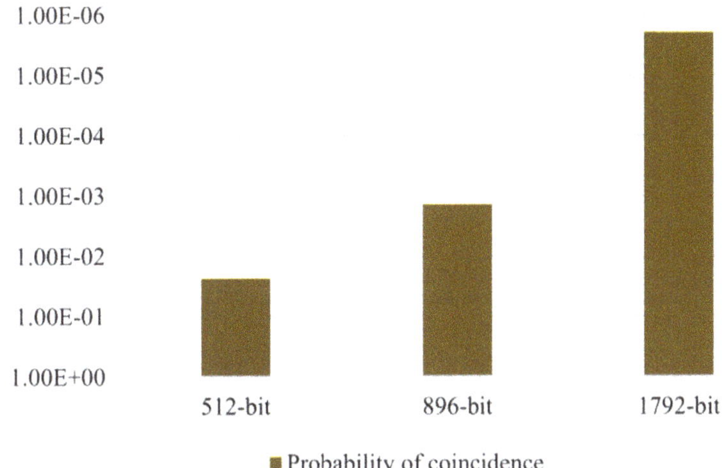

Figure 4.15 Variation in probability of coincidence with varying watermarking signature size for mathematical watermarking technique (Sengupta and Anshul, 2024)

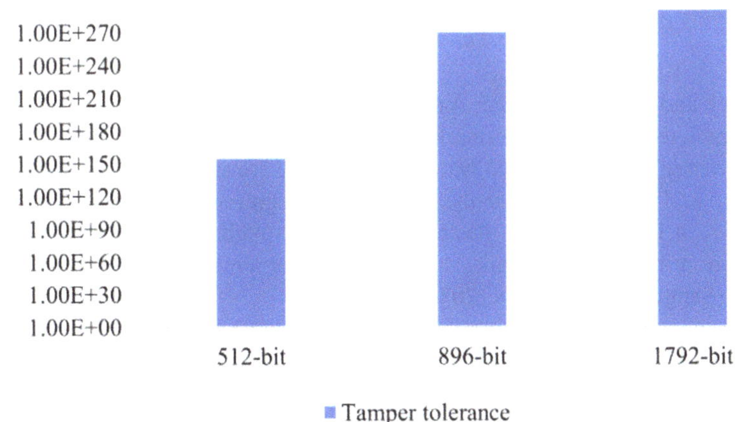

Figure 4.16 Variation in tamper tolerance with varying watermarking signature size for mathematical watermarking technique (Sengupta and Anshul, 2024)

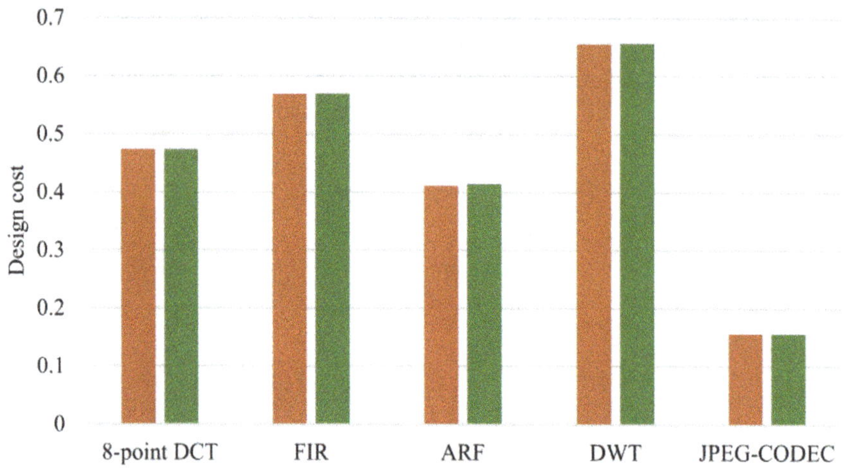

Figure 4.17 Comparison of design cost before and after embedding mathematical watermark (Sengupta and Anshul, 2024)

4.5.2 Design cost analysis

The design cost of the mathematical watermarking approach is evaluated using the following design cost function (Sengupta and Anshul, 2024; Anshul and Sengupta, 2023b; Anshul and Sengupta, 2023c).

$$Design\ \ cost = x1 * (((Design\_Area))/A_{\max}) + x2* (((Design\_Latency))/L_{\max}) \tag{4.3}$$

Here, $x1$ and $x2$ are set to 0.5, indicating equal importance for both area and latency parameters during watermark embedding process. "*Design_Area*" and "*Design_Latency*" denote the final watermarked hardware area and latency (delay). Next, '$A_{\max}$' and '$L_{\max}$' represent the maximum design area and latency, respectively. Figure 4.17 presents the design cost for the mathematical watermarking approach on different benchmarks before and after embedding the generated watermark (Sengupta and Anshul, 2024).

4.6 Conclusion

A mathematical watermarking framework has been discussed in this chapter for offering robust protection against IP piracy and false IP ownership assertion. A reader learns the following concepts by reading this chapter:

(a) A mathematical hardware watermarking framework using HLS that incorporates watermarking constraints derived from the characteristics of hardware design parameters and design space.
(b) Usage of statistical model to extract the characteristics of design parameters (such as area, delay, etc.) and the design space (resource configurations) as watermarking constraints.
(c) Exploitation of dispersion matrix and eigen decomposition framework to generate a robust mathematical watermark.

Comparative analysis of various IP watermarking techniques against existing attacks of ghost insertion search attach, brute force attack, tampering attack, and false IP ownership claim.

4.7 Questions and exercise

1. Explain the growing importance of application-specific computing hardware systems within the context of the modern technological landscape.
2. Discuss the primary security threats associated with the global design supply chain process of hardware systems.
3. Explain the role of the defender in the context of the global design supply chain process.
4. Explain the overview of the encrypted mathematical watermarking-based security approach.
5. List the primary inputs and output of the encrypted mathematical watermarking approach.
6. Briefly explain the mathematical watermark signature generation process along with the details of the dispersion matrix and eigen decomposition framework.
7. Explain the advantages or motivations of the mathematical watermarking approach.
8. Briefly explain the dispersion matrix generation process with a suitable example.
9. Briefly explain the determination of eigen roots from the corresponding design apace with a suitable example
10. Explain the encrypted mathematical watermark generation process using AES and IP seller's concatenation rule.
11. Explain the resolution of false IP ownership claim and IP piracy detection process.
12. Discuss the various security parameters of mathematical watermarking approach that make it forgery proof.
13. What are the security metrics used to compare and analyze the security strength of the discussed security approaches?
14. Discuss the advantages of the mathematical watermarking approach over contemporary approaches in the literature that help in achieving robust security.

References

A. Anshul and A. Sengupta, (2022) "IP Core Protection of Image Processing Filters with Multi-Level Encryption and Covert Steganographic Security Constraints," *2022 IEEE International Symposium on Smart Electronic Systems (iSES)*, Warangal, India, pp. 83–88.

A. Anshul and A. Sengupta, (2023a) "A Survey of High Level Synthesis Based Hardware Security Approaches for Reusable IP Cores [Feature]," *IEEE Circuits and Systems Magazine*, vol. 23, no. 4, pp. 44–62.

A. Anshul and A. Sengupta, (2023b) "Exploration of Optimal Crypto-chain Signature Embedded Secure JPEG-CODEC Hardware IP During High Level Synthesis," *Microprocessors and Microsystems*, vol. 102, 104916.

A. Anshul and A. Sengupta, (2023c) "PSO Based Exploration of Multi-phase Encryption Based Secured Image Processing Filter Hardware IP Core Datapath During High Level Synthesis," *Expert Systems with Applications*, vol. 223, 119927.

A. Anshul and A. Sengupta, (2023d) "Low-Cost Hardware Security of Laplace Edge Detection and Embossment Filter Using HLS Based Encryption and PSO," *2023 IEEE International Symposium on Smart Electronic Systems (iSES)*, Ahmedabad, India, pp. 135–140.

A. Anshul, K. Bharath and A. Sengupta, (2022), "Designing Low Cost Secured DSP Core Using Steganography and PSO for CE systems," *2022 IEEE International Symposium on Smart Electronic Systems (iSES)*, Warangal, India, 2022, pp. 95–100.

E. Castillo, L. Parrilla, A. Garcia, U. Meyer-Baese, G. Botella and A. Lloris, (2008) "Automated Signature Insertion in Combinational Logic Patterns for HDL IP Core Protection," *2008 4th Southern Conference on Programmable Logic*, Bariloche, Argentina, pp. 183–186.

R. Chaurasia, A. Anshul, A. Sengupta and S. Gupta, (2022) "Palmprint Biometric Versus Encrypted Hash Based Digital Signature for Securing DSP Cores Used in CE Systems," *IEEE Consumer Electronics Magazine*, vol. 11, no. 5, pp. 73–80.

B. Colombier and L. Bossuet, (2015) "Survey of Hardware Protection of Design Data for Integrated Circuits and Intellectual Properties," *IET Computers & Digital Techniques*, vol. 8, no. 6, pp. 274–287.

W. Hu, C. Chang, A. Sengupta, S. Bhunia, R. Kastner and H. Li, (2021) "An Overview of Hardware Oriented Security and Trust: Threats, Countermeasures and Design Tools," *IEEE Transactions on Computer-Aided Design of Integrated Circuits and Systems*, Invited Paper, vol. 40, no. 6, pp. 1010–1038.

S. A. Islam, L. K. Sah and S. Katkoori, (2020) "High-Level Synthesis of Key-Obfuscated RTL IP with Design Lockout and Camouflaging." *ACM Transactions on Design Automation of Electronic Systems*, vol. 26, no. 1, pp. 1–35.

F. Koushanfar, I. Hong and M. Potkonjak, (2005) "Behavioral Synthesis Techniques for Intellectual Property Protection," *ACM Transactions on Design Automation of Electronic Systems*, vol. 10, no. 3, pp. 523–545.

B. Le Gal and L. Bossuet, (2012) "Automatic Low-cost IP Watermarking Technique Based on Output Mark insertions," *Design Automation for Embedded Systems*, vol. 16, no. 2, pp. 71–92.

M. Potkonjak, (2006) "Methods and Systems for the Identification of Circuits and Circuit Designs," United State Patent, US7017043B1.

M. Rathor, A. Sengupta, R. Chaurasia and A. Anshul, (2023a) "Exploring Handwritten Signature Image Features for Hardware Security," *IEEE Transactions on Dependable and Secure Computing*, vol. 20, no. 5, pp. 3687–3698.

M. Rathor, A. Anshul, K. Bharath, R. Chaurasia and A. Sengupta, (2023b) "Quadruple Phase Watermarking During High Level Synthesis for Securing Reusable Hardware Intellectual Property Cores," *Computers and Electrical Engineering*, vol. 105, 108476.

M. Rathor, A. Anshul and A. Sengupta, (2024) "Securing Reusable IP Cores Using Voice Biometric Based Watermark," *IEEE Transactions on Dependable and Secure Computing*, vol. 21, no. 4, pp. 2735–2749.

S. Rizzo, F. Bertini, and D. Montesi, (2019) "Fine-grain Watermarking for Intellectual Property Protection," *EURASIP Journal on Information Security*, vol. 2019, Article no. 10.

M. Rostami, F. Koushanfar and R. Karri, (2014) "A Primer on Hardware Security: Models, Methods, and Metrics," *Proceedings of the IEEE*, vol. 102, no. 8, pp. 1283–1295.

R. Schneiderman, (2010) "DSPs Evolving in Consumer Electronics Applications," *IEEE Signal Processing Magazine*, vol. 27, no. 3, pp. 6–10.

A. Sengupta and A. Anshul, (2024) "Watermarking Hardware IPs Using Design Parameter Driven Encrypted Dispersion Matrix with Eigen Decomposition Based Security Framework," *IEEE Access*, vol. 12, pp. 47494–47507, doi:10.1109/ACCESS.2024.3382202.

A. Sengupta and M. Rathor, (2020) "Securing Hardware Accelerators for CE Systems Using Biometric Fingerprinting," *IEEE Transactions on Very Large Scale Integration (VLSI) Systems*, vol. 28, no. 9, pp. 1979–1992.

A. Sengupta and M. Rathor, (2021) "Facial Biometric for Securing Hardware Accelerators," *IEEE Transactions on Very Large Scale Integration (VLSI) Systems*, vol. 29, no. 1, pp. 112–123.

A. Sengupta, R. Chaurasia and A. Anshul, (2023a) "Hardware Security of Digital Image Filter IP Cores against Piracy Using IP Seller's Fingerprint Encrypted Amino Acid Biometric Sample," *2023 Asian Hardware Oriented Security and Trust Symposium (AsianHOST)*, Tianjin, China, pp. 1–6.

A. Sengupta, A. Anshul and R. Chaurasia, (2023b) "Exploration of Optimal Functional Trojan-Resistant Hardware Intellectual Property (IP) Core Designs During High Level Synthesis," *Microprocessors and Microsystems*, vol. 103, 104973.

A. Sengupta, R. Chaurasia and A. Anshul, (2023c) "Robust Security of Hardware Accelerators Using Protein Molecular Biometric Signature and Facial Biometric Encryption Key," *IEEE Transactions on Very Large Scale Integration (VLSI) Systems*, vol. 31, no. 6, pp. 826–839.

M. Yasin, J. J. Rajendran, O. Sinanoglu and R. Karri, (2016) "On Improving the Security of Logic Locking," *IEEE Transactions on Computer-Aided Design of Integrated Circuits and Systems*, vol. 35, no. 9, pp. 1411–1424.

Chapter 5
High-level synthesis-based watermarking using multimodal biometric

Anirban Sengupta[1] *and Aditya Anshul*[1]

This chapter discusses hardware security methodologies for securing the transient fault-secured hardware intellectual property (IP) designs using unified multimodal biometric and an encoded dictionary. The explained framework leverages a high-level synthesis (HLS) process to create transient fault-secured hardware IPs. Further, it integrates encoded unified biometric-based hardware security constraints into the hardware design to protect the designs from IP piracy and false IP ownership claim. This ensures the creation of watermarked fault-secured hardware designs, thereby enhancing IP seller's and end consumer's safety by mitigating the risk of using pirated/unreliable designs. The discussed watermarking methodologies provide stronger digital evidence for piracy detection and greater resilience against tampering attack.

5.1 Introduction

Ensuring robust security of reusable intellectual property (IP) cores is paramount to mitigate risks associated with IP piracy, counterfeiting, and false ownership claim. These challenges pose significant threats to the integrity and exclusivity of valuable hardware IP assets, potentially leading to financial losses, reputational damage for IP sellers, and threats to end consumers. In the contemporary consumer electronics (CE) system design process, there is a reliance on multiple third-party IP sellers (Anshul and Sengupta, 2023a). This exposes the design process to potential piracy, as a malicious actor within the system on chip (SoC)/fabrication houses could unlawfully replicate the design without the original seller's/designer's awareness. From a different perspective, an IP core sourced from an untrustworthy third-party seller must undergo thorough verification to confirm its authenticity before integration into a CE system (Rostami *et al.*, 2014; Rathor *et al.*, 2023a). These precautions are essential because an adversary may attempt to replicate the authentic watermark of the genuine IP design to generate fake IP versions that may impersonate the original IP

[1]Department of Computer Science and Engineering, Indian Institute of Technology Indore, India

design. By doing so, they may seek to circumvent piracy detection measures by embedding the copied watermark into counterfeit and unreliable hardware IP cores (Rizzo *et al.*, 2019). Such pirated IP cores pose significant risks to end consumers as they bypass rigorous testing and quality assurance protocols, making them inherently unreliable and unsafe. Additionally, these counterfeit hardware IPs may contain hardware Trojans, which may cause multiple hazards for the IP seller and end consumers (Anshul and Sengupta, 2023b). Therefore, safeguarding reusable IP cores against piracy, counterfeiting, and false ownership claim is crucial (Schneiderman, 2010; Le Gal and Bossuet, 2012; Anshul and Sengupta, 2023b; Sengupta *et al.*, 2023a; Rathor *et al.*, 2023b; Colombier and Bossuet, 2015; Rathor *et al.*, 2024; Anshul and Sengupta, 2023c; Islam *et al.*, 2020; Anshul *et al.*, 2022; Yasin *et al.*, 2016; Sengupta *et al.*, 2023b; Anshul and Sengupta, 2022).

5.1.1 Why biometric-based watermark over conventional watermark techniques

To address the challenges posed by IP piracy, counterfeiting, and false claims of ownership, techniques involving hardware watermarking and steganography have been proposed in the form of detective control mechanisms and detective countermeasures. These techniques involve selecting a seller's signature and translating it into hardware security constraints implanted within the design of the hardware IP (Sengupta and Rathor, 2019; Sengupta and Bhadauria, 2016). Merely selecting a signature or using randomly generated binary sequences does not accurately represent the seller's unique identity (natural identity). Additionally, physical parameters of integrated circuits (ICs) or images can serve as sources of randomness for generating secret signatures, but they do not inherently reflect the seller's natural identity and can lead to ownership disputes if compromised. Adversaries may attempt to imitate/replicate the watermark to evade piracy detection or falsely claim it as their own covert mark. Therefore, there is a demand for a covert mark derived from the IP seller's biometric (natural) traits to facilitate seamless detection of IP piracy and verification of genuine ownership during conflict. Biometric-based watermarking provides a distinctive natural identity of the IP seller within the design (Sengupta *et al.*, 2021; Chaurasia *et al.*, 2022). Compared to conventional watermarking techniques, biometric-based watermarking offers numerous advantages for securing hardware IP cores, including uniqueness, generation of tamper-resistant signatures, seamless authentication, stronger entropy, and robustness (Rathor *et al.*, 2024; Sengupta *et al.*, 2023a; Sengupta *et al.*, 2023c; Sengupta *et al.*, 2023d; Sengupta and Rathor, 2020; Sengupta and Rathor, 2021).

5.1.2 Why multimodal biometric over unimodal biometric-based watermark for hardware security

The integration of multimodal biometric-based watermark security results in the creation of a highly unique hybrid feature set that cannot be duplicated and compromised/impersonated. This robust hybrid feature set is utilized to generate an imperceptible unified biometric security identifier, offering superior resistance to tampering and a lower likelihood of coincidental matches (probability of coincidence) compared to unimodal biometric hardware watermarking methods. Multimodal biometrics-based IP watermark, which relies on multiple physiological or behavioral attributes, presents numerous advantages over unimodal biometric-based watermarks, which depend on a single characteristic (Rathor *et al.*, 2024; Sengupta *et al.*, 2023a; Sengupta *et al.*, 2023c; Sengupta *et al.*, 2023d; Sengupta and Rathor, 2020; Sengupta and Rathor, 2021; Sengupta *et al.*, 2021; Chaurasia *et al.*, 2022). These advantages include enhanced watermark strength, increased credibility of authorship proof, lower false positive, improved robustness, lower chances of escaping IP piracy detection, etc. This chapter discusses the generation and embedding of unified multimodal biometrics (*viz.,* IP seller's fingerprints, facial and palmprint biometrics) based hardware watermarking (security constraints) within the hardware IP design. These covert security constraints serve as digital evidence and can be utilized to effectively provide detective countermeasure against IP piracy and nullify false IP ownership claim (Sengupta *et al.*, 2023c).

5.1.3 Why hardware security (watermarking) of transient fault-detectable hardware IP cores

Fault-secured hardware designs are crucial for ensuring reliability and security in electronic systems by mitigating the risks associated with potential faults and vulnerabilities. Hardware IP cores designed to be fault-secured are also susceptible to the threat of IP piracy (threats have been discussed earlier in this section), wherein counterfeit fault-secured IP cores may be illicitly incorporated into consumer electronic systems. Consequently, safeguarding these fault-secured IP designs against the risk of piracy is essential (Sengupta *et al.*, 2023c).

This chapter discusses a hardware watermarking-based security approach that utilizes unified biometrics with an encoded dictionary concept to safeguard fault-secured hardware IP designs (Sengupta *et al.*, 2023c). The discussed method leverages unified biometrics to generate hardware watermark (security constraints), enabling detective countermeasure and detective control against pirated IP cores (Sengupta *et al.*, 2023c).

5.2 Hardware watermarking-based security framework for handling IP piracy

This chapter discusses a watermarking-based security methodology for safeguarding fault-secured hardware IP designs against piracy and false IP ownership claim threats by employing a unified multimodal biometric-based hardware security approach coupled with an encoded dictionary (Sengupta *et al.*, 2023c). This technique can be applied not only to fault-secured hardware IP designs but also to regular ones. Figure 5.1 illustrates the overview of the unified multimodal biometric-based security approach, where the primary inputs are as follows: the control data flow graph (CDFG) of the hardware application, resource constraints, module library, fingerprint, facial, and palmprint biometric data of the original IP

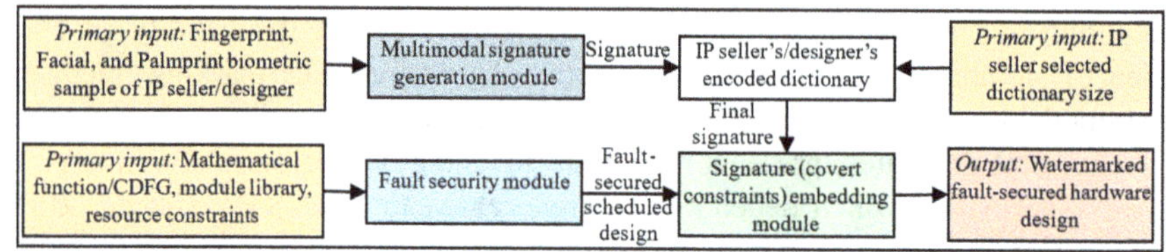

Figure 5.1 Overview of fault-secured multimodal biometric-based hardware watermarking methodology

seller, along with the designated dictionary size. The output is a protected fault-secured hardware IP design using multimodal biometrics and an encoded dictionary (Sengupta *et al.*, 2023c).

As depicted in Figure 5.1, the process of creating the protected fault-secured design involves four main processing modules: (a) the multimodal signature generation module, (b) the transient fault-security module, (c) the encoded dictionary module, and (d) the signature (covert constraints) embedding module. The first module uses captured images of the IP seller's fingerprint, facial, and palmprint to generate a unified biometric binary signature serving as digital evidence within the hardware design. The second module takes inputs such as the CDFG, resource constraints, transient fault strength (K_C), and module library, producing the fault-secured hardware design. Subsequently, the third module processes the obtained signature to produce covert security constraints based on the chosen combination and strength (with the help of the IP seller's selected encoded dictionary). It's worth noting that the number of covert security constraints embedded can be increased by incorporating more biometric features, each with its respective encoding. The strength of embedded security information (i.e., digital evidence) can be adjusted by varying the number of biometric features. Additionally, the specific set of security constraints depends on the size of the encoded dictionary and the corresponding encoded bits selected by the IP seller. A larger encoded dictionary provides more options for selecting the biometric signature (watermark) to generate covert security constraints (Sengupta *et al.*, 2023c).

The fourth module takes the generated transient fault-secured schedule and encoded dictionary-based unified biometric signature to produce a watermarked register transfer level (RTL) datapath using high-level synthesis (HLS) framework. This approach effectively safeguards generated fault-secured hardware designs against IP piracy and false IP ownership claim (Sengupta *et al.*, 2023c). Detailed discussions on each module are provided in the subsequent subsections.

5.2.1 Multimodal biometric-based hardware watermarking approach

The unified multimodal biometric-based watermarking approach involves the fusion of three distinct biometric approaches: IP seller's fingerprint, facial, and palmprint biometrics-based hardware security watermarks. This amalgamation

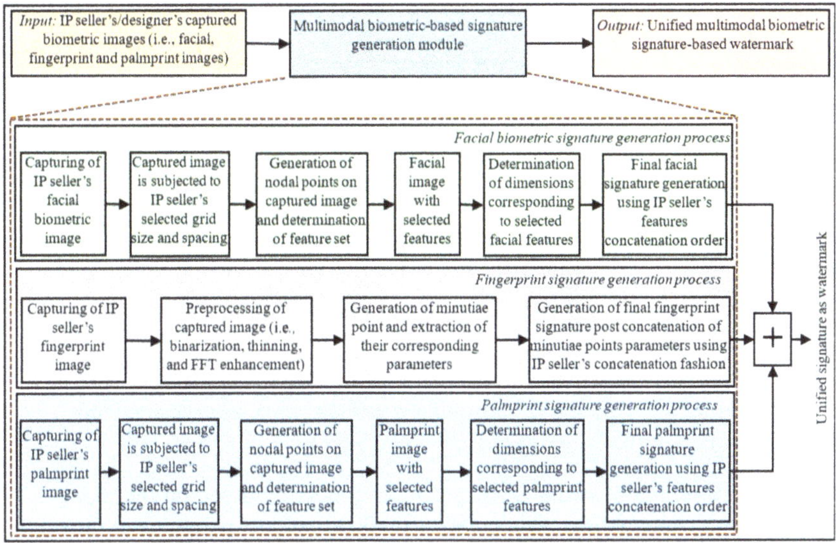

Figure 5.2 Generation process of unified multimodal biometric signature (watermark)

generates a covert unified multimodal biometric-based security signature (i.e., watermark). Figure 5.2 provides an overview of the process for generating signatures based on fingerprint, facial, and palmprint biometrics. The primary input to the multimodal biometric-based signature generation module comprises the IP seller's captured biometric images, and the output is a unified multimodal biometric signature-based watermark. In the fingerprint signature generation process, the captured fingerprint is subjected to preprocessing steps, and a further preprocessed image is used for minutiae points generation (Sengupta *et al.*, 2023c). Next, the overall process of facial and palmprint biometrics is almost similar, where the input image is subjected to the IP seller's specified grid size and spacing (Sengupta *et al.*, 2023c). After this, the nodal points and corresponding feature sets are extracted, forming the foundation of signature generation (Sengupta *et al.*, 2023c). The details of the signature generation corresponding to each biometric and final unified multimodal-based watermark are discussed in the next subsections.

5.2.2 Generating transient fault-detectable dual modular redundant (DMR) schedule

As motivated earlier in the introduction section regarding the importance of fault-detectable hardware IP designs, this subsection presents the generic design process of creating fault-detectable hardware IP designs (Sengupta *et al.*, 2023c).

Figure 5.3 depicts the process for generating K_C-cycle transient fault-secured hardware designs, which involves accepting the CDFG of an input hardware application, the transient fault strength (K_C), resource constraints, and the module

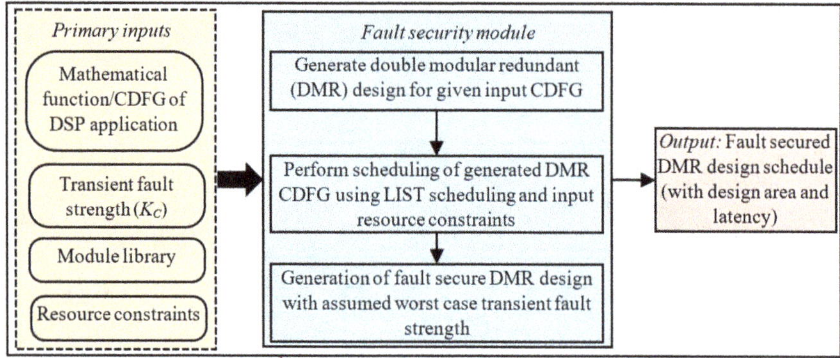

Figure 5.3 Process of designing transient fault secured schedule

library as inputs. Initially, the DMR design is generated based on the input CDFG of the hardware application. This involves duplicating the operations of the original unit (M_{OG}) to obtain a duplicate unit (M_{DP}) and combining both units to form the DMR design of the hardware application (Sengupta *et al.*, 2017). Following this, the obtained DMR design is scheduled using input resource constraints represented as $X_C = \{HX_1, HX_2 \ldots\ldots HX_d\}$, where '$HX_n$' denotes the number of hardware/resource units and 'd' represents the hardware resource types. The scheduling of the DMR design utilizes the LIST scheduling algorithm. Following this, once the scheduled DMR design ($SDFG_{DMR}$) is obtained, the K_C-cycle fault security rules are implemented, and adjustments are made to the scheduled DMR design locally (Sengupta *et al.*, 2017). These fault security rules govern operation assignment to hardware components based on fault resiliency conditions. The following rules have been used. Assign operation *opn* (*P*) to M_{OG} and operation *opn* (P') to M_{DP} using separate operators if available. Here, M_{OG} and U_{DP} denote the original and duplicate units, respectively. If the distinct operators (hardware sellers) are unavailable, the same assignment is maintained for operations in the duplicate unit based on the following condition mentioned in (5.1) (SEU, 2024; Sengupta, 2016; Sengupta and Kachave, 2017; Yuce *et al.*, 2019):

$$D(P') - D(P) > K_C \tag{5.1}$$

where $D(P')$ and $D(P)$ are the respective timestamps of sister operations P' and P respectively. However, if any of the above conditions are violated, it may result in transient fault hazard between similar operations assigned to similar hardware units, leading to undetectability. A transient fault hazard occurs between similar operations within the same hardware when (Sengupta, 2015):

$$D(P') - D(P) \leq K_C \tag{5.2}$$

where $(P) \in U_{OG}$ and $(P') \in U_{DP}$. To address these hazards, affected operations and their successors are pushed to the duplicate unit in later control steps, ensuring a sufficient time stamp interval between operations in the original and duplicate units (Sengupta *et al.*, 2023c).

5.2.3 Case study of fault-detectable FIR IP

In the previous subsection, we discussed the process of generating transient fault-secured (detectable) IP design. In this subsection, we subsequently employ a finite impulse response (FIR) filter as a case study to explain the process. Figure 5.4(a) depicts the scheduled data flow graph (SDFG) of the FIR filter (Sengupta *et al.*, 2023c; Sengupta and Mohanty, 2019; EXP BEN, 2024) scheduled using three adders and four multipliers. Next, Figure 5.4(b) illustrates the fault-secured design derived from the scheduled DMR associated with the DFG of the FIR filter (with worst-case transient fault strength K_C = 2). The worst-case pulse width of this temporal effect is considered to be up to 2000ps, 1000ps per control step (Sengupta and Kachave, 2017; Rezgui *et al.*, 2008; Lisboa and Carro, 2007). Through the configuration of K_C = 2T and hardware reallocation after every two control steps (T), the methodology effectively considers the impact of the transient fault, ensuring transient fault detection (Sengupta *et al.*, 2023c). In this representation, J_0 to J_{61} indicate storage variables utilized in the design for managing input and intermediate values, T0 to T14 represent the required control steps for scheduling the fault-secured design, and M1, M2, A1, A2, and A3 symbolize multipliers and adders, respectively.

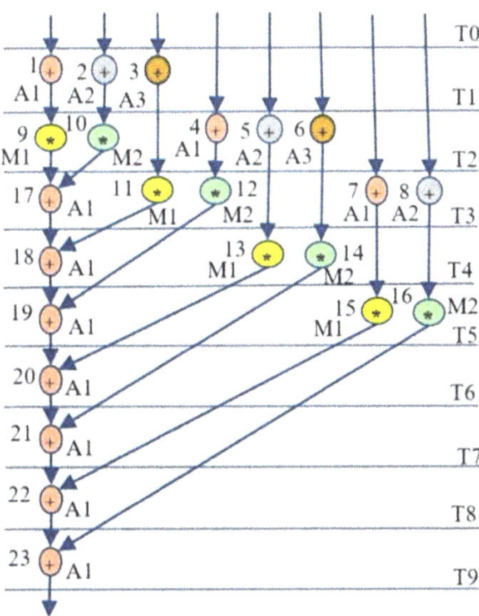

Figure 5.4 (a) Scheduled DFG of FIR filter using three adders and two multipliers. (b) Fault-secured scheduled DFG of FIR filter embedded with encoded dictionary-based unified signature.

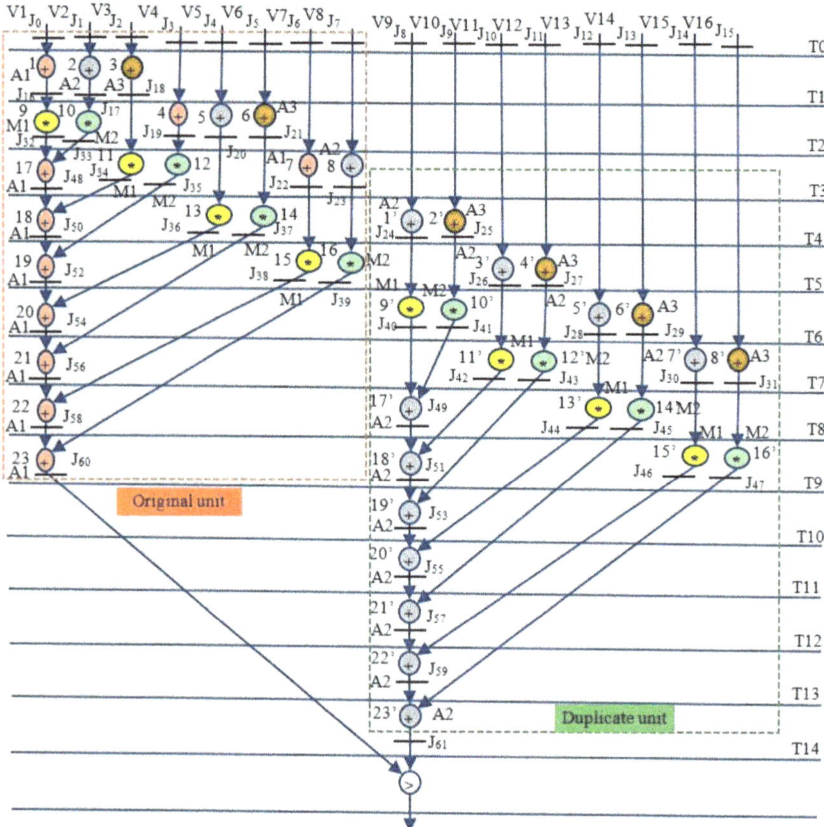

Figure 5.4 (Continued)

5.2.4 Fingerprint biometric signature generation process

The complete fingerprint signature generation process is illustrated in Figure 5.5 (Sengupta *et al.*, 2023c). The initial step involves capturing the IP seller's fingerprint impression using a fingerprint optical scanner. The initial captured fingerprint is depicted in Figure 5.5(a). Subsequently, several preprocessing steps are employed to accurately extract the minutiae points from the fingerprint sample. This preprocessing stage encompasses three key sub-processes: (i) image enhancement utilizing fast Fourier transform (FFT) to amplify and reconnect broken ridges, thereby improving the quality of the image. (ii) Binarization reduces the image to two intensity levels ('0' representing low intensity and '255' representing high intensity) by comparing pixel intensities with a threshold value. The binarized fingerprint image is illustrated in Figure 5.5(b). Next, (iii) thinning is a process that reduces the thickness of ridge lines, resulting in the thinned image shown in Figure 5.5(c). After the preprocessing steps, the thinned fingerprint image is analyzed to extract minutiae points, which are crucial in

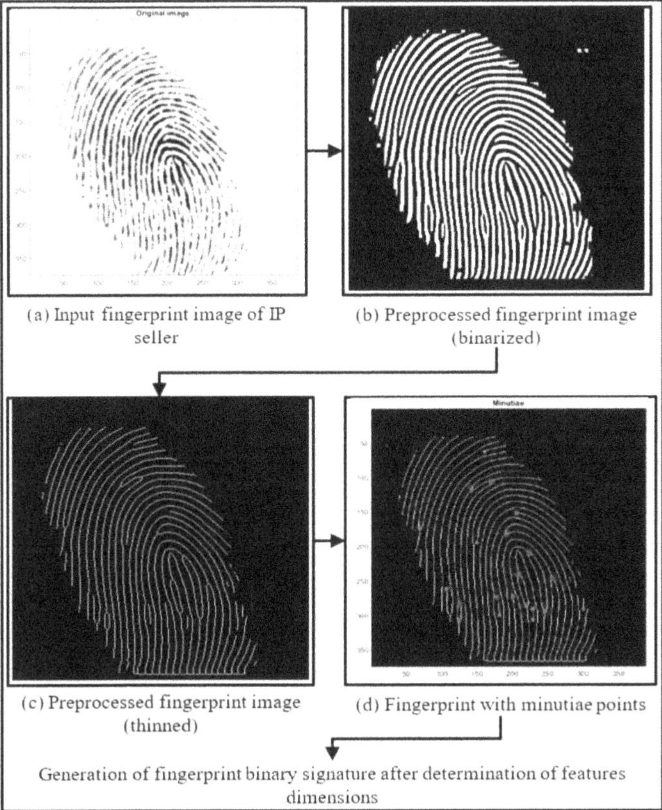

Original image
(a) Input fingerprint image of IP seller
(b) Preprocessed fingerprint image (binarized)
Minutiae
(c) Preprocessed fingerprint image (thinned)
(d) Fingerprint with minutiae points
Generation of fingerprint binary signature after determination of features dimensions

Figure 5.5 Fingerprint feature extraction and its corresponding signature generation process

identifying unique features of an IP seller's/designer's fingerprint. Minutiae points are categorized into two types: ridge endings (depicted in red), where ridge lines terminate abruptly, and ridge bifurcations (depicted in blue), where a ridge line branches into two. The fingerprint image with generated minutiae points is shown in Figure 5.5(d). Next, each minutiae point is represented in the corresponding binary form, incorporating minutiae point parameters, such as coordinates (x,y), minutiae type, and the ridge angle in degrees to construct the fingerprint signature. Subsequently, a final fingerprint signature is formed by concatenating the binary representations of all minutiae point parameters. The strength and characteristics of the fingerprint signature can vary depending on the number of minutiae points considered and the order of concatenation, offering flexibility to the IP designer (Sengupta *et al.*, 2023c). Table 5.1 depicts the coordinates of selected fingerprint minutiae points chosen by the IP seller, along with their corresponding minutiae type, ridge angles, and binary representations.

Table 5.1 Minutiae coordinates, type, angle, and its respective binary equivalent post concatenation of all minutiae parameters

X-axis	Y-axis	Minutiae type	Angle (in degree)	Binary equivalent
161	63	1	153	1010 0001111111110011001
171	106	3	337	1010101111010101110101000 1
143	118	3	130	1000111111101101110000010
207	152	1	187	1100111110011000110111011
70	174	3	99	1000110101011101111100011
191	181	3	131	1011111110110101110000011
150	195	1	95	1001011011000011110111111
224	210	3	234	1110000011010010111101010
210	241	3	252	1101001011110001111111100
257	247	3	247	10000000111110111111110111
107	262	1	262	1101011100000110110000011 0
201	272	1	255	1100100110001000011111111
179	274	1	259	10110011100010010110000001 1

5.2.5 Facial biometric signature generation process

The complete facial biometric signature generation process is illustrated in Figure 5.6 (Sengupta *et al.*, 2023c). The initial step involves capturing the facial biometric data of the IP seller using an image-capturing device (such as a high-resolution camera). Subsequently, the captured facial biometric image undergoes a process where a specific grid size/spacing is applied. The facial biometric image with grid size and spacing is depicted in Figure 5.6(b). After this, nodal points are generated on the facial image based on a predetermined set of facial features chosen by the IP seller. A total of 16 nodal points (X1 to X16, indicated in red) are generated to identify these facial features (shown in Figure 5.6(c)). Next, a facial image is generated where the distance between two nodal points represents each feature. Eight facial features are selected in the facial biometric image, as illustrated in Figure 5.6(e). Subsequently, the dimensions corresponding to each facial feature are determined using Manhattan distance, resulting in a decimal value for each feature. These decimal values (magnitude of each feature) are then transformed into their binary equivalents. Finally, the facial biometric signature is generated by combining the binarized signature of each facial feature through concatenation. The IP seller/designer can customize the concatenation sequence to obtain the desired strength's facial biometric signature (Sengupta *et al.*, 2023c). Table 5.2 provides details of the selected facial features by the IP seller, along with corresponding nodal points, coordinates, feature dimensions, and their binary representations.

5.2.6 Palmprint biometric signature generation process

The complete palmprint signature generation process is illustrated in Figure 5.7 (Sengupta *et al.*, 2023c). The initial step involves capturing the palmprint image of the IP seller using a high-resolution camera. Subsequently, the captured palmprint

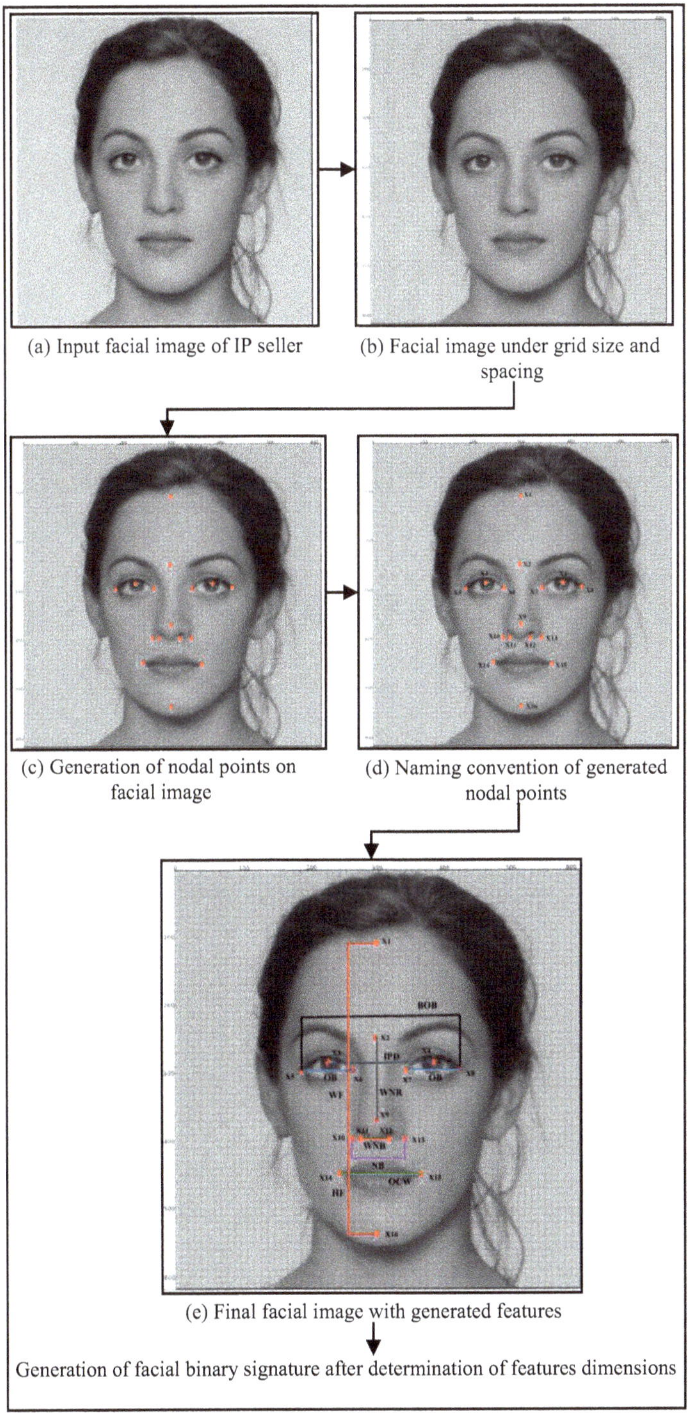

Figure 5.6 Facial feature extraction and its corresponding signature generation process

Table 5.2 IP seller selected facial features, corresponding nodal points, coordinates, feature dimensions, and its binary equivalent representation

Feature No.	Facial feature abbreviation	Facial feature name	Nodal points name	Co-ordinates *(s1,s2)-(t1,t2)*	Feature dimension (Manhattan distance)	Binary equivalent representation
1	*WNR*	Width of nasal ridge	(X2) – (X9)	(305, 250)–(305, 370)	120	1111000
2	*OCW*	Oral commissure width	(X14) – (X15)	(250, 450)–(370, 450)	120	1111000
3	*OB*	Ocular breadth	(X5) – (X6)	(190,300)–(190, 270)	30	11110
4	*BOB*	Bio-ocular breadth	(X5) – (X8)	(190,300)–(430, 300)	240	11110000
5	*NB*	Nasal breadth	(X10) – (X13)	(270, 400)–(350, 400)	80	1010000
6	*IPD*	Inter pupillary distance	(X3) – (X4)	(230, 285)–(390, 285)	160	10100000
7	*WNB*	Width of nasal base	(X11) – (X12)	(280, 400)–(325, 400)	45	101101
8	*HF*	Height of face	(X1) – (X16)	(305, 110)–(305,540)	430	110101110

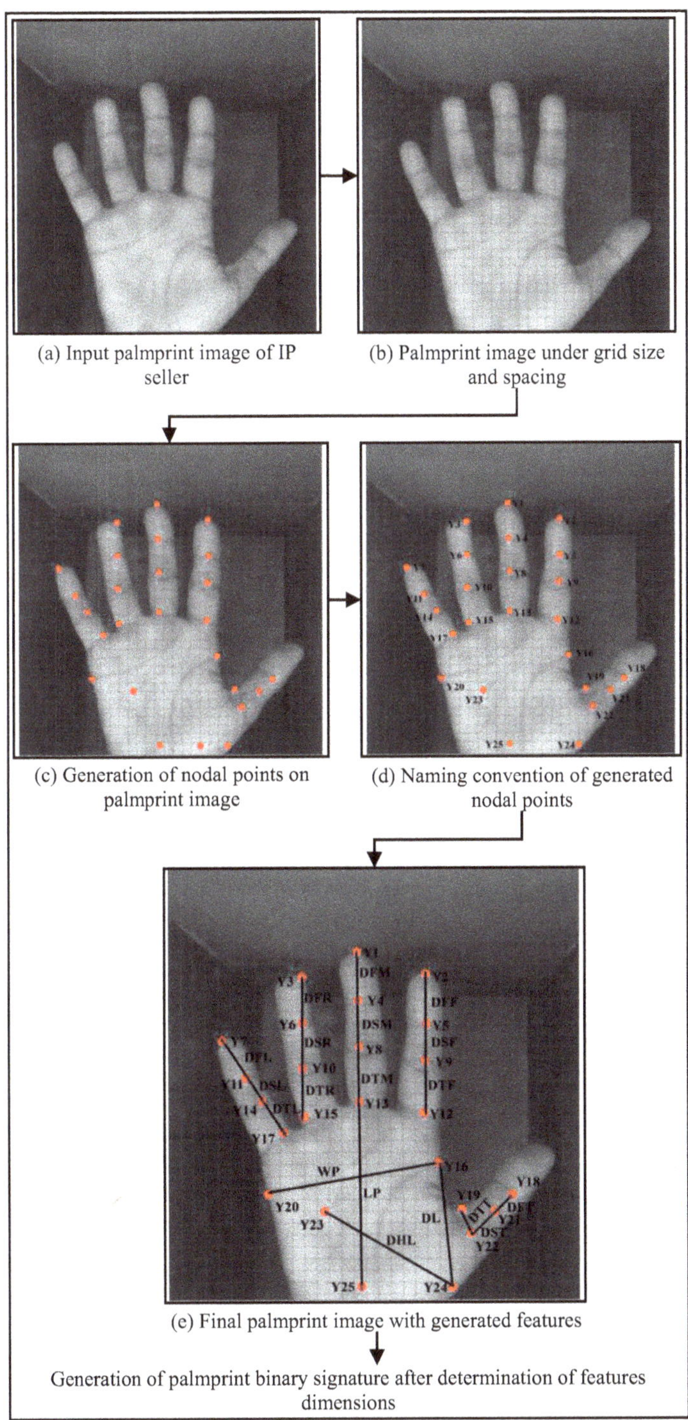

Figure 5.7 Palmprint feature extraction and its corresponding signature generation process

biometric image is subjected to specific grid size/spacing. The palmprint biometric image with grid size and spacing is depicted in Figure 5.7(b). Subsequently, nodal points are generated on the palmprint image based on a predetermined set of palmprint features chosen by the IP seller, as illustrated in Figure 5.7(c). Next, Figure 5.7(d) depicts the naming assignment of the nodal points. Each palm feature is measured as a function of the distance between two nodal points (marked in red). A palm image with the selected palm features is then generated (as depicted in Figure 5.7(e)), containing all the necessary details for generating the palmprint signature. Next, the feature dimensions of all selected features are determined using the Manhattan distance. Each feature dimension (decimal value) is then converted into its corresponding binary equivalent. Finally, the palmprint signature is obtained by concatenating the binary string of each palm feature. It's important to note that various signature combinations are possible based on different concatenation orders chosen by the IP seller (Sengupta *et al.*, 2023c). Table 5.3 provides details of the palmprint features selected by the IP seller, along with corresponding nodal points, coordinates, feature dimensions, and their binary representations.

Consequently, a unified multimodal biometric signature (L) based on the encoded dictionary rule (discussed in the next subsection) is generated by concatenating the signatures corresponding to the fingerprint, facial, and palmprint biometrics of the IP seller/designer (Sengupta *et al.*, 2023c). The resulting signatures for each biometric modality are presented below.

The fingerprint biometric signature is as follows: 101000011111 11110011001101010111101010111010100011000111111011011100000101100 111110011000110111011100011010101110111100011101111111011010111100 00011100101101100001111011111111100000011010010111110101011010010111 100011111111100100000001111101111111110111110101110000011011000001 1011001001100010 0 0011111111111011 0011100010010110000 00011.

The facial biometric signature is as follows: 11110001111000111110 1111000010100001010000010110111010 1110.

The palmprint biometric signature is as follows: 10011101110101011 0000011011111101 0101101000011010011010000101010110100001001 11011.

5.2.7 *Encoded dictionary (IP seller/designer selected)*

As previously explained, the primary objective of the encoded dictionary module is to determine the final unified signature (used for embedding in the hardware design) with the selected strength and combination chosen by the IP seller. The encoded dictionary (created by the IP seller/designer) contains details regarding encoding bits and their respective encoding rules. For the sake of brevity, an example of 3-bit encoded dictionary is illustrated in Figure 5.8, showing eight different encoding rules (2^X, where X = dictionary size; e.g., X = 3 bit) for choosing unique combinations of the unified multimodal signature. Similarly, multiple encoding rules corresponding to encoding bits can be devised to construct an expandable encoded dictionary. Moreover, the IP seller has the flexibility to choose

Table 5.3 Selected 11 palmprint features, corresponding nodal points, coordinates, feature dimensions, and corresponding binary representation

Feature. No.	**Facial feature abbreviation**	**Facial feature name**	**Nodal points name**	**Co-ordinates *(s1,s2)–(t1,t2)***	**Feature dimension**	**Binary equivalent representation**
1	*LP*	Length of palm	(Y13) – (Y25)	(325,415)–(325,730)	315	100111011
2	*DFF*	Distance between first consecutive intersection points of forefinger	(Y2) – (Y5)	(445,185)–(445,270)	85	1010101
3	*DSF*	Distance between second consecutive intersection points of forefinger	(Y5) – (Y9)	(445,270)–(445,335)	65	1000001
4	*DTF*	Distance between third consecutive intersection points of forefinger	(Y9) – (Y12)	(445,335)–(445,430)	95	1011111
5	*DFM*	Distance between first consecutive intersection points of middle finger	(Y1) – (Y4)	(325,145)–(325,230)	85	1010101
6	*DSM*	Distance between second consecutive intersection points of middle finger	(Y4) – (Y8)	(325,230)–(325,310)	80	1010000
7	*DTM*	Distance between third consecutive intersection points of middle finger	(Y8) – (Y13)	(325,310)–(325,415)	105	1101001
8	*DFR*	Distance between first consecutive intersection points of ring finger	(Y3) – (Y6)	(230,190)–(230,270)	80	1010000
9	*DSR*	Distance between second consecutive intersection points of ring finger	(Y6) – (Y10)	(230,270)–(230,355)	85	1010101
10	*DTR*	Distance between third consecutive intersection points of ring finger	(Y10) – (Y15)	(230,355)–(230,435)	80	1010000
11	*LP*	Length of palm	(Y13) – (Y25)	(325,415)–(325,730)	315	100111011

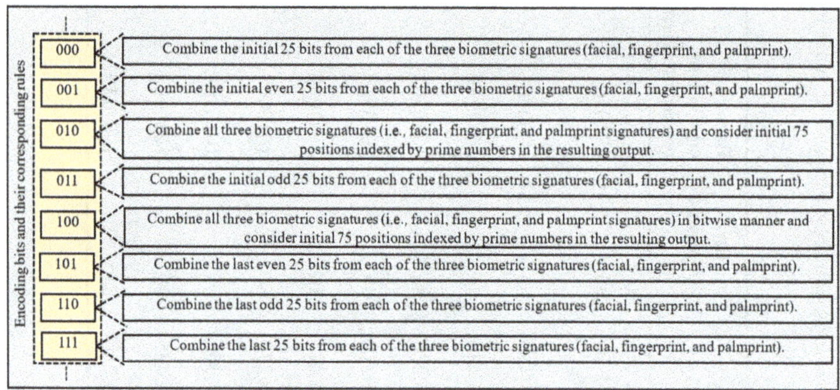

Figure 5.8 Example of an encoded dictionary with 3 bits ($X = 3$), which can expand up to 2^X encoding rules

the target unified multimodal biometric signature of desired strength and combination based on the chosen encoding bits (specific value of dictionary size X), thereby enhancing the robustness of the unified multimodal biometric signature. Subsequently, the obtained multimodal biometrics signature is implanted in the hardware design. For example, the resulting IP designer selected unified signature is obtained as $L = 75$ bits according to a specific encoded dictionary. The resulting sequence is: "0001100101111111 1010111010001011010000001000110010 1100010111100011001011010", consisting of 38 *zeros* and 37 *ones* (Sengupta *et al.*, 2023c). The watermark signature embedding process is elaborated in the following subsection.

5.2.8 *Unified multimodal biometric signature (watermark) generation and its embedding*

The unified multimodal biometric watermark generation process has already been discussed in Section 5.2.6. The procedure for embedding the signature comprises two stages. Firstly, the IP designer determines covert hardware security constraints from the chosen final watermark (discussed in the previous subsection). The generation of covert security constraints corresponding to the chosen signature relies on the SDFG of the fault-secured DMR design and the IP seller's chosen encoding rule. The SDFG of the target transient fault-secured hardware design helps in the determination of the number of storage variables utilized to construct pairs of hardware security constraints. The IP seller selected encoding rule is as follows:

For a signature bit '*0*', the security constraints are embedded between even-even storage variable pairs (J_X, J_Y) of the fault-secured SDFG, where '*X*' and '*Y*' denote the storage variable numbers. Thus, the resulting security constraints for 38 '0's in the unified multimodal biometric signature (derived earlier) are *<J0, J2>,< J0, J4>,<J0, J6>,<J0,J8>,<J0,J10>,<J0,J12>,<J0,J14>,<J0,J16>,<J0,J18>,*

<J0,J20>,<J0,J22>,<J0,J24>,<J0,J26>,<J0,J28>,… …..,<J0,J56>,<J0, J58>,<J0,J60>,<J2,J4>,<J2,J6>,<J2,J8>,<J2,J10>,<J2,J12>,<J2,J14>,<J2, J16>,<J2,J18>.

For a signature bit '*1*', the security constraints are embedded between odd-odd storage variable pairs of the scheduled DFG. Consequently, the resulting security constraints for 37 '1's in the unified multimodal biometric signature are *<J1, J3>,<J1,J5>,<J1,J7>,<J1,J9>,<J1,J11>,<J1,J13>,<J1,J15>,<J1,J17>, <J1,J19>,<J1,J21>,<J1,J23>,<J1,J25>,<J1,J27>,<J1,J29>,… ……, <J1,J57>,<J1,J61>,<J3,J5>,<J3,J7>,<J3,J9>,<J3,J11>,<J3,J13>,<J3, J15>,<J3,J17>.*

Then, the obtained covert security constraints are integrated into the hardware design during the register allocation phase of the HLS process without incurring any additional design costs. The register allocation table (RAT) for the transient fault-secured DMR design is generated (providing details of storage variables, control steps, and register allocation information). Subsequently, the register allocation information and covert security constraints are fed into the constraints embedding module, which generates the unified biometric watermark embedded register transfer level (RTL) datapath for the transient fault-secured hardware design. To embed the covert security constraints into the fault-secured design (i.e., the corresponding RAT), local adjustments are made among registers to accommodate the security constraints, ensuring that storage variables within incoming constraint pairs do not share the same register. Thus, adjustments among registers are made accordingly. If any incoming security constraints cannot be accommodated within existing registers (due to register conflict as mentioned above), either a local swapping between registers is made, or a new register is allocated (Sengupta *et al.*, 2023c; Anshul and Sengupta, 2023b; Anshul and Sengupta, 2023c). However, security constraints that inherently satisfy the distinct register assignment rule do not require such local adjustments (Sengupta *et al.*, 2023c). Table 5.4 shows the register allocation information after embedding the generated unified watermark, along with the required number of control steps for obtaining the watermark-embedded transient fault-secured schedule design.

5.2.9 How IP piracy is detected and how false IP ownership claim is resolved?

The security constraints for the discussed unified multimodal biometric approach are regenerated during piracy detection. From the suspected chip, the layout file is reverse-engineered to retrieve the RTL information of the IP design. Based on the regenerated multimodal biometric information, matching is performed with the retrieved RTL IP design. In case of complete match of the regenerated unified watermark constraints with the security constraints of the retrieved RTL IP design from the suspected chip, IP piracy is detected (Sengupta *et al.*, 2023c). *Note: The regeneration process does not require the reacquisition of the genuine IP seller's multimodal biometric unified watermark. Instead, the original pre-stored (in an*

Table 5.4 Register allocation table before and after embedding crypto-chain-based security constraints corresponding to FIR

Control steps	T0	T1	T2	T3	T4	T5	T6	T7	T8	T9	T10	T11	T12	T13	T14
V1	J0	J17	J19	J23	J23	J27	J27	J31	J31	J47	J47	J47	J47	J47	J61
V2	J1	J16	J20	J20	J24	J24	J28	J28	J44	J44	J44	—	—	—	—
V3	J2	—	J21	J21	J25	J25	J29	J29	J45	J45	J45	J45	J57	J59	—
V4	J3	J3	J32	J22	J22	J26	J26	J30	J30	J46	J46	J46	J46	—	—
V5	J4	J4	J33	J34	J36	J36	J40	J40	J49	J51	J53	J55	—	—	—
V6	J5	J5	—	J35	J35	J38	J38	J38	J58	J60	—	—	—	—	—
V7	J6	J6	J6	J48	J37	J37	J37	J42	J42	—	—	—	—	—	—
V8	J7	J7	J7	—	J50	J39	J39	J39	J39	—	—	—	—	—	—
V9	J8	J8	J8	J8	—	J52	J41	J41	—	—	—	—	—	—	—
V10	J9	J9	J9	J9	—	—	J54	J43	J43	J43	—	—	—	—	—
V11	J10	J10	J10	J10	J10	—	—	J56	—	—	—	—	—	—	—
V12	J11	J11	J11	J11	J11	—	—	—	—	—	—	—	—	—	—
V13	J12	J12	J12	J12	J12	J12	—	—	—	—	—	—	—	—	—
V14	J13	J13	J13	J13	J13	J13	—	—	—	—	—	—	—	—	—
V15	J14	J14	J14	J14	J14	J14	J14	—	—	—	—	—	—	—	—
V16	J15	J15	J15	J15	J15	J15	J15	—	—	—	—	—	—	—	—
V17	—	J18	J18	—	—	—	—	—	—	—	—	—	—	—	—

encrypted format in a safe database) multimodal biometric data of the genuine IP seller is utilized for regeneration during the matching process. The same biometric features, their respective signature, and corresponding security constraints can be precisely regenerated from the pre-stored fingerprint, facial, and palmprint images during IP piracy detection.

Furthermore, the inclusion of the unified multimodal biometric-based watermark within the IP design also safeguards against the assertion of false IP ownership claim. Instances of fraudulent IP ownership claim may arise in the SoC design house and foundry. In the case of ownership conflict resolution, the original security constraints are regenerated and then compared against the extracted security constraints from the register allocation of the IP design under test. Ownership rights are awarded to the entity that can establish complete matching. Thus, the discussed security algorithm effectively resolves false IP ownership assertion (Sengupta *et al.*, 2023c).

5.3 Analysis of case studies

5.3.1 Security analysis

The security analysis of the discussed unified multimodal biometric-based security methodology employs the probability of coincidence and tamper tolerance (Sengupta *et al.*, 2023c). The probability of coincidence refers to the likelihood of detecting identical covert security information (constraints) in an unsecured design.

In other words, the probability of coincidence is also a measure of false positives. It is represented as (Sengupta *et al.*, 2023c; Koushanfar *et al.*, 2005; Sengupta and Bhadauria, 2016; Hu *et al.*, 2021; Potkonjak, 2006):

$$\text{Probability of coincidence} = (1 - 1/u)^k \tag{5.3}$$

where 'u' represents the number of registers in the schedule design before incorporating covert security constraints, and 'k' denotes the total implanted covert security constraints. A lower probability of coincidence value denotes more robust security, thus providing stronger digital evidence and credible authorship proof. The incorporation of distinct unified biometric-based watermark into the hardware IP significantly contributes to effective IP piracy detection process and legal proof of IP ownership. Figure 5.9 presents a comparative analysis of the probability of coincidence between the unified multimodal biometric-based approach (Sengupta *et al.*, 2023c), palmprint biometric (Sengupta *et al.*, 2021), (Chaurasia *et al.*, 2022), steganography (Sengupta and Rathor, 2019), dynamic watermarking (Koushanfar *et al.*, 2005), fingerprint biometric (Sengupta and Rathor, 2020), and chromosomal DNA (Sengupta and Chaurasia, 2022). The discussed unified biometric-based watermarking approach outperforms all contemporary approaches with a lower probability of coincidence value (Sengupta *et al.*, 2023c).

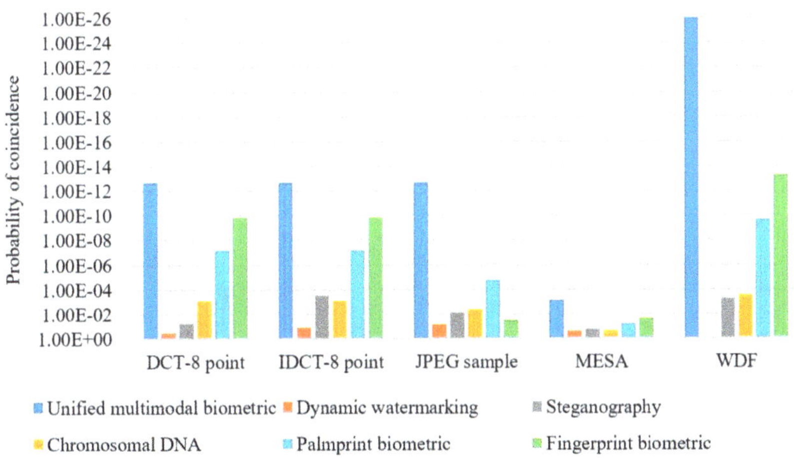

Figure 5.9 Comparison of probability of coincidence between unified multimodal biometric (Sengupta et al., 2023d), dynamic watermarking (Koushanfar et al., 2005), steganography (Sengupta and Rathor, 2019), chromosomal DNA (Sengupta and Chaurasia, 2022), palmprint biometric (Sengupta et al., 2021), (Chaurasia et al., 2022), and fingerprint biometric (Sengupta and Rathor, 2020)

Further, tamper tolerance refers to the ability of a security mechanism to withstand and resist unauthorized tampering using brute-force attack. It is represented as (Sengupta *et al.*, 2023c; Koushanfar *et al.*, 2005; Sengupta and Bhadauria, 2016; Hu *et al.*, 2021; Potkonjak, 2006):

$$\text{Tamper tolerance} = v^s \tag{5.4}$$

where 's' denotes the total implanted covert security constraints, and 'v' represents the IP seller's selected distinct mapping variables. A higher tamper tolerance signifies a larger signature space, indicating more robust security. With a higher tamper tolerance value, it is possible to generate various signature combinations. With a larger signature combination, it becomes difficult for attackers to extract the exact watermark (covert security constraints corresponding to the watermark signature). An adversary aims to tamper or remove the embedded watermark. Further, an adversary may attempt to regenerate the exact watermark signature to insert in fake IPs, such that IP evasion of IP piracy detection can be performed. Thus, a higher tamper tolerance value impedes an adversary from realizing his malicious intent. Figure 5.10 presents a comparative analysis of the tamper tolerance between the unified multimodal biometric-based approach (Sengupta *et al.*, 2023c), palmprint biometric (Sengupta *et al.*, 2021),

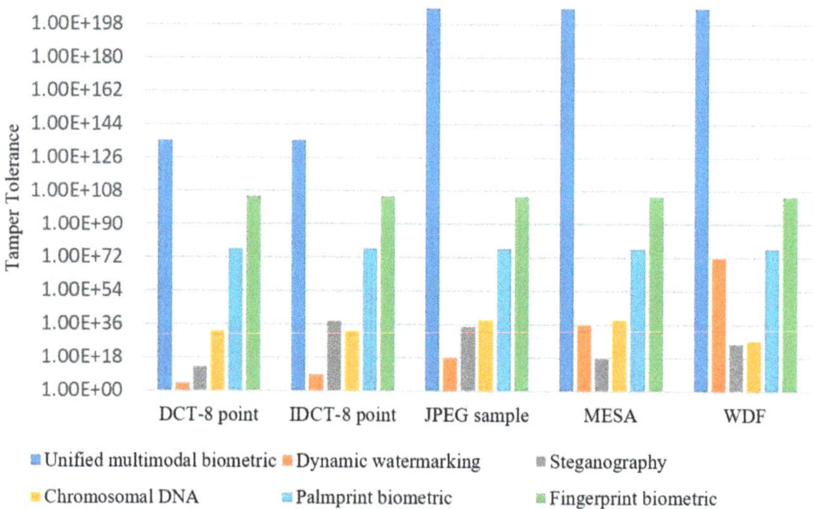

Figure 5.10 Comparison of tamper tolerance between unified multimodal biometric (Sengupta et al., 2023d), dynamic watermarking (Koushanfar et al., 2005), steganography (Sengupta and Rathor, 2019), chromosomal DNA (Sengupta and Chaurasia, 2022), palmprint biometric (Sengupta et al., 2021; Chaurasia et al., 2022), and fingerprint biometric (Sengupta and Rathor, 2020)

(Chaurasia *et al.*, 2022), steganography (Sengupta and Rathor, 2019), dynamic watermarking (Koushanfar *et al.,* 2005), fingerprint biometric (Sengupta and Rathor, 2020), and chromosomal DNA (Sengupta and Chaurasia, 2022). The discussed watermarking approach outperforms all contemporary approaches with a higher tamper tolerance value. Furthermore, Figures 5.11 and 5.12

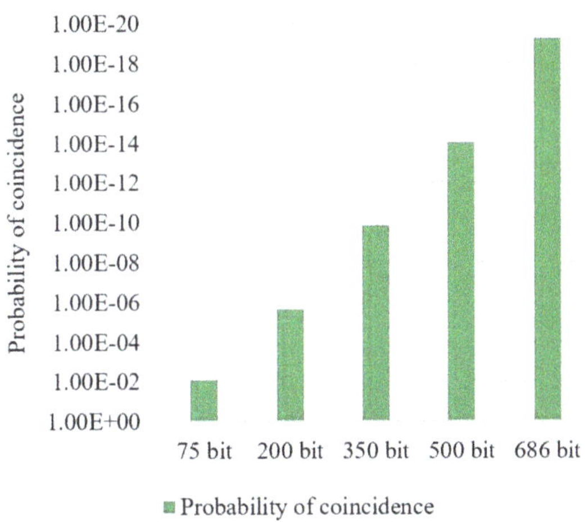

Figure 5.11 Variation in probability of coincidence with varying signature sizes for discussed approach for DCT-8 point application

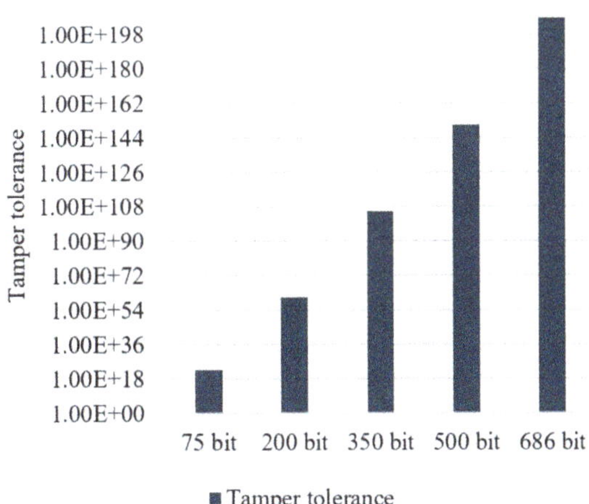

Figure 5.12 Variation in tamper tolerance with varying signature sizes for discussed approach for DCT-8 point application

illustrate the variation in the probability of coincidence and tamper tolerance values of the unified watermark signature for 8-point DCT hardware design. As the number of embedded signature bits increases, the probability of coincidence decreases, whereas the value of tamper tolerance escalates (Sengupta *et al.*, 2023c).

5.3.2 Design cost analysis

The design cost of the discussed unified biometric-based watermarking approach is evaluated using a design cost function mentioned in (5.5) (Sengupta *et al.*, 2023c; Sengupta *et al.*, 2023d; Anshul and Sengupta, 2023b; Anshul and Sengupta, 2023c).

$$\text{Designcost} = d1 * \left(\frac{(Area)}{A\max}\right) + d2 * \left(\frac{(Latency)}{L\max}\right) \tag{5.5}$$

Here, *d1* and *d2* are set to 0.5, indicating equal importance for both area and latency parameters. "*Area*" and "*Latency*" denote the final hardware design area and latency (delay). Next, "$A_{\max}$" and "$L_{\max}$" represent the maximum achievable design area and latency, respectively. Evaluation of area and latency for the selected hardware benchmarks is conducted on a 15 nm scale using the NanGate library (OCL, 2024). Figure 5.13 presents the design cost for the discussed unified biometric-based watermark approach on different benchmarks before and after incorporating the watermark. It is evident from Figure 5.13 that the unified biometric-based watermark imposes zero design

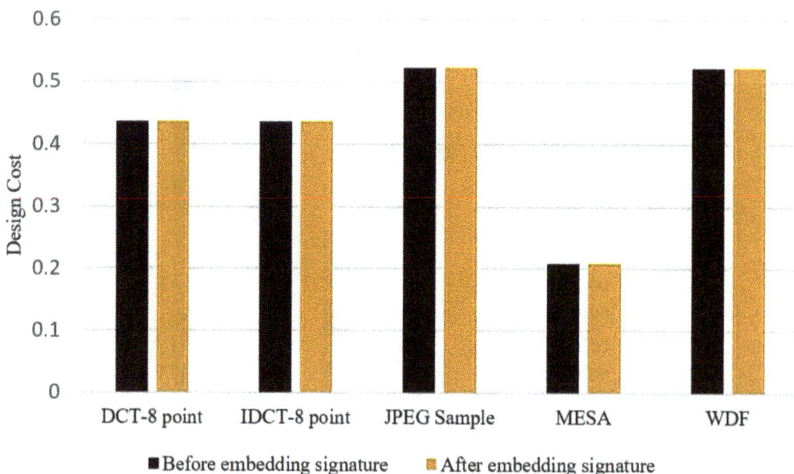

Figure 5.13 Design cost comparison for the discussed approach before and after embedding unified multimodal signature

cost overhead while ensuring the robust security of the hardware IPs (Sengupta *et al.*, 2023c).

5.4 Conclusion

In this chapter, IP piracy and false IP ownership assertion were discussed as the primary threats in the hardware global design supply chain process. A unified multimodal biometric-based watermarking approach has been discussed in this chapter for offering robust protection against the mentioned threats. A reader learns the following concepts by reading this chapter:

(a) Contemporary approaches for IP watermarking that address threats of IP piracy and false IP ownership assertion.
(b) Generation of unified IP watermark using multimodal biometric framework.
(c) Encoded dictionary framework for selecting the watermark signature strength (size).

Comparative analysis of various IP watermarking techniques against existing attacks of ghost insertion search attack, brute force attack, tampering attack, false IP ownership claim using probability of coincidence metric, and tamper tolerance metric.

5.5 Questions and exercise

1. Briefly explain the different hardware security threats associated with the global design supply chain process of reusable hardware IP cores.
2. Discuss the importance of biometric-based watermarking techniques over conventional watermarking techniques for securing hardware IP cores.
3. What are the advantages of multimodal biometric over unimodal biometric for hardware security?
4. Explain the importance of transient fault-secured hardware design and why it is important to secure them.
5. Explain the overview of unified multimodal biometric-based hardware security approach.
6. List primary inputs and output of the unified multimodal biometric-based hardware security approach.
7. Briefly explain the overview of the multimodal biometric-based security approach.
8. Explain the generation process of transient fault-secured hardware design with a suitable example.
9. Explain the importance of dual modular redundancy (DMR) based technique to generate fault-secured hardware designs.
10. Discuss the fingerprint biometric signature generation process along with the primary inputs and output.

11. Discuss the facial biometric signature generation process along with the primary inputs and output.
12. Discuss the palmprint biometric signature generation process along with the primary inputs and output.
13. Explain the generation process of unified multimodal biometric signature and the importance of encoded dictionary in the context of the final signature (watermark) generation process.
14. What are some of the advantages/merits of the unified multimodal biometric-based security module that helps in achieving robust security?
15. Demonstrate the unified multimodal biometric watermark embedding process with a suitable example.
16. What are the security metrics used to compare and analyze the security strength of the discussed security approaches?

References

A. Anshul and A. Sengupta, (2022) "IP Core Protection of Image Processing Filters with Multi-Level Encryption and Covert Steganographic Security Constraints," *2022 IEEE International Symposium on Smart Electronic Systems (iSES)*, Warangal, India, pp. 83–88.

A. Anshul and A. Sengupta, (2023a) "A Survey of High Level Synthesis Based Hardware Security Approaches for Reusable IP Cores [Feature]," *IEEE Circuits and Systems Magazine*, vol. 23, no. 4, pp. 44–62.

A. Anshul and A. Sengupta, (2023b) "Exploration of Optimal Crypto-chain Signature Embedded Secure JPEG-CODEC Hardware IP During High Level Synthesis," *Microprocessors and Microsystems*, vol. 102, 104916.

A. Anshul and A. Sengupta, (2023c) "PSO Based Exploration of Multi-phase Encryption Based Secured Image Processing Filter Hardware IP Core Datapath During High Level Synthesis," *Expert Systems with Applications*, vol. 223, 119927.

A. Anshul, K. Bharath and A. Sengupta, (2022) "Designing Low Cost Secured DSP Core using Steganography and PSO for CE Systems," *2022 IEEE International Symposium on Smart Electronic Systems (iSES)*, Warangal, India, 95–100.

R. Chaurasia, A. Anshul, A. Sengupta and S. Gupta, (2022) "Palmprint Biometric Versus Encrypted Hash Based Digital Signature for Securing DSP Cores Used in CE Systems," *IEEE Consumer Electronics Magazine*, vol. 11, no. 5, pp. 73–80.

B. Colombier and L. Bossuet, (2015) "Survey of Hardware Protection of Design Data for Integrated Circuits and Intellectual Properties," *IET Computers and Digital Techniques*, vol. 8, no. 6, pp. 274–287.

EXP BEN, University of California Santa Barbara Express Group, accessed on Jan. 2024. [Online]. Available: http://express.ece.ucsb.edu/benchmark/.

W. Hu, C. Chang, A. Sengupta, S. Bhunia, R. Kastner and H. Li, (2021) "An Overview of Hardware Oriented Security and Trust: Threats, Countermeasures and Design

Tools," *IEEE Transactions on Computer-Aided Design of Integrated Circuits and Systems*, Invited Paper, vol. 40, no. 6, pp. 1010–1038.

S. A. Islam, L. K. Sah and S. Katkoori, (2020) "High-Level Synthesis of Key-Obfuscated RTL IP with Design Lockout and Camouflaging," *ACM Transactions on Design Automation of Electronic Systems*, vol. 26, no. 1, pp. 1–35.

F. Koushanfar, I. Hong and M. Potkonjak, (2005) "Behavioral Synthesis Techniques for Intellectual Property Protection," *ACM Transactions on Design Automation of Electronic Systems*, vol. 10, no. 3, pp. 523–545.

B. Le Gal and L. Bossuet, (2012) "Automatic Low-cost IP Watermarking Technique Based on Output Mark Insertions," *Design Automation for Embedded Systems*, vol. 16, no. 2, pp. 71–92.

C.A. Lisboa and L. Carro, (2007) "System Level Approaches for Mitigation of Long Duration Transient Faults in Future Technologies," *12th IEEE European Test Symposium – ETS 2007*, in Freiburg, Germany, pp. 165–172.

OCL, (2024) 15 nm open cell library. [Online], Available: https://si2.org/open-cell-library/, last accessed on Jan. 2024.

M. Potkonjak, (2006) "Methods and Systems for the Identification of Circuits and Circuit Designs," United States Patent, US7017043B1.

M. Rathor, A. Sengupta, R. Chaurasia and A. Anshul, (2023a) "Exploring Handwritten Signature Image Features for Hardware Security," *IEEE Transactions on Dependable and Secure Computing*, vol. 20, no. 5, pp. 3687–3698.

M. Rathor, A. Anshul, K. Bharath, R. Chaurasia and A. Sengupta, (2023b) "Quadruple Phase Watermarking During High Level Synthesis for Securing Reusable Hardware Intellectual Property Cores," *Computers and Electrical Engineering*, vol. 105, 108476.

M. Rathor, A. Anshul and A. Sengupta, (2024) "Securing Reusable IP Cores Using Voice Biometric Based Watermark," *IEEE Transactions on Dependable and Secure Computing*, vol. 21, no. 4, pp. 2735–2749.

S. Rezgui, J. J. Wang, Y. Sun, B. Cronquist and J. McCollum, (2008) "Configuration and Routing Effects on the SET Propagation in Flash-Based FPGAs," *IEEE Transactions on Nuclear Science,* vol. 55, no. 6, pp. 3328–3335.

S. Rizzo, F. Bertini, and D. Montesi, (2019) "Fine-grain Watermarking for Intellectual Property Protection," *EURASIP Journal on Information Security*, vol. 2019, Article no. 10.

M. Rostami, F. Koushanfar and R. Karri, (2014) "A Primer on Hardware Security: Models, Methods, and Metrics," *Proceedings of the IEEE*, vol. 102, no. 8, pp. 1283–1295.

SEU, "Single Event Upsets," Intel [online], (2024) Available: https://www.intel.com/content/www/us/en/support/programmable/supportresources/quality/seu.html.

R. Schneiderman, (2010) "DSPs Evolving in Consumer Electronics Applications," *IEEE Signal Processing Magazine*, vol. 27, no. 3, pp. 6–10.

A. Sengupta, (2015) "Exploration of kc-Cycle Transient Fault-Secured Datapath and Loop Unrolling Factor for Control Data Flow Graphs During High-Level Synthesis," *Electronics Letters*, vol. 51, pp. 562–564.

A. Sengupta, (2016) "Resilient Soft IP-Core Design Against Terrestrial Transient Faults for CE Products," *IEEE Consumer Electronics Magazine,* vol. 5, no.4, pp. 129–131.

A. Sengupta and S. Bhadauria, (2016) "Exploring Low Cost Optimal Watermark for Reusable IP Cores During High Level Synthesis," *IEEE Access*, vol. 4, pp. 2198–2215.

A. Sengupta and R. Chaurasia, (2022) "Securing IP Cores for DSP Applications Using Structural Obfuscation and Chromosomal DNA Impression," *IEEE Access*, vol. 10, 50903–50913.

A. Sengupta and D. Kachave, (2017) "Low Cost Fault Tolerance against kc-cycle and km-unit Transient for Loop Based Control Data Flow Graphs during Physically Aware High Level Synthesis," *Elsevier Journal on Microelectronics Reliability*, vol. 74, pp. 88–99.

A. Sengupta and S. P. Mohanty, (2019) "IPCore and Integrated Circuit Protection Using Robust Watermarking", *IP Core Protection and Hardware-Assisted Security for Consumer Electronics*. Stevenage: The Institution of Engineering and Technology, Chap. 4, pp. 123–170.

A. Sengupta and M. Rathor, (2019) "IP Core Steganography for Protecting DSP Kernels Used in CE Systems," *IEEE Transactions on Consumer Electronics*, vol. 65, no. 4, pp. 506–515.

A. Sengupta and M. Rathor, (2020) "Securing Hardware Accelerators for CE Systems Using Biometric Fingerprinting," *IEEE Transactions on Very Large Scale Integration (VLSI) Systems*, vol. 28, no. 9, pp. 1979–1992.

A. Sengupta and M. Rathor, (2021) "Facial Biometric for Securing Hardware Accelerators," *IEEE Transactions on Very Large Scale Integration (VLSI) Systems*, vol. 29, no. 1, pp. 112–123.

A. Sengupta, S. Bhadauria and S. P. Mohanty, (2017) "TL-HLS: Methodology for Low Cost Hardware Trojan Security Aware Scheduling with Optimal Loop Unrolling Factor during High Level Synthesis," *IEEE Transactions on Computer-Aided Design of Integrated Circuits and Systems*, vol. 36, no.4, pp. 655–668.

A. Sengupta, R. Chaurasia and T. Reddy, (2021) "Contact-Less Palmprint Biometric for Securing DSP Coprocessors Used in CE Systems," *IEEE Transactions on Consumer Electronics*, vol. 67, no. 3, pp. 202–213.

A. Sengupta, R. Chaurasia and A. Anshul, (2023a) "Hardware Security of Digital Image Filter IP Cores against Piracy Using IP Seller's Fingerprint Encrypted Amino Acid Biometric Sample," *2023 Asian Hardware Oriented Security and Trust Symposium (AsianHOST)*, Tianjin, China, pp. 1–6.

A. Sengupta, A. Anshul and R. Chaurasia, (2023b) "Exploration of Optimal Functional Trojan-Resistant Hardware Intellectual Property (IP) Core Designs during High Level Synthesis," *Microprocessors and Microsystems*, vol. 103, 104973.

A. Sengupta, R. Chaurasia and K. Bharath, (2023c) "Exploring Unified Biometrics with Encoded Dictionary for Hardware Security of Fault Secured IP Core Designs," *Computers and Electrical Engineering Part A*, vol. 111, 108928.

A. Sengupta, R. Chaurasia and A. Anshul, (2023d) "Robust Security of Hardware Accelerators Using Protein Molecular Biometric Signature and Facial Biometric Encryption Key," *IEEE Transactions on Very Large Scale Integration (VLSI) Systems*, vol. 31, no. 6, pp. 826–839.

M. Yasin, J. J. Rajendran, O. Sinanoglu and R. Karri, (2016) "On Improving the Security of Logic Locking," *IEEE Transactions on Computer-Aided Design of Integrated Circuits and Systems*, vol. 35, no. 9, pp. 1411–1424.

B. Yuce, C. Deshpande, M. Ghodrati, A. Bendre, L. Nazhandali and P. Schaumont, (2019) 'A Secure Exception Mode for Fault-Attack-Resistant Processing," *IEEE Transactions on Dependable and Secure Computing*, vol. 16, no. 3, pp. 388–401.

Chapter 6

High-level synthesis-based watermarking using crypto-chain signature framework

Anirban Sengupta[1] and Aditya Anshul[1]

Designing reusable hardware intellectual property (IP) core (such as JPEG-CODEC) for multimedia/electronic systems involves managing competing design goals like latency and area during high-level synthesis (HLS). Additionally, it requires addressing the critical aspect of hardware security to prevent IP piracy and false IP ownership claim. This chapter discusses the generation of low-cost, secure IP design using the firefly algorithm (FA) based space exploration or design space exploration (DSE) and crypto-based watermarking (security) framework. The discussed crypto (key-driven) watermarking approach leads to the generation of watermarking constraints, which are further implanted into the hardware design through an HLS-based embedding process. This discussed low-cost security methodology facilitates IP piracy detection and resolution of false IP ownership assertion.

6.1 Introduction

The adoption of digital/computing gadgets, such as laptops, smartphones, several sensors, etc., has experienced a substantial increase owing to the notable advancements and automation in contemporary modern specialized society. These gadgets usually incorporate general-purpose computing systems or processors to manage user applications alongside specialized (application-specific) hardware intellectual property (IP) cores customized for particular data and computation-intensive functions that include the execution of complex mathematical calculations, data filtering, data compression-decompression, etc. These IPs are developed to attain efficient performance within real-world scenarios/constraints (Anshul and Sengupta, 2023a; Schneiderman, 2010; Colombier and Bossuet, 2015).

Further, the swift growth of integrated circuit (IC) manufacturing units, accompanied by fierce rivalry, and the need for brisk time-to-market demands have resulted in the participation of various entities in the design process of hardware computing systems/gadgets. IP sellers, integration house (system-on-chip (SoC)

[1]Department of Computer Science and Engineering, Indian Institute of Technology, Indore, India

design house), and manufacturing unit are the key units in the global design supply chain process of hardware computing gadgets/systems. Nevertheless, the engagement of multiple entities/parties exposes hardware IP designs to risks such as IP piracy, false assertion of IP ownership, and IP counterfeiting (Rostami *et al.*, 2014). The concern arises from the possibility that a malicious actor within the SoC integrator house can unlawfully pirate or fraudulently claim a design acquired from a legitimate IP seller. Further, a rouge foundry may also try to perform IP piracy/counterfeiting (Rathor *et al.*, 2023a). This can result in unpredictable behavior and potential issues like improper functioning/computation of the device, information leakage, excessive heat dissipation, a revenue loss of the original IP seller, etc. Pirated/counterfeited IPs may also incorporate malicious logic because they lack comprehensive/rigorous testing (Rizzo *et al.*, 2019). This form of IP piracy causes financial losses and also maligns genuine IP seller's reputations. Hence, prioritizing security, along with optimizing the design, is crucial to produce an efficient and secure/watermarked IP design (Islam *et al.*, 2020; Yasin *et al.*, 2016; Le Gal and Bossuet, 2012; Anshul and Sengupta, 2022; Sengupta *et al.*, 2023a; Rathor *et al.*, 2023b; Rathor *et al.*, 2024).

Hardware IP cores have become more prevalent nowadays. They meet the strict constraints of the present global design supply chain operation of integrated circuits, where trust and security are crucial design constraints. Therefore, designing reusable IP cores involves careful considerations of security and optimization. The HLS framework plays a pivotal role in generating optimized, secure hardware IP designs customized for data and computation-intensive applications. Design space exploration (DSE) is a significant component of HLS that actively contributes to the determination of the optimal/low-cost secure architectural configuration (Anshul *et al.*, 2022). DSE performs tradeoff between conflicting parameters such as latency, power, security, area, etc., with the primary objective of securing and optimizing while selecting the most appropriate solution or design from the possible solutions within the search space (Anshul *et al.*, 2022; Sengupta *et al.*, 2023b).

Furthermore, among data and computation-intensive functions, compression and decompression of image data are particularly prevalent. The joint photographic expert group compression-decompression (JPEG-CODEC) distinguishes itself as a commonly utilized application in almost every camera-enabled multimedia device. Its effectiveness in compressing and decompressing multimedia files (containing images or videos) by minimizing their size without compromising any crucial information while making storage and transportation convenient (Koff and Shulman, 2006). JPEG-CODEC also holds considerable importance in healthcare applications. Designing JPEG-CODEC as a dedicated hardware IP proves valuable in electronic devices used for Several crucial computing devices such as medical imaging modalities (like computed tomography (CT) scans and magnetic resonance imaging (MRI) systems) work on high-resolution images, where JPEG-CODEC as a dedicated hardware IP. These modalities capture higher-quality images of patient organs for precise disease detection. The images produced by CT and MRI scanners are significantly larger (Gokturk *et al.*, 2001). For example, a solitary higher-quality image obtained from a CT scan of stomach may comprise 512×512 pixels (where one pixel requires 16 bits). A complete abdomen CT scan could include anywhere between 300 and 500 higher-quality images, leading to approximate data of around 100 MB. In situations like remote healthcare services (such as telemedicine, telepathology, teleconsultation, etc.), the captured images are first stored on a local server/on the cloud before transporting them to healthcare professionals for further evaluations. Efficient storage and transportation of large-sized medical images (often generated by CT and MRI scanners) are facilitated by compressing them using JPEG-CODEC before transmission. Various methods have been implemented to compress higher-quality medical images. One common strategy involves utilizing lossy compression, offering a suitable compression ratio range (Chen, 2007). Furthermore, hybrid compression using JPEG-CODEC is applied, where diagnostically crucial areas in a medical image requiring higher quality undergo minimal or no compression. At the same time, other regions maintain an acceptable image quality with lossy compression (Agarwal *et al.*, 2019). Besides medical usage, the JPEG-CODEC finds application in several camera-based gadgets/systems like digital cameras, laptops, smartphones, etc. Therefore, JPEG-CODEC can be useful if designed as a dedicated secure reusable IP core using the HLS framework.

This chapter discusses the generation of crypto-signature embedded JPEG-CODEC IP design using the HLS framework for the security of such important design IPs (Anshul and Sengupta, 2023b; Sengupta *et al.*, 2018). The presence of a crypto-signature, i.e., the IP seller's unique digital evidence in the design, provides a detective countermeasure against IP piracy and false assertion of IP ownership. Further, it also helps in the detection of counterfeited/fake hardware IPs or products.

6.2 Different hardware security threats and discussion on the threat actors

The entire IP design supply chain process is vulnerable to various security risks due to the participation of multiple units (Rostami *et al.*, 2014; Le Gal and Bossuet, 2012; Anshul and Sengupta, 2022). The dedicated reusable IP core (such as the JPEG-CODEC) faces threat of piracy when an IP seller distributes it to the customer (i.e., system-on-chip (SoC) integrator). The potential threat arises when an attacker in the SoC integrator house may attempt to unlawfully replicate/pirate the genuine IP design and resell it under the same/different brand, making it difficult to distinguish between the authentic and pirated versions. This not only poses a challenge in identifying genuine products but also leads to revenue loss for the original IP seller (Anshul and Sengupta, 2023a). Besides IP piracy, an adversary might falsely assert the ownership rights of the IP (Anshul and Sengupta, 2023a). Additionally, an attacker within a rogue foundry may unlawfully pirate the IP without the designer's knowledge/consent (Rathor *et al.*, 2023b; Anshul and Sengupta, 2023b).

Furthermore, IP counterfeiting occurs when an attacker in the rogue foundry and a new/secondary IP seller illicitly replicate or imitate the original IP design (Rostami *et al.*, 2014). Counterfeit semiconductor components result in financial losses for the original IP sellers and cause safety hazards for end consumers. The subpar performance of counterfeit products (typically of inferior quality or older generation) negatively affects overall system performance and reliability. It also tarnishes the reputation of authentic IP sellers and poses risks to critical applications, such as military applications, aircraft, cars, etc., that rely on these components. Moreover, the ease of incorporating intentional hardware Trojans/malicious logic in fake/counterfeited IPs poses a significant security threat to the entire integrated design supply chain system (Islam *et al.*, 2020; Yasin *et al.*, 2016; Sengupta *et al.*, 2023a; Rathor *et al.*, 2024).

6.3 Summary of crypto-chain signature embedding method

This section discusses the summary of crypto-chain signature embedding HLS methodology used for IP piracy detection (Anshul and Sengupta, 2023b; Sengupta *et al.*, 2018). Figure 6.1 illustrates the summary of low-cost crypto-chain-based secure design flow for JPEG-CODEC hardware using HLS. As shown in Figure 6.1, the low-cost crypto-chain-based secure design flow accepts the control data flow graph (CDFG) of hardware application (such as JPEG-CODEC) and module libraries as the primary input. It yields a secure low-cost register transfer level (RTL) datapath/IP design corresponding to input hardware application. Further, the discussed secure design flow consists of two primary blocks: (a) architecture exploration block and (b) crypto-chain-based hardware security block. Next, Figure 6.2 depicts the outline of the architecture exploration block and generation of scheduling information corresponding to JPEG-CODEC hardware design. The architecture exploration block accepts CDFG of JPEG-CODEC and module library as primary input along with various design space exploration (DSE) control parameters. It yields a low-cost/optimal architecture and its scheduling information corresponding to the input hardware design. Further, the scheduling

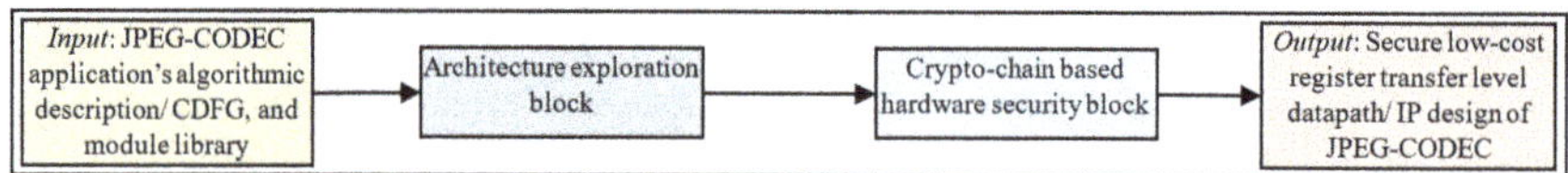

Figure 6.1 Overview of low-cost crypto-chain-based secure design flow for JPEG-CODEC hardware using HLS

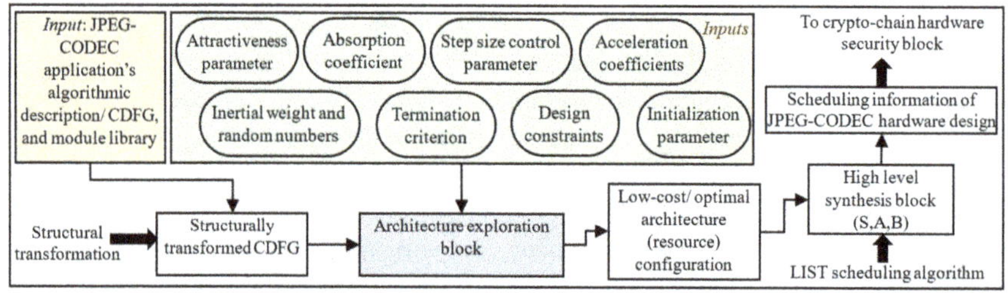

Figure 6.2 Overview of architecture exploration block and generation of scheduling information corresponding to JPEG-CODEC hardware design

information based on optimal architecture configuration is passed to the crypto-based hardware watermarking (security) block. Subsequently, Figure 6.3 highlights the overview of crypto-based watermarking block and generation of watermarked JPEG-CODEC hardware IP. This block accepts the scheduling information and crypto-chain security algorithms' parameters as primary input and yields a secure low-cost register transfer level (RTL) datapath/IP design corresponding to hardware design. The details corresponding to architecture exploration and crypto-chain-based hardware security block are discussed in the next sections. Additionally, Figure 6.4. depicts the step-wise flow diagram of the low-cost crypto-signature watermarking methodology (Anshul and Sengupta, 2023b).

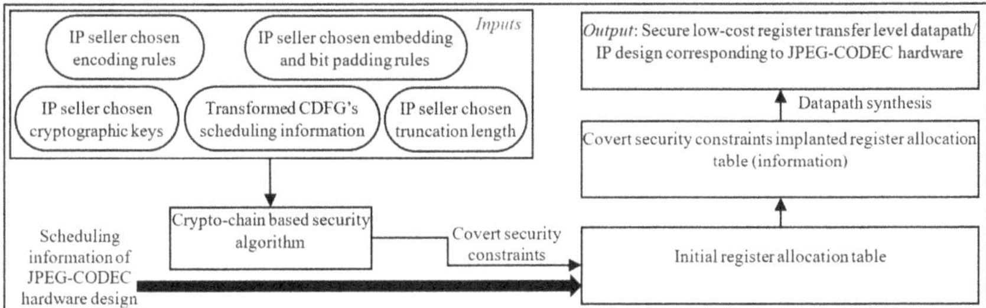

Figure 6.3 Overview of crypto-chain-based hardware security block and generation of secure JPEG-CODEC hardware IP design

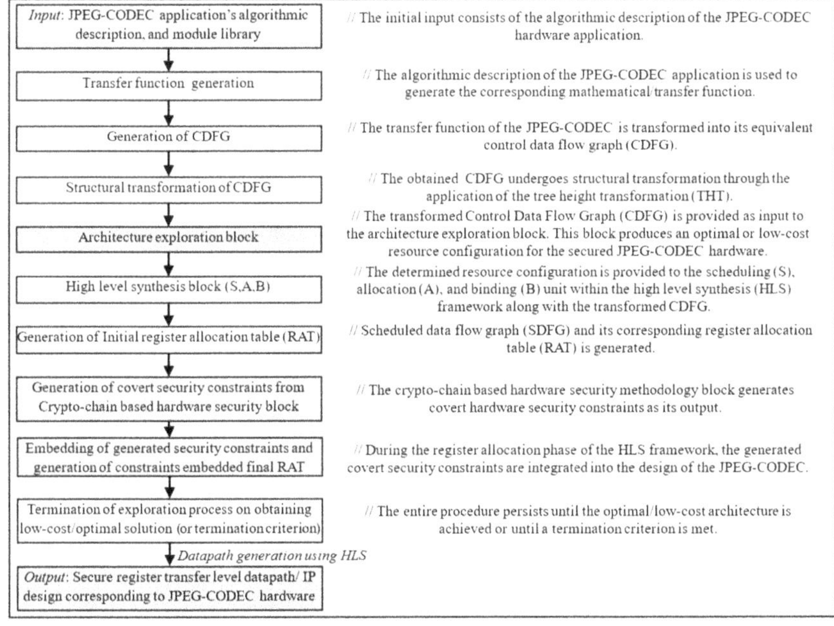

Figure 6.4 Step-wise flow diagram of the low-cost crypto-chain signature-based hardware security methodology

6.4 Details of low-cost crypto-chain signature embedding method

6.4.1 Architecture exploration block

This section discusses the architecture exploration process in detail. Heuristic-based design search space algorithms can be employed to perform architecture exploration (Anshul and Sengupta, 2023b). This section explains the details of firefly (Anshul and Sengupta, 2023b) and particle swarm optimization (Anshul and Sengupta, 2023c) based architecture exploration process corresponding to hardware application.

6.4.1.1 Overview of firefly algorithm (FA)-based design space exploration (DSE) block

The firefly-driven design search exploration (FA-DSE) effectively eliminates undesired designs (those with higher cost/lesser fitness value) based on IP seller chosen design constraints (such as latency and area). This subsection explains the exploration/identification of the optimal design architecture for the hardware application to be designed (such as JPEG-CODEC). The flow of the architectural exploration based on the firefly algorithm is depicted in Figure 6.5. The key inputs for this methodology include (a) firefly initialization parameter, (b) control data flow graph (CDFG) of the hardware application, (c) absorption coefficient ("γ"), (d) design constraints, (e) terminating criterion, (f) step size control parameter ("α"), (g) module library, and (h) attractiveness

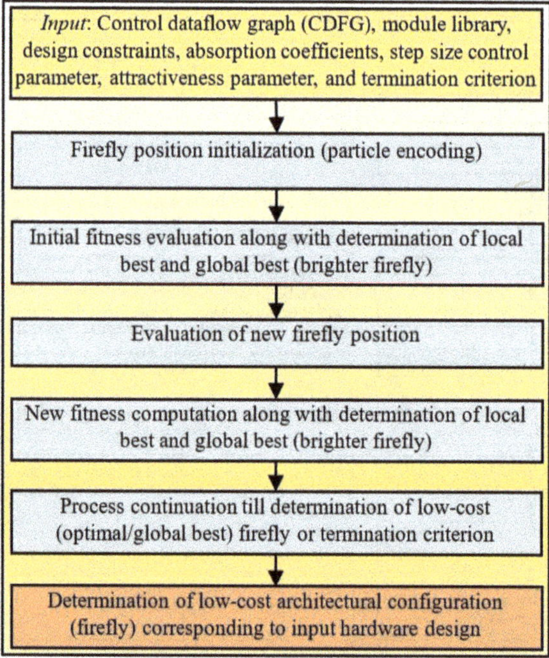

Figure 6.5 FA-DSE-based architectural exploration process

parameter ("β"). The initial positions of the fireflies are determined, and following this, an initial fitness value/cost is computed for every firefly with the help of fitness/cost function. Subsequently, the local and global best positions are updated based on the determined initial design cost. The cost value corresponding to the initial firefly positions is termed as their local best, and the firefly with lowest cost is considered as global best. New positions are then evaluated using parameters such as the control step size parameter ("α"), attractiveness parameter ("β"), and absorption coefficient ("γ"). After determining these new positions, their corresponding design costs are computed. Following the computation, if the new cost for any firefly is lower than the previously computed cost, the local best positions for those fireflies are updated. The firefly with lowest cost is considered as global best. This entire process is iteratively repeated until the termination. Finally, the output of the process is an optimal architecture configuration for the hardware application (Anshul and Sengupta, 2023b).

6.4.1.2 Overview of particle swarm optimization (PSO)-based DSE block

The flow of the architectural exploration based on the particle swarm optimization (PSO) is depicted in Figure 6.6. The key inputs for this methodology include

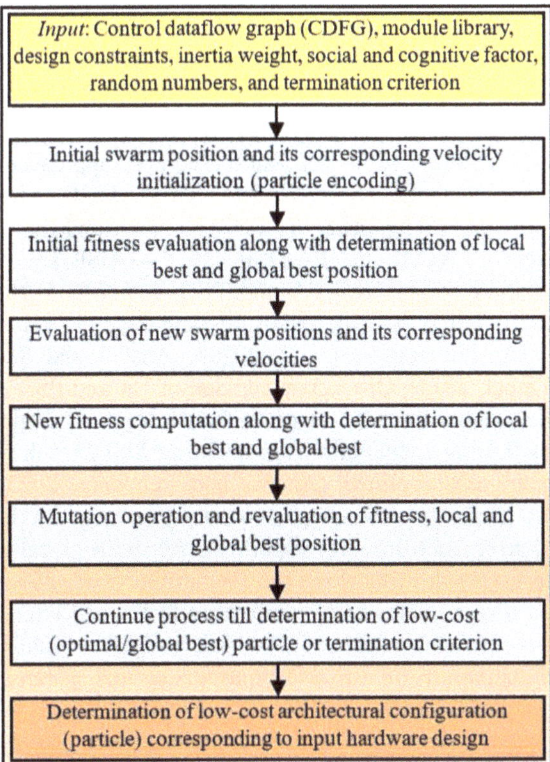

Figure 6.6 PSO-DSE-based architectural exploration process

(a) PSO initialization parameter, (b) control data flow graph (CDFG) of the hardware application, (c) inertia weight, (d) random number, (e) terminating criterion, (f) social and cognitive factor, (g) module library, and (h) design constraints. Initially, the population of the swarm, along with their velocities and dimensions, is initialized. Subsequently, the primary cost/fitness, considering both latency and area, is calculated for each particle, and their corresponding local and global best positions are determined. The particle with the lowest cost is identified as the global best. Following this, new positions are generated based on the computation of updated velocity using inertia weight. The cost for all the newly generated positions is recalculated. If the new cost is lower than the previously determined cost, the corresponding local and global best positions are updated. Following this, velocity clamping and the application of adaptive end terminal perturbation are executed to maintain the computed velocity and particle positions within the desired range. Finally, a mutation process is introduced among particle positions to enhance diversity in the search across the design space. After the mutation, the design cost is re-evaluated, and the respective global best is updated if a lower design cost is achieved. This entire process is iteratively repeated until the termination. Finally, the output of the process is an optimal architecture configuration for the hardware application (Anshul and Sengupta, 2023c).

6.4.2 Benefits of using FA/PSO for DSE over other heuristic-based design space exploration algorithms

The firefly algorithm (FA) and particle swarm optimization (PSO) based DSE offer distinct improvements over various heuristic-based DSE approaches, including ant colony optimization (ACO) (Zukhri *et al.*, 2013; Ge *et al.*, 2016; Kao *et al.*, 2012), Genetic Algorithm (GA) (Sengupta *et al.*, 2012; Krishnan and Katkoori, 2006), etc. In comparison to GA, ACO, etc., FA-DSE and PSO-DSE architectures exhibit lower implementation complexity. FA and PSO demonstrate the capability to achieve an optimal solution in fewer rounds, making them more time-efficient than other heuristics (Anshul and Sengupta, 2023b; Anshul and Sengupta, 2023c; Mishra and Sengupta, 2014). One key advantage of FA and PSO over alternative approaches is their ability to strike a balance between intensification (exploitation time) and diversification (exploration time) during DSE, resulting in a quicker convergence to a global optimal (Anshul and Sengupta, 2023b; Anshul and Sengupta, 2023c; Mishra and Sengupta, 2014). Moreover, GA faces challenges with premature convergence, limiting its ability to reach the global optimal solution and causing it to get stuck in local optima (Wihartiko *et al.*, 2018). Furthermore, GA exhibits poor time complexity in comparison to FA and PSO, requiring more iterations to converge under higher constraints/variables (Zukhri *et al.*, 2013; Wihartiko *et al.*, 2018). FA's variable control step size parameter and PSO's inertial weight parameter help overcome this limitation of GA. Similarly, ACO, similar to GA, experiences higher implementation complexity due to a supplementary pheromone laying action (that acts as a message channel between ants) (Ge *et al.*, 2016). ACO also has the risk of getting stuck in a local optimum

(Ge *et al.*, 2016; Kao *et al.*, 2012). The firefly algorithm needs lesser rounds and is particularly appropriate for non-linear real-world problems (Sarkar *et al.*, 2017).

Primary advantages of using the firefly for conducting DSE in HLS (Anshul and Sengupta, 2023b):

a) FA incorporates hyperparameters such as the absorption coefficient and step-size control variables, influencing randomness during design search exploration, leading to efficient convergence to the global optimal. The values of both variables decrease linearly as the algorithm progresses, adapting to the changing distance between the global best and current and firefly positions.
b) FA operates on a divide-and-conquer approach, using an attraction parameter where it depends inversely on the distance between fireflies. This enables FA to create subgroups, each swarming around different local optimums, contributing to the exploration of a final optimal solution.
c) The values of the absorption coefficient and step-size control variables decrease linearly, which enables FA to maintain a stable balance between exploration and exploitation, facilitating the escape from local optima and achieving an optimal solution in fewer rounds.

Primary advantages of using particle swarm optimization for conducting design space exploration in HLS (Anshul and Sengupta, 2023c):

a) PSO incorporates the momentum of its prior iteration through the inertia variable, ensuring a smooth exploration by avoiding falling into local optima.
b) PSO uses an inertia weight value that linearly decreases from 0.9 to 0.1, allowing more significant steps in the beginning and smaller steps later, maintaining a balanced exploration and exploitation process.
c) The inclusion of various hyperparameters in PSO, such as cognitive factor, social factor, and random numbers, enhances its ability to generate an optimal solution in fewer iterations by avoiding trapping into local optima

6.4.3 Overview of crypto-chain signature generation algorithm

As illustrated in Figure 6.7, this module accepts the hardware application's scheduled design flow graph (SDFG), encoded rule, mapping rule, and truncation length as its primary inputs. The encoded template is created from the SDFG and the IP sellers' chosen encoding rule. Once generated, the encoded templates undergo initial pre-processing and bit-filling before being sent as input to primary hash slices (HS_1–HS_N). The first input (generated using encoding 1) is expanded using the IP seller chosen pre-processing rule and then sent as input to SHA-512 hash slice (*HS*)-*1* to generate a 512-bit encrypted hash digest. The output from the first hash slice is further expanded based on the IP seller-specified bit-filling rule. Subsequently, the expanded output from hash slice-1 is inputted to hash slice-2 to produce its respective 512-bit encrypted output. This process continues iteratively until "HS_N." Subsequently, the final 512-bit digest from the hash slice (*HS*)-*N* is inputted to the first block, i.e., H_N+1, of the hash slice (HS_{N+1}–HS_{2N}) block. The additional "*N*" hash slices accept encoded templates as

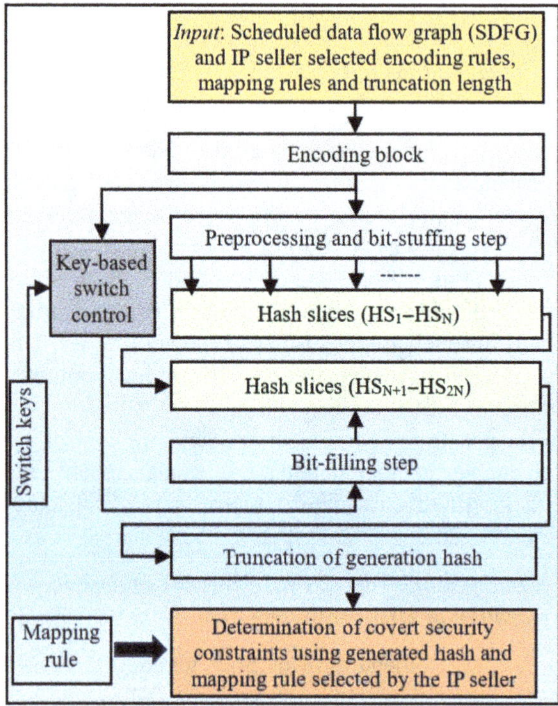

Figure 6.7 Crypto-chain signature generation algorithm

inputs regulated by a switch control component (key-based), made up of multiplexers (Muxes). These Muxes select the encoded templates from the encoding block using IP seller-selected switch keys (X_1–X_N). The selected encoded templates are then expanded according to the IP seller's chosen bit-filling rule. The complete process is reiterated until the hash slice (HS_{2N}) ultimately yields a 512-bit digest as output. This obtained digest output is subsequently truncated to the length specified by the IP seller. Following truncation, the output is transformed into watermarking constraints based on the IP seller's selected mapping rule (discussed in the subsequent subsection). The resulting watermarking constraints are sent as input to the embedding block to generate a secure hardware IP design (Anshul and Sengupta, 2023b).

6.4.4 Details of crypto-chain signature generation and embedding process

6.4.4.1 Details of firefly algorithm to generate low-cost secure architectural configuration corresponding to hardware design

This subsection explains the detailed process of firefly algorithm-based architectural exploration. The encoding rules selected by the IP seller and the workflow of FA-driven exploration are illustrated in Figure 6.8. The fundamental inputs for

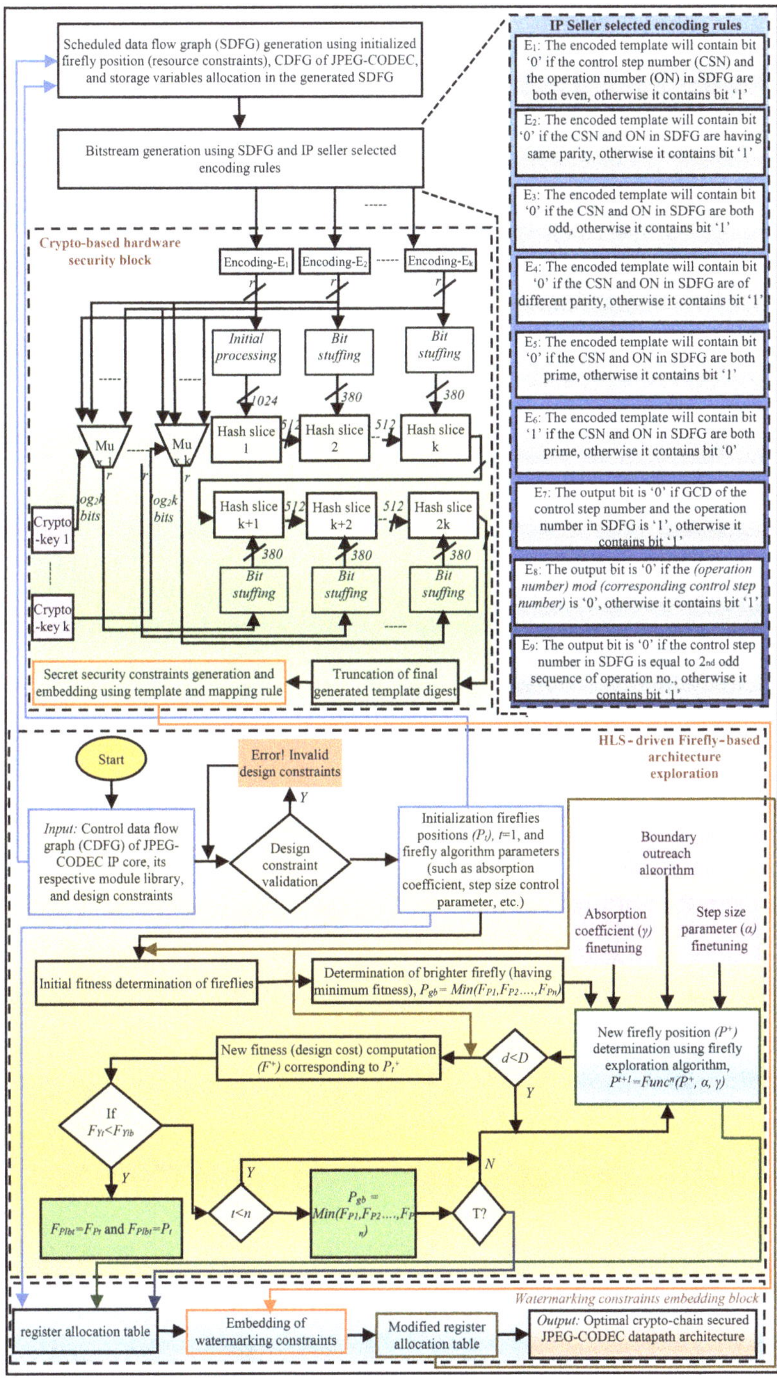

Figure 6.8 Details of low-cost crypto-chain signature generation and embedding process

the FA-DSE have already been discussed in the overview Section 6.4.1. At first, the population (P) size of the firefly algorithm and the respective dimensions (z) are set as three and two, respectively. It is also possible to assume a larger population size. "z" signifies the different functional units, such as adder, multiplier, etc. For instance, when the hardware resource types include only an adder and multiplier, then d is two. The value of "d" varies depending on the number of hardware resource types utilized for a specific application. In FA-DSE, the design constraints corresponding to latency ($Latency_{con}$) and area ($Area_{con}$) are confirmed to fall in the range of lower and higher values of *latency* and *area* corresponding to the hardware design (such as JPEG-CODEC) (i.e., $L_{max} > Latency_{con} > L_{min}$ and $A_{max} > Area_{con} > A_{min}$). The firefly positions are then initialized post-design constraints confirmation, as shown in Figure 6.8. The first position (P_1) is assigned a higher value of functional units, while the second position is assigned a lower number of functional units. The third position represents the average of the first and second. Subsequent positions for the remaining fireflies, if required, are initialized based on the (Anshul and Sengupta, 2023b; Anshul and Sengupta, 2023c; Mishra and Sengupta, 2014). Subsequently, an initial fitness/cost is computed for each firefly position using the fitness/cost function. It is important to note that this chapter considers area and latency specifications as parameters for evaluating the design cost. The corresponding local (P_{lbi}) and global best (P_{gb}) positions are determined. The position with the lowest cost is identified as the global best. The functions for determining area, latency, and fitness/cost are provided in (6.1)–(6.3), respectively (Anshul and Sengupta, 2023b; Anshul and Sengupta, 2023c).

$$Area = \sum_{i=1}^{2} (A(D_i) * (D_i)) \tag{6.1}$$

Here, $A(D_i)$ represents the area occupied by resource type (D_i), and (D_i) indicates the number of instances utilized for that particular resource type.

$$\text{Latency} = (CON_{\text{Mul}} * T_{\text{Mul}}) + (CON_{\text{Add}} * T_{\text{Add}}) \tag{6.2}$$

Here, "CON_{Mul}" *and* "CON_{Add}" denote the control step (CS) needed in the schedule for the multiplier and adder, while "T_{Add}" and "T_{Mul}" denote the latency of an adder and a multiplier, respectively.

$$\text{Fitness/cost} = s1 * \left(\frac{(Area - Area_{\text{con}})}{Area_{\text{max}}}\right) + s2 * \left(\frac{(Latency - Latency_{\text{con}})}{Latency_{\text{max}}}\right) \tag{6.3}$$

Here, $s1$ and $s2$ are set to 0.5, indicating equal importance for latency and area. "$Area_{\text{con}}$" and "$Latency_{\text{con}}$" denote the design constraints set by the IP seller. At the same time, "$Latency_{\text{max}}$" and "$Area_{\text{max}}$" represent the maximum achievable latency and area, respectively.

Furthermore, new firefly positions (P^{t+1}) are evaluated using the attractiveness parameter (β), control step size parameter (α), and absorption coefficient (γ). The boundary outreach algorithm adjusts that firefly within its limits in case of a boundary violation (Anshul and Sengupta, 2023b). The determination of new positions is done using equations (6.4), (6.5), and (6.6), respectively (Anshul and Sengupta, 2023b; Anshul and Sengupta, 2023c).

$$P_i^{t+1} = P_i^t + (\beta(P_j - P_i) + \alpha(rand - ½) \tag{6.4}$$

Here, the new positions are calculated by incorporating a drift factor into the initial position.

$$c = (P_{ij})^2 \tag{6.5}$$

$$\beta = \beta_0 e^{-\gamma c} \tag{6.6}$$

$$Pij = |Pi - Pj| = \sqrt{\sum_{k=0}^{d} (P_{i,k} - P_{j,k})} \tag{6.7}$$

Here, "P_i^{t+1}" represents the updated firefly position, "P_i^t" indicates the previous position of the firefly, while "P_i" and "P_j" correspond to the positions of the "i^{th}" and "j^{th}" fireflies, respectively. The parameters "γ", "α", and "β" are the hyperparameters defined earlier. "β_0" signifies the attractiveness at zero distance, and "P_{ij}" represents the cartesian distance between the fireflies "P_i" and "P_j". In cases where the generated positions exceed the boundary limits, the boundary outreach algorithm (BOA) is implemented to bring the firefly particles into the desirable range. After evaluation of the new firefly positions, their corresponding design costs are calculated. Subsequent to design cost computation, the local best positions for all such fireflies are updated, where the newly calculated cost is lower than the prior cost. Following this, the position with the lowest cost is identified as the global best. This entire process continues until the termination criterion is met. The termination criterion (T) considers that either the algorithm continues for fifty rounds, or it will terminate in case of constant cost value for 15 iterations (Anshul and Sengupta, 2023b). Ultimately, an optimal architecture configuration for the specific hardware IP is determined. This optimal resource configuration is subsequently forwarded to the watermarking constraints embedding block (as shown in Figure 6.8), facilitating the integration of watermarking constraints to yield a secure low-cost hardware IP design (Anshul and Sengupta, 2023b).

Customizing the firefly algorithm (parameter tuning) for DSE in HLS: The essential conditions or specific parameter values necessary for the fireflies to achieve equilibrium, aiding in the convergence of the firefly algorithm, are derived from (Yang and He, 2013) and (Yang, 2009). More precisely, the parameters "γ", "β_0", "$rand$", "α_x", and "α_y" are assigned the values of 0.5, 1, 1.5, the first dimension's upper value, and the second dimension's upper value, respectively (Sarkar *et al.*, 2017; Anshul and Sengupta, 2023b). It is to be noted that the absorption coefficient ("γ") and the control step size parameter ("α") should not remain static/constant; specifically, they should follow a linearly decreasing pattern, as discussed earlier in this section (Anshul and Sengupta, 2023b).

6.4.4.2 Details of crypto-chain signature generation process (Anshul and Sengupta, 2023b)

The core objective of the crypto-chain watermarking block is to generate watermarking (security) constraints using scheduling details of the hardware design, encoding rules specified by the IP seller, and crypto-keys chosen by the IP seller. The key inputs of crypto-chain-based security mechanism have been previously outlined in Section 6.4.3. The resultant watermarking constraints are discreetly incorporated into the IP design *via* the HLS framework. This incorporation of secret digital evidence protects (as a detective countermeasure) the design against risks such as IP piracy/theft and false assertion of IP ownership. Figure 6.8 depicts the crypto-based watermarking approach and its fusion with the firefly algorithm. The step-by-step process for generating these watermarking constraints is elaborated below:

- *Initial template generation using IP seller chosen encoding mechanism:* Initially, a template is created using the scheduling information (SDFG) of the IP design (we use JPEG-CODEC as a case study) and the encoding mechanisms specified by the IP seller. Table 6.1 depicts the details of scheduling for JPEG-CODEC, while the IP seller's encoding rules are highlighted in Figure 6.8. The selection of encoding rules for generating the initial templates is determined by the legitimate IP seller, remaining entirely non-cognizant to potential adversaries/attackers. The JPEG-CODEC's SDFG comprises 136 (r) operations (as indicated in Table 6.1), which are scheduled across various control steps based on architectural constraint (for instance, scheduled using three multipliers and adders, respectively). The three adders and multipliers-based schedule is only for demonstration. However, in each round of the security approach based on the FA, scheduling is done using the resource configuration obtained through FA-DSE. This specific detail of scheduling is then utilized to create the primary templates. Consequently, these primarily created templates vary in each round for each resource configuration. Additionally, for clarity and simplicity, there are nine IP sellers specified encoding rules (E), as depicted in Figure 6.8. (Note: "E" can be chosen to higher value also, as the IP seller can potentially devise numerous encoding algorithms.) Hence, an initial r-bit template is produced for each nine encodings. For example, in Figure 6.8, for encoding 1 (E_1), the template bit is 0 when both the operation number and control step number in SDFG are even; otherwise, the template bit is "1". Thus, for the first operation in SDFG, the control step is odd (as indicated in Table 6.1), therefore the output is "1". The output for all other operations of the JPEG-CODEC application is determined in a similar manner.
- *Transformation of the initially produced encoded template into 1024 bits and explanation of the crypto-algorithm:* The crypto-based watermarking methodology (driven through IP seller selected crypto-keys) utilizes "*2n*" hash slices, with each slice incorporating an SHA-512 based crypto-chain module to generate the encrypted template. Given that SHA-512 operates on 1024-bit inputs, the initial r-bit bitstream undergoes conversion/expansion into a 1024-bit template using a pre-processing technique specified by the IP seller. This pre-processing involves the following steps: first, the r-bit input is extended to 896 bits through bit stuffing.

Table 6.1 Scheduling information corresponding to JPEG-CODEC (scheduled using three multiplier and three adder)

Control step	Operation on Multi-plier 1	Operation on Multi-plier 2	Operation on Multi-plier 3	Operation on Adder 1	Operation on Adder 2	Operation on Adder 3
1	1	2	3	–	–	–
2	4	5	6	9	–	–
3	7	8	17	10	11	–
4	18	19	20	12	13	–
5	21	22	23	25	26	14
6	24	33	34	27	29	15
7	35	36	37	28	41	–
8	38	39	40	42	30	–
9	49	50	51	43	44	45
10	52	53	65	57	46	31
11	55	56	65	58	59	47
12	66	67	68	60	61	–
13	69	70	71	73	74	62
14	72	81	82	75	77	63
15	83	84	85	76	89	–
16	86	87	88	90	78	–
17	97	98	99	91	92	93
18	100	101	102	105	94	79
19	103	104	113	106	107	95
20	114	115	116	108	109	–
21	117	118	119	121	122	110
22	120	16	32	123	125	111
23	48	64	80	124	129	–
24	96	112	–	126	130	–
25	–	–	–	127	131	134
26	–	–	128	–	–	–
27	–	–	–	132	–	–
28	–	–	–	–	134	–
29	–	–	–	–	–	135
30	–	–	136	–	–	–

"1" is appended to the end of the initial *r*-bits of the template, followed by the addition of "0"s to achieve a total of 896 bits. Next, the initial template's length (*r*) is transformed into 128-bit by first converting "*r*" into binary and then appending "0"s until it reaches a length of 128 bits before the binary template. Subsequently, these 128 bits are concatenated with the 896 bits to yield a 1024-bit input template for the first slice of the crypto-based watermarking methodology. The incorporation of nine distinct IP seller-chosen encodings enhances the watermarking approach's robustness. Each encoding produces a unique *r*-bit encoded template, which serves as input for various hash slices in the watermarking methodology.

- *Crypto-chain*: The watermarking (security) approach involves multiple hash slices linked together in a sequential manner, forming a chain. Each hash slice's output feeds into the subsequent one, creating a cascading arrangement as depicted in Figure 6.8. The initial "*n*" hash slices, where "*n*" (here, set to nine), serve as the core cryptographic components. They receive input from an encoded block-generated template following the IP seller chosen pre-processing step. However, the remaining "*n*" hash slices are managed *via* multiplexers, with IP seller selected cryptographic keys utilized to regulate their outputs.

- *Bit stuffing*: The process of expanding the "*r*" bits template into a 1024-bit template (e.g., here 136 bits) for the initial slice has been previously described in this section. Subsequently, the 136-bit output from the remaining encoding blocks (excluding the first one) is transformed into 380 bits. This involves appending *(380-r)* continuous "0" bits as a suffix to each "*r*"-bit template. These algorithms for appending bits are proprietary to the IP seller, making it difficult for an adversary to replicate the exact output.

  Moreover, the generation of 1024 bits, which will be inputted to second slice from 512 bits of the preceding slice and 380 bits of the primary encoded template input (post-bit-stuffing) is as follows: the 512-bit output of the prior hash slice is concatenated with "1000" followed by 380 bits to yield 896 bits. Subsequently, the preceding slice output length (i.e., 512) is transformed into 128-bit, a process similar to that discussed earlier in this section. After obtaining the 128 bits, they are added to the 896 bits to create a total of 1024 bits. Likewise, each hash slice's output (512 bits) is extended to 1024 bits by incorporating 380 bits (obtained after bit-stuffing the input "*r*"-bit template) and 128 bits (generated from the length of the preceding slice's output, i.e., 512 bits). The output of the $(n-1)_{\text{th}}$ slice serves as the input to the n_{th} slice.
- *Hash Slice (slice)*: The discussed security methodology incorporates a total of "*2n*" hash slices. Every slice is invoked once to produce an encrypted value as output (512 bits), subsequently serving as the input to its next hash slice. Moreover, the execution of the round function is determined based on iterations specified by the IP seller. As illustrated in Figure 6.8, the initial *n* hash slices receive pre-processed templates from encoding blocks as input. However, the remaining "*n*" slice's input is governed by Muxes. The input selection for the additional "*n*" slices is determined based on the IP seller's chosen cryptographic keys. Here, the maximum number of slices for the "*n*" encodings is "*2n*".

Further, the resulting output (512-bit) is translated into its corresponding watermarking constraints using the IP seller chosen mapping rule. Subsequently, these watermarking constraints are implanted into the JPEG-CODEC application's design during the register allocation phase of the HLS framework. Following this, FA-DSE explores the low-cost architecture configuration for the secured hardware design through a fitness/cost function. The next subsection demonstrates the embedding process of generated watermarking constraints.

6.4.4.3 Details of low-cost crypto-chain signature embedding process

The generation of 512-bit output and its conversion into its corresponding hardware watermarking constraints is discussed in the previous subsection. Figure 6.9 depicts

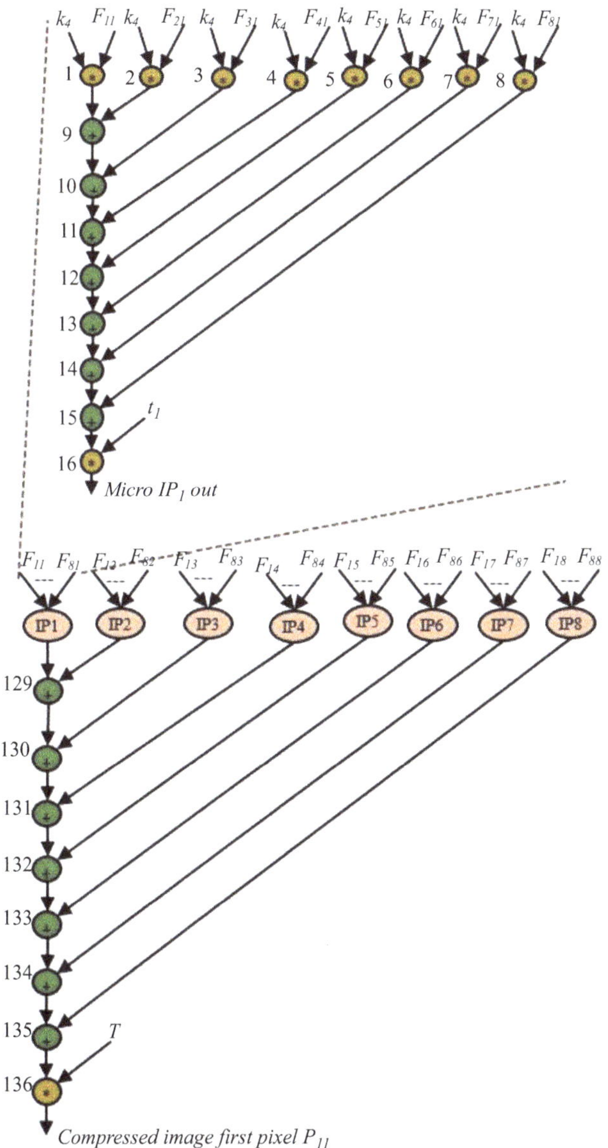

Figure 6.9 CDFG of JPEG-CODEC to determine the compressed image first pixel P_{11}

the CDFG of the JPEG-CODEC, respectively, derived from (6.8)–(6.11) (Anshul and Sengupta, 2023b; Sengupta *et al.*, 2018)

$$P_{11} = (k_4 * D_{11} + k_4 * D_{12} + k_{4*} * D_{13} + k_4 * D_{14} + k_4 * D_{15} + k_4 * D_{16} + k_4 * D_{17} + k_4 * D_{18}) \tag{6.8}$$

where, D_{11}, D_{12}, D_{13}, $\ldots\ldots D_{18}$ is evaluated as follows:

$$D_{11} = (k_4 * F_{11} + k_4 * F_{21} + k_{4*} * F_{31} + k_4 * F_{41} + k_4 * F_{51} + k_4 * F_{61} + k_4 * F_{71} + k_4 * F_{81}) \tag{6.9}$$

$$D_{12} = (k_4 * F_{12} + k_4 * F_{22} + k_{4*} * F_{32} + k_4 * F_{42} + k_4 * F_{52} + k_4 * F_{62} + k_4 * F_{72} + k_4 * F_{82}) \tag{6.10}$$

Similarly,

$$D_{18} = (k_4 * F_{18} + k_4 * F_{28} + k_{4*} * F_{38} + k_4 * F_{48} + k_4 * F_{58} + k_4 * F_{68} + k_4 * F_{78} + k_4 * F_{88}) \tag{6.11}$$

where "k_4" signifies the row elements of standard 2D Discrete Cosine Transform (DCT) matrix "*K*" utilized for the manipulation of grayscale images. "*F*" denotes a generic 8 × 8 matrix block representing a portion of the input image. Additionally, "P_{11}" indicates the computed compressed image's first pixel data.

Next, Figure 6.10 depicts the structurally transformed (using tree height transformation) CDFG of JPEG-CODEC (Anshul and Sengupta, 2023b). The

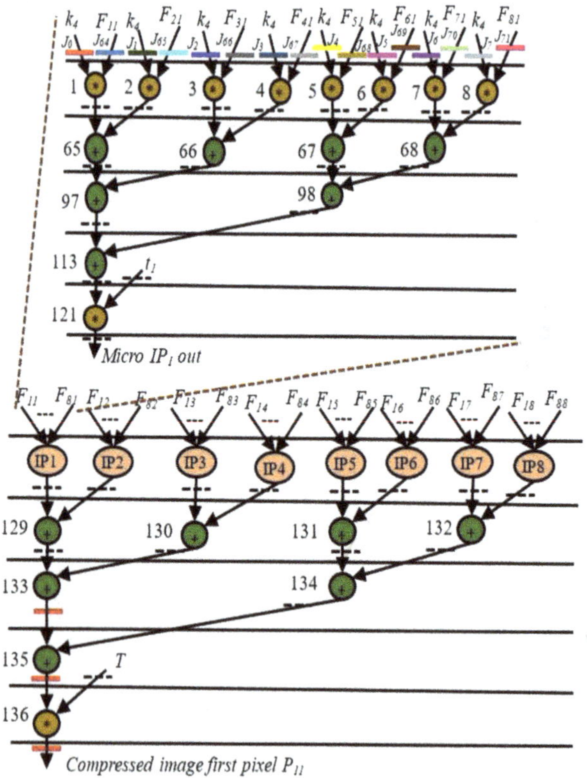

Figure 6.10 Scheduled CDFG of structurally transformed JPEG-CODEC for determining compressed image first pixel P_{11} with registers at input, output, and intermediate storage points

structurally transformed CDFG is scheduled using low-cost architecture configuration generated through the FA-DSE block. After SDFG generation, its corresponding RAT is created. The IP seller-defined mapping rule for the determination of watermarking constraints (indicated using storage variable pairs as highlighted in Figure 6.10) involves the following steps: For bit "0", add an artificial edge between pairs of storage variables in the RAT framework where both variables are even, and for any other bit value, add an edge between pairs of odd storage variables. For example, the derived watermarking constraints include pairs such *as* <*J*0, *J*2>, <*J*0, *J*4>—<*J*0, *J*262>—<*J*244, *J*128>, and <*J*1, *J*3>—<*J*1, *J*263>—<*J*3, *J*261>. Subsequently, the determined crypto-based watermarking constraints are integrated into the JPEG-CODEC design within the RAT framework (*i.e.*, register allocation information) of the HLS process.

The core concept underlying the incorporation of watermarking constraints (represented as artificial edges) is to ensure that storage variables associated with these edges cannot share the same register (Anshul and Sengupta, 2023b; Anshul and Sengupta, 2023c) in the RAT. Failure to adhere to this requirement prompts either register swapping or the allocation of a new register to satisfy hardware watermarking constraints, effectively enforcing distinct register allocation (Anshul and Sengupta, 2023b). Moreover, Table 6.2 presents the register allocation table before and after the incorporation of the generated crypto-signature-based watermarking constraints into the JPEG-CODEC. The table denotes the initial (normal) and final positions of registers (with bolded entries) indicating post-embedding positions. The added watermarking constraints manifest as alterations

Table 6.2 Register allocation table before and after embedding crypto-chain-based watermarking constraints corresponding to JPEG-CODEC

Control steps (C0–C16)/ Registers (R1–R129)	***R*1**	***R*2**	***R*3**	***R*4**	***R*5**	***R*6**	—	***R*129**
*C*0	J0	J1	J2	J3	J4	J5	—	J135
*C*1	J128/**J129**	J129/**J128**	J130/**J131**	J131/**J130**	J132	J133	—	J135
*C*2	J136/**J137**	J137/**J136**	J138/**J139**	J139/**J138**	J140	J141	—	J135
*C*3	J144/**J145**	J145/**J144**	J146/**J147**	J147/**J146**	J148	J149	—	J135
*C*4	J152/**J153**	J153/**J152**	J154/**J155**	J155/**J154**	J156	J157	—	J135
*C*5	J160/**J161**	J161/**J160**	J162/**J163**	J163/**J162**	J164	J164	—	J135
*C*6	J168/**J169**	J169/**J168**	J170/**J171**	J171/**J170**	J172	J173	—	J135
*C*7	J176/**J177**	J177/**J176**	J178/**J179**	J179/**J178**	J180	J181	—	J135
*C*8	J184/**J185**	J185/**J184**	J186/**J187**	J187/**J186**	J188	J189	—	J135
*C*9	J220/**J221**	J221/**J220**	J222/**J223**	J223/**J222**	J236	J237	—	J135
*C*10	J238/**J239**	J239/**J238**	J240/**J241**	J241/**J240**	J236	J237	—	J135
*C*11	J244/**J245**	J245/**J244**	J246	J247/**J248**	J248/**J247**	J249	—	J135
*C*12	J252/**J253**	J253/**J252**	J254	J255	J256	J257	—	J135
*C*13	J258/**J259**	J259/**J258**	J260	—	—	—	—	J135
*C*14	J261	—	J260	—	—	—	—	J135
*C*15	—	**J262**	—	—	—	—	—	J135
*C*16	J263	—	—	—	—	—	—	—

in register positions within the SDFG of the JPEG-CODEC. Notably, this low-cost security algorithm guarantees robust detective countermeasure against IP piracy and false assertion of IP ownership without incurring additional design costs (as no extra registers are required post-embedding watermarking constraints).

6.5 IP piracy detection and nullification of false assertion of IP ownership

IP piracy detection involves the use of crypto-signature as hardware watermarking constraints. During piracy detection, the crypto-signature-based watermarking constraints are regenerated using the discussed algorithm and checked in the suspected chip. The watermarking constraints are extracted from the suspected chip's register transfer level information (after reverse engineering). In case of a match with the suspected chip, IP piracy is detected (Anshul and Sengupta, 2023b).

In the case of attempting to evade IP piracy detection, an attacker would try to reproduce the watermarking constraints of the IP seller. However, he/she would be unable to reproduce the unique signature and implant it in the pirated version (to escape piracy detection). This is because the regeneration of original crypto-based watermarking constraints depends upon various security factors discussed in prior sections (that are only decodable by the original IP seller). Therefore, these watermarking constraints make it complex for an adversary to accurately regenerate the signature (Anshul and Sengupta, 2023b).

Furthermore, the inclusion of the crypto-based watermarking constraints within the IP design also safeguards against the assertion of false IP ownership claim. Instances of fraudulent IP ownership claim may arise in the SoC design house and foundry. In the case of ownership conflict resolution, the original watermarking constraints are regenerated and then compared against the extracted register allocation information of the IP design under test. Ownership rights are awarded to the entity that is able to establish complete matching of the security information. Thus, the discussed security algorithm effectively cancels false IP ownership assertion (Anshul and Sengupta, 2023b; Anshul and Sengupta, 2023c).

6.6 Discussion and analysis on case studies

The discussed approach uses specific parametric values, such as a firefly population size (*P*) of 3, 5, and 7, with $s1$ and $s2$ set to 0.5, γ linearly decreasing from 0.5 to 0.1, β_0 set to 1, *rand* set to 1.5, and α_x and α_y linearly decreasing from the maximum value of the first and second dimensions respectively (Anshul and Sengupta, 2023b). Evaluation of area and latency for the JPEG-CODEC is conducted on a 15 nm scale using the NanGate library (EXP BEN, 2024).

6.6.1 Design cost analysis

The design cost of the discussed low-cost crypto-chain security approach is evaluated using a design cost function mentioned in (6.3). Figure 6.11 presents the

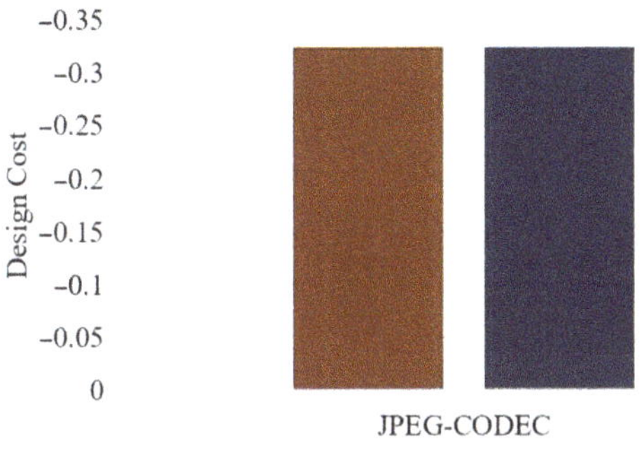

Figure 6.11 Design cost comparison before and after embedding crypto-chain signature for JPEG-CODEC

design cost for the low-cost secure JPEG-CODEC IP, both before and after incorporating the watermarking constraints. The determined FA-DSE-based low-cost architectural solution for the secured JPEG-CODEC hardware IP consists of six adders and eight multipliers (Anshul and Sengupta, 2023b). It is evident from Figure 6.11 that the discussed low-cost security approach imposes no design cost overhead while ensuring the robust security of the JPEG-CODEC hardware IP core.

6.6.2 Security analysis

The security analysis of the discussed low-cost crypto-based watermarking methodology employs the probability of coincidence and tamper tolerance (Anshul and Sengupta, 2023b). The probability of coincidence refers to the likelihood of detecting identical covert security information (constraints) in an unsecured design. In other words, the probability of coincidence is also a measure of false positives. It is represented by the formula (Anshul and Sengupta, 2023b; Koushanfar *et al.*, 2005; Sengupta and Bhadauria, 2016; Hu *et al.*, 2021; Potkonjak, 2006):

$$\text{Probability of coincidence} = (1 - 1/b)^k \tag{6.12}$$

where "b" represents the number of registers in the JPEG-CODEC's SDFG before incorporating watermarking constraints, and "k" denotes the total implanted watermarking constraints. A lower probability of coincidence value denotes more robust security, thus providing stronger digital evidence. The incorporation of distinct crypto-based watermarking constraints into the JPEG-CODEC IP design significantly contributes to the precise and reliable identification of pirated IP cores from authentic ones. Figure 6.12 presents a comparative analysis of the probability of coincidence between the discussed approach, palmprint biometric (Sengupta *et al.*, 2021; Chaurasia *et al.*, 2022), steganography (Sengupta and Rathor, 2019), multi-variable watermarking (Sengupta and Bhadauria, 2016), encrypted signature

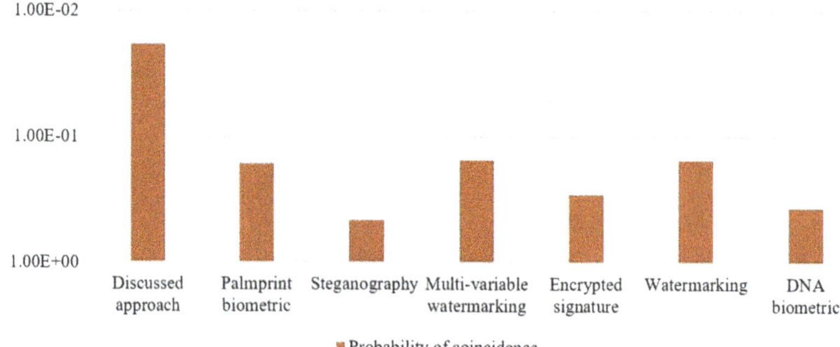

Figure 6.12 Comparison of probability of coincidence between discussed approach, palmprint biometric (Sengupta et al., 2021), (Chaurasia et al., 2022), steganography (Sengupta and Rathor, 2019), multi-variable watermarking (Sengupta and Bhadauria, 2016), encrypted signature (Castillo et al., 2008), watermarking (Koushanfar et al., 2005), and DNA biometric (Sengupta and Chaurasia, 2022)

(Castillo *et al.*, 2008), watermarking (Koushanfar *et al.*, 2005), and DNA biometric (Sengupta and Chaurasia, 2022). The discussed FA-based crypto-chain security approach outperforms all similar prior approaches with a lower probability of coincidence value. The discussed approach facilitates the determination and incorporation of higher covert constraints, thus making the occurrence of identical security information in an unsecured design highly improbable for the attackers.

Next, tamper tolerance refers to the ability of a security mechanism to withstand and resist unauthorized tampering/brute-force attack (Anshul and Sengupta, 2023b). It is represented by the formula (Anshul and Sengupta, 2023b; Koushanfar *et al.*, 2005; Sengupta and Bhadauria, 2016; Hu *et al.*, 2021; Potkonjak, 2006):

$$\text{Tamper tolerance} = v^s \tag{6.13}$$

where "s" denotes the total implanted watermarking constraints, and "v" represents the IP seller's selected distinct mapping variables. A higher tamper tolerance signifies a larger signature space, indicating more robust security. With a higher tamper tolerance value, it is possible to generate various signature combinations. With a larger signature combination, it becomes difficult for attackers to extract exact covert constraints (or to decode the precise signature combination). Adversaries aim to regenerate the exact signature to evade IP piracy detection. Thus, a higher tamper tolerance value impedes adversaries from tampering with the secured JPEG-CODEC IP design. Figure 6.13 presents a comparative analysis of the tamper tolerance between the discussed approach, palmprint biometric (Sengupta *et al.*, 2021; Chaurasia *et al.*, 2022), multi-variable watermarking (Sengupta and Bhadauria, 2016), encrypted signature (Castillo *et al.*, 2008), watermarking (Koushanfar *et al.*, 2005), and DNA biometric (Sengupta and

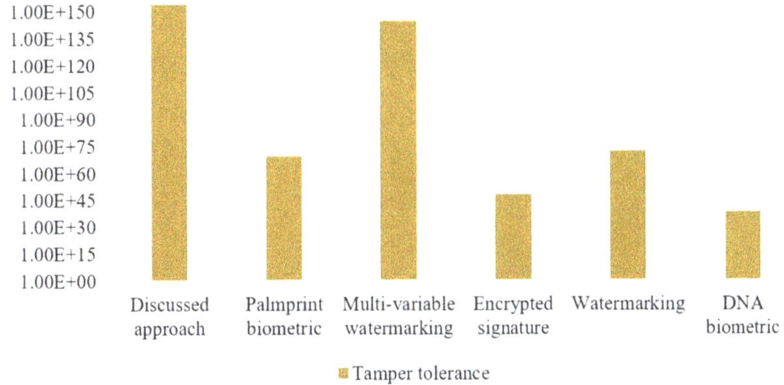

Figure 6.13 Comparison of tamper tolerance between discussed approach, palmprint biometric (Sengupta et al., 2021), (Chaurasia et al., 2022), multi-variable watermarking (Sengupta and Bhadauria, 2016), encrypted signature (Castillo et al., 2008), watermarking (Koushanfar et al., 2005), and DNA biometric (Sengupta and Chaurasia, 2022)

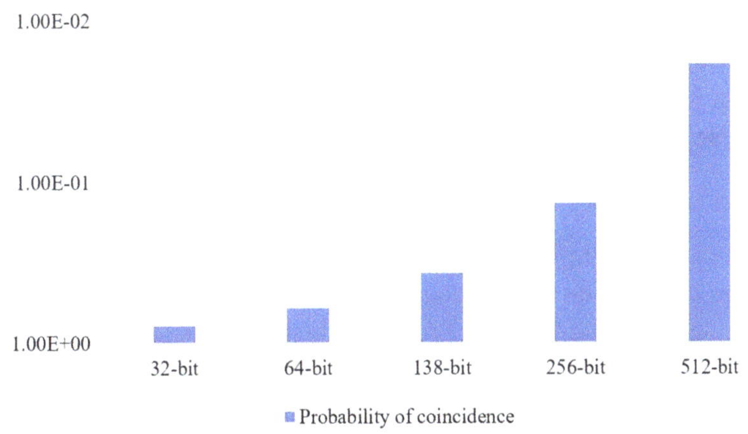

Figure 6.14 Variation in probability of coincidence with increase in signature size

Chaurasia, 2022). The discussed low-cost security approach outperforms all mentioned approaches with a higher tamper tolerance value. With a large tamper tolerance value, extracting the precise watermarking constraints from the multitude of combinations within the signature space becomes exceedingly difficult. Furthermore, Figures 6.14 and 6.15 illustrate the variation in the probability of coincidence and tamper tolerance values regarding the incorporation of crypto-chain signature into the JPEG-CODEC hardware IP design. As the number of

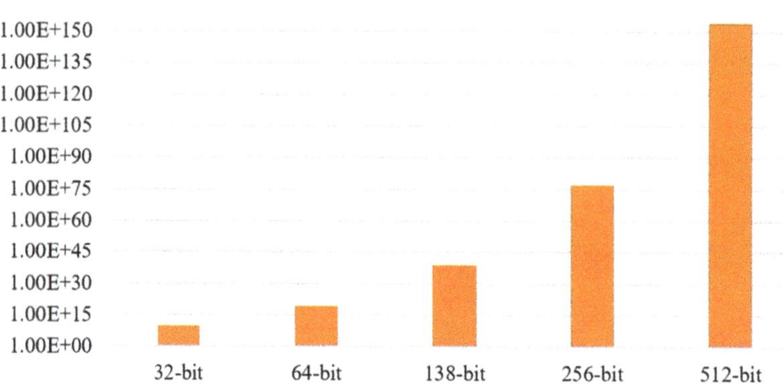

Figure 6.15 Variation in tamper tolerance with increase in signature size

embedded signature bits increases, the probability of coincidence decreases, whereas the value of tamper tolerance rises.

6.6.3 Entropy analysis

Entropy refers to the measure of uncertainty/randomness within the security system, indicating an adversary's difficulty in predicting or deducing covert information (NCSRC, 2024). It is represented by the formula (Anshul and Sengupta, 2023b; Sengupta *et al.*, 2023c):

$$Entropy = \left((1/2^s) * (1/E_n) * (1/R) * \left(1/2^{64}\right)\right) \tag{6.14}$$

where "*s*" represents the strength of the obtained IP seller's signature, "*En*" denotes the IP seller selected encoding rules, "*R*" indicates the maximum value of the round computation, and $(1/2^{64})$ signifies the probability of encountering the exact key hash buffer initialized value in the SHA-512 cryptographic module (each hash buffer being initialized with a predefined 64-bit value). Figure 6.16 presents a comparison of the entropy between the discussed approach, palmprint biometric (Sengupta *et al.*, 2021; Chaurasia *et al.*, 2022), multi-variable watermarking (Sengupta and Bhadauria, 2016), encrypted signature (Castillo *et al.*, 2008), watermarking (Koushanfar *et al.*, 2005), and DNA biometric (Sengupta and Chaurasia, 2022). The discussed approach demonstrates a stronger entropy value (lower probability value) in comparison to the previously mentioned approaches. More explicitly, an attacker needs to decode the following resistive parameters to overcome the uncertainty in regenerating the exact watermarking constraints, which offers stronger entropy than existing works: (a) IP seller's selected encoding mechanism, (b) IP seller's selected mapping rule, and (c) IP seller's selected number of hash slices and round function computation value.

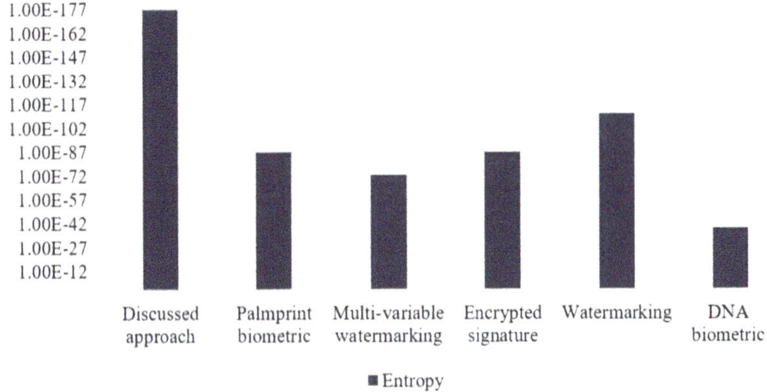

Figure 6.16 Comparison of entropy between discussed approach, palmprint biometric (Sengupta et al., 2021; Chaurasia et al., 2022), multi-variable watermarking (Sengupta and Bhadauria, 2016), encrypted signature (Castillo et al., 2008), watermarking (Koushanfar et al., 2005), and DNA biometric (Sengupta and Chaurasia, 2022)

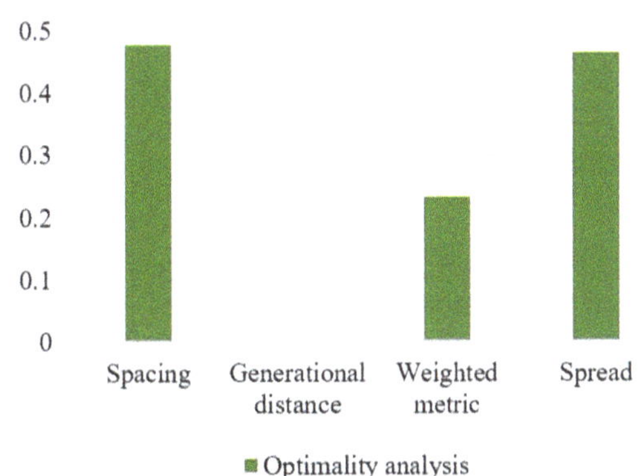

Figure 6.17 Optimality analysis of the discussed low-cost crypto-chain-based hardware security approach

6.6.4 Optimality analysis

The optimality analysis for the discussed low-cot security approach, regarding the determination of explored resource (architectural) configurations for the secured JPEG-CODEC IP, is conducted using the following metrics: (i) generational distance, (ii) spacing, (iii) weighted sum, and (iv) spreading. Figure 6.17 shows the evaluated optimality metrics values corresponding to the discussed approach. A zero generational distance value signifies that the list of solutions obtained aligns with the true Pareto front. Similarly, a zero value (or slightly higher) for the spacing parameter indicates an even distribution of Pareto points on the curve. Furthermore, the spread metric evaluates how thoroughly the true Pareto front is covered. Weighted sum is evaluated by considering generation distance and spread metric (both given equal weightage) (Anshul and Sengupta, 2023b).

6.7 Conclusion

This chapter discusses an optimal crypto-based hardware watermarking approach employing the firefly algorithm and crypto-chain-based signature generation process for generating low-cost secure hardware IP design. The primary aim is to provide a detective measure against IP piracy and false assertion of IP ownership within SoC design houses or foundries. Here, the IP seller is classified as the defender, and the foundry/SoC integrator is the potential attacker. The discussed security approach incorporates an encoding mechanism specified by the IP seller, SHA-512 hash slices, cryptographic keys, and mapping rules to determine watermarking constraints. These components collectively lead to the generation of tamper-resistant signature, which are then translated into watermarking constraints and implanted into the low-cost hardware design. This incorporation of IP seller's watermarking constraints serves as digital evidence, safeguarding the hardware IPs from piracy and false assertion of ownership rights.

6.8 Questions and exercise

1. Briefly explain the importance of application-specific integrated circuits in the context of multimedia and electronic systems.
2. What factors affect the security of hardware IP cores? What is the importance of safeguarding hardware IP cores from various threats to hardware?
3. Discuss the advantages of JPEG-CODEC with respect to electronic and multimedia systems with some practical examples.
4. Define the inputs of the crypto-chain-based security approach for hardware security.
5. Define the inputs of the firefly-based design space exploration module.
6. Explain the design flow of crypto-chain-based hardware security approach.
7. What is the importance of design space exploration in a high-level synthesis process?
8. Explain the generation process of low-cost resource configuration using firefly-based design space exploration.
9. List the advantages of particle swarm optimization and firefly-based design space exploration.
10. Illustrate the generation process of crypto-chain-based watermarking constraints in brief.

11. Describe the embedding of the generated covert constraints with a suitable example and determine its register allocation table corresponding post embedding.
12. What is the importance of a register allocation table while performing the embedding of crypto-chain-based signature?
13. Briefly explain the detection and validation for detective control against fraud claim of IP ownership and IP piracy.
14. What are some of the advantages/merits of the crypto-chain-based security module that helps in achieving robust security?
15. What are the security metrics used to compare and analyze the security strength of the discussed security approaches?
16. Comment on the optimality analysis of firefly-based design space exploration.

References

R. Agarwal, C. S. Salimath and K. Alam, (2019) "Multiple Image Compression in Medical Imaging Techniques Using Wavelets for Speedy Transmission and Optimal Storage," *Biomed and Pharmacology Journal*, vol. 12, no.1, pp. 183–198.

A. Anshul and A. Sengupta, (2022) "IP Core Protection of Image Processing Filters with Multi-Level Encryption and Covert Steganographic Security Constraints," *2022 IEEE International Symposium on Smart Electronic Systems (iSES)*, Warangal, India, pp. 83–88.

A. Anshul and A. Sengupta, (2023a) "A Survey of High Level Synthesis Based Hardware Security Approaches for Reusable IP Cores [Feature]," *IEEE Circuits and Systems Magazine*, vol. 23, no. 4, pp. 44–62.

A. Anshul and A. Sengupta, (2023b) "Exploration of Optimal Crypto-chain Signature Embedded Secure JPEG-CODEC Hardware IP during High Level Synthesis," *Microprocessors and Microsystems*, vol. 102, 104916.

A. Anshul and A. Sengupta, (2023c) "PSO Based Exploration of Multi-phase Encryption Based Secured Image Processing Filter Hardware IP Core Datapath during High Level Synthesis," *Expert Systems with Applications*, vol. 223, 119927.

A. Anshul, K. Bharath and A. Sengupta, (2022) "Designing Low Cost Secured DSP Core using Steganography and PSO for CE systems," *2022 IEEE International Symposium on Smart Electronic Systems (iSES)*, Warangal, India, pp. 95–100.

E. Castillo, L. Parrilla, A. Garcia, U. Meyer-Baese, G. Botella and A. Lloris, (2008) "Automated Signature Insertion in Combinational Logic Patterns for HDL IP Core Protection," *2008 4th Southern Conference on Programmable Logic, Bariloche*, Argentina, pp. 183–186.

R. Chaurasia, A. Anshul, A. Sengupta and S. Gupta, (2022) "Palmprint Biometric Versus Encrypted Hash Based Digital Signature for Securing DSP Cores Used in CE Systems," *IEEE Consumer Electronics Magazine*, vol. 11, no. 5, pp. 73–80.

Y. Chen, (2007) "Medical Image Compression Using DCT-Based Subband Decomposition and Modified SPIHT Data Organization," *International Journal of Medical Informatics*, vol. 76, no. 10, pp. 717–725.

B. Colombier and L. Bossuet, (2015) "Survey of Hardware Protection of Design Data for Integrated Circuits and Intellectual Properties," *IET Computers & Digital Techniques*, vol. 8, no. 6, pp. 274–287.

EXP BEN, (2024) University of California Santa Barbara Express Group, accessed on Jan. 2024. [Online]. Available: http://express.ece.ucsb.edu/benchmark/.

B. Ge, Y. Han and C. Bian, (2016) "Hybrid Ant Colony Optimization Algorithm for Solving the OpenVehicle Routing Problem," *Journal of Computers*, vol. 27, no.4, pp. 41–54.

S. B. Gokturk, C. Tomasi, B. Girod and C. Beaulieu, (2001) "Medical Image Compression Based on Region of Interest, with Application to Colon CT Images," *2001 Conference Proceedings of the 23rd Annual International Conference of the IEEE Engineering in Medicine and Biology Society*, Istanbul, Turkey, vol. 3, pp. 2453–2456.

W. Hu, C. Chang, A. Sengupta, S. Bhunia, R. Kastner and H. Li, (2021) "An Overview of Hardware Oriented Security and Trust: Threats, Countermeasures and Design Tools," *IEEE Transactions on Computer-Aided Design of Integrated Circuits and Systems*, Invited Paper, vol. 40, no. 6, pp. 1010–1038.

S. A. Islam, L. K. Sah and S. Katkoori, (2020) "High-Level Synthesis of Key-Obfuscated RTL IP with Design Lockout and Camouflaging," *ACM Transactions on Design Automation of Electronic Systems (TODAES)*, vol. 26, no. 1, pp. 1–35.

Y. Kao, M.-H. Chen and Y.-T. Huang, (2012) "A Hybrid Algorithm Based on ACO and PSO for Capacitated Vehicle Routing Problems," *Mathematical Problems in Engineering*, vol. 2012, no. 1, 726564.

D. A. Koff and H. Shulman, (2006) "An Overview of Digital Compression of Medical Images: Can We Use Lossy Image Compression in Radiology?" *Canadian Association of Radiologists Journal*, vol. 57, no.4, pp. 211–217.

F. Koushanfar, I. Hong and M. Potkonjak, (2005) "Behavioral Synthesis Techniques for Intellectual Property Protection," *ACM Transactions on Design Automation of Electronic Systems (TODAES)*, vol. 10, no. 3, pp. 523–545.

V. Krishnan and S. Katkoori, (2006) "A Genetic Algorithm for the Design Space Exploration of Datapaths during High-level Synthesis," *IEEE Transactions on Evolutionary Computation*, vol. 10, no. 3, pp. 213–229.

B. Le Gal and L. Bossuet, (2012) "Automatic Low-cost IP Watermarking Technique Based on Output Mark Insertions," *Design Automation for Embedded Systems*, vol. 16, no. 2, pp. 71–92.

V. K. Mishra and A. Sengupta, (2014) "MO-PSE: Adaptive Multi-objective Particle Swarm Optimization Based Design Space Exploration in Architectural Synthesis for Application Specific Processor Design," *Advances in Engineering Software*, vol. 67, pp. 111–124.

NCSRC, (2024) NIST Computer Security Resource Center, Glossary, https://csrc.nist.gov/glossary/term/entropy#:~:text=A%20measure%20of%20the%20amount, is%20usually%20stated%20in%20bits, accessed on Jan 2024.

M. Potkonjak, (2006) "Methods and systems for the identification of circuits and circuit designs," United State Patent, US7017043B1.

M. Rathor, A. Sengupta, R. Chaurasia and A. Anshul, (2023a) "Exploring Handwritten Signature Image Features for Hardware Security," *IEEE Transactions on Dependable and Secure Computing*, vol. 20, no. 5, pp. 3687–3698.

M. Rathor, A. Anshul, K. Bharath, R. Chaurasia and A. Sengupta, (2023b) "Quadruple Phase Watermarking during High Level Synthesis for Securing Reusable Hardware Intellectual Property Cores," *Computers and Electrical Engineering*, vol. 105, 108476.

M. Rathor, A. Anshul and A. Sengupta, (2024) "Securing Reusable IP Cores Using Voice Biometric Based Watermark," *IEEE Transactions on Dependable and Secure Computing*, vol. 21, no. 4, pp. 2735–2749.

S. Rizzo, F. Bertini, and D. Montesi, (2019) "Fine-grain Watermarking for Intellectual Property Protection," *EURASIP Journal on Information Security*, vol. 2019, Article no. 10.

M. Rostami, F. Koushanfar and R. Karri, (2014) "A Primer on Hardware Security: Models, Methods, and Metrics," *Proceedings of the IEEE*, vol. 102, no. 8, pp. 1283–1295.

P. Sarkar, A. Sengupta, S. Rathlavat and M. K. Naskar, (2017) "Designing Low-Cost Hardware Accelerators for CE Devices," *IEEE Consumer Electronics Magazine*, vol. 6, no. 4, 140–149.

R. Schneiderman, (2010) "DSPs Evolving in Consumer Electronics Applications," *IEEE Signal Processing Magazine*, vol. 27, no. 3, pp. 6–10.

A. Sengupta and S. Bhadauria, (2016) "Exploring Low Cost Optimal Watermark for Reusable IP Cores during High Level Synthesis," *IEEE Access*, vol. 4, pp. 2198–2215.

A. Sengupta and R. Chaurasia, (2022) "Securing IP Cores for DSP Applications Using Structural Obfuscation and Chromosomal DNA Impression," *IEEE Access*, vol. 10, 50903–50913.

A. Sengupta and M. Rathor, (2019) "IP Core Steganography for Protecting DSP Kernels Used in CE Systems," *IEEE Transactions on Consumer Electronics*, vol. 65, no. 4, pp. 506–515.

A. Sengupta, R. Sedaghat and P. Sarkar, (2012) "A Multi structure Genetic Algorithm for Integrated Design Space Exploration of Scheduling and Allocation in High Level Synthesis for DSP Kernels," *Swarm and Evolutionary Computation*, vol. 7, pp. 35–46.

A. Sengupta, D. Roy, S. P. Mohanty and P. Corcoran, (2018) "Low-Cost Obfuscated JPEG CODEC IP Core for Secure CE Hardware," *IEEE Transactions on Consumer Electronics*, vol. 64, no. 3, pp. 365–374.

A. Sengupta, R. Chaurasia and T. Reddy, (2021) "Contact-Less Palmprint Biometric for Securing DSP Coprocessors Used in CE Systems," *IEEE Transactions on Consumer Electronics*, vol. 67, no. 3, pp. 202–213.

A. Sengupta, R. Chaurasia and A. Anshul, (2023a) "Hardware Security of Digital Image Filter IP Cores against Piracy using IP Seller's Fingerprint Encrypted Amino Acid Biometric Sample," *2023 Asian Hardware Oriented Security and Trust Symposium (AsianHOST)*, Tianjin, China, pp. 1–6.

A. Sengupta, A. Anshul and R. Chaurasia, (2023b) "Exploration of Optimal Functional Trojan-Resistant Hardware Intellectual Property (IP) Core Designs during High Level Synthesis," *Microprocessors and Microsystems*, vol. 103, 104973.

A. Sengupta, R. Chaurasia and A. Anshul, (2023c) "Robust Security of Hardware Accelerators Using Protein Molecular Biometric Signature and Facial Biometric Encryption Key," *IEEE Transactions on Very Large Scale Integration (VLSI) Systems*, vol. 31, no. 6, pp. 826–839.

F. Wihartiko, H. Wijayanti and F. Virgantari, (2018) "Performance Comparison of Genetic Algorithms and Particle Swarm Optimization for Model Integer Programming Bus Timetabling Problem," *IOP Conference Series: Materials Science and Engineering*, vol. 332, 012020.

X.-S. Yang, (2009) "Firefly Algorithms for Multimodal Optimization," in *Proceedings of 5th International Conference Stochastic Algorithms: Foundations and Applications (SAGA '09)*, pp. 169–178.

X.-S. Yang and X. He, (2013) "Firefly Algorithm: Recent Advances and Applications," *International Journal of Swarm Intelligence*, vol. 1, no. 1, pp. 36–50.

M. Yasin, J. J. Rajendran, O. Sinanoglu and R. Karri, (2016) "On Improving the Security of Logic Locking," *IEEE Transactions on Computer-Aided Design of Integrated Circuits and Systems*, vol. 35, no. 9, pp. 1411–1424.

Z. Zukhri, U. Islam and Zukhri, (2013) "A Hybrid Optimization Algorithm based on Genetic Algorithm and Ant Colony Optimization," *International Journal of Artificial Intelligence and Application*, vol. 4, pp. 63–75.

Chapter 7
HLS-based fingerprinting

Anirban Sengupta[1] *and Aditya Anshul*[1]

The utilization of hardware intellectual property (IP) cores within system-on-chip computing architectures offers a distinct advantage by enhancing design productivity while reducing the overall design cycle time. However, it is necessary to secure these IP designs against potential threats from the perspective of both, the seller and the buyer within the global design supply chain process. This chapter discusses an IP fingerprinting and symmetrical IP protection mechanism for securing IP buyer's and IP seller's rights. This chapter demonstrates the embedding of both a buyer's fingerprint and a seller's watermark simultaneously using the high-level synthesis (HLS) process. By integrating the buyer's fingerprint and seller's watermark into the IP design, robust protection against IP piracy/unauthorized usage and false IP ownership claim is ensured.

7.1 Introduction

The growing demands of modern technological society underscore the significance of application-specific computing hardware in driving innovation and addressing the diverse computational challenges of the digital era. As technology continues to evolve, the development and adoption of specialized hardware solutions play a pivotal role in shaping the future of computing, enabling breakthroughs in areas ranging from artificial intelligence (AI) and machine learning (ML) to the Internet of Things (IoT) and beyond. This surge is primarily fueled by the ever-evolving nature of technological requirements and the need for optimized performance across various domains. Unlike traditional and generalized computing solutions, application-specific hardware caters to specific tasks or functions, offering superior efficiency and performance. These application-specific computing systems are designed as reusable intellectual property (IP) cores using a high-level synthesis (HLS) framework (Anshul and Sengupta, 2023a). There has been widespread adoption of dedicated hardware IP cores (like FIR, ML, AI accelerators, DCT, JPEG, etc.) to meet the modern demands of higher performance at low development cost (Rathor *et al.*, 2023a; Le Gal and Bossuet 2012; Anshul *et al.*, 2022).

[1]Department of Computer Science and Engineering, Indian Institute of Technology Indore, India

Protection of IP buyer and seller: In the design process of reusable hardware IP cores, both IP buyers and IP sellers play crucial roles. Buyers seek exclusive user rights to prevent unauthorized redistribution of purchased IPs. On the other hand, sellers must protect their designs from piracy and false IP ownership claim. This demand arises when a buyer procures an IP based on custom specifications from a seller, establishing a unique mapping between the two entities. Buyer fingerprinting enables the tracing of illegally sold copies of an IP core by dishonest sellers. Similarly, IP core sellers must safeguard their work from IP piracy and false IP ownership claims. Conventional intellectual property protection measures like copyrights, trademarks, and patents are insufficient to address these concerns effectively (Roy and Sengupta, 2017; Sengupta *et al.*, 2019). Hence, a robust protection scheme is necessary from both perspectives. Buyer fingerprinting emerges as a solution to safeguard exclusive buyer rights by embedding user-specific signature without altering functionality, thereby deterring illegal redistribution. Additionally, IP sellers must safeguard their design from piracy, and false IP ownership claim. This underscores the importance of a robust symmetrical protection mechanism for hardware IPs from both the buyer's and seller's perspectives. There are two primary objectives in the field of hardware IP fingerprinting: (i) from the angle of the IP buyer, to trace illegally redistributed and resold copies by dishonest IP sellers, and (ii) from the angle of the IP seller to secure the design against false IP ownership claims and IP piracy (Rostami *et al.*, 2014; Rizzo *et al.*, 2019; Anshul and Sengupta, 2023b; Sengupta *et al.*, 2023a; Rathor *et al.*, 2023b; Colombier and Bossuet, 2015; Rathor *et al.*, 2024; Anshul and Sengupta, 2023c; Islam *et al.*, 2020; Anshul *et al.*, 2022; Yasin *et al.*, 2016; Sengupta *et al.*, 2023b; Anshul and Sengupta, 2023d; Sengupta *et al.*, 2023c; Chaurasia *et al.*, 2022; Anshul and Sengupta, 2022).

While various signature integration schemes have been proposed in different multimedia contexts over the past few decades, they are often not applicable for IP core protection as they may alter functionality and accuracy during signature insertion (Roy and Sengupta, 2017; Sengupta *et al.*, 2019). Therefore, buyer fingerprinting, and seller watermark insertions must not only provide necessary security to both entities but also preserve the IP design's correct functionality. Additionally, despite the importance of security, minimizing design overhead remains a priority. The objective is to minimize the design overhead without compromising the security of the IP design. Certain criteria must be met for the fingerprint to be effective. It should be permanently embedded into the design, resistant to easy tracing, unique for each buyer, non-interfering with the seller's IP rights protection scheme, incurring minimal overhead post-embedding, and preserving the IP core's functional behavior. This chapter discusses two recent IP fingerprinting methods.

The first approach (Sengupta *et al.*, 2019) discusses a cryptographic multivariable fingerprinting approach employing a four-variable signature (incurring zero design cost). This approach utilizes a cryptographic technique (secure hashing and unique processing of the hashed digest) to enhance the security of the design by employing a distinct encoding mechanism for the intermediate representation of the

IP core (known as control data flow graph, CDFG). On the other hand, the second approach (Roy and Sengupta, 2017) discusses low-overhead security methodology (symmetrical security methodology) for reusable IP cores, integrating robust buyer fingerprinting and seller watermarking during the high-level synthesis process.

7.2 Crypto multi-variable fingerprinting

7.2.1 Details of the technique (Sengupta et al., 2019)

The overview of the cryptographic multi-variable fingerprinting algorithm is depicted in Figure 7.1. As shown in Figure 7.1, it accepts the control data flow graph (CDFG) or transfer function of the respective hardware application, IP seller's/designer's resource constraints, encoding mechanism, and fingerprint signature combination as the primary input. The cryptographic fingerprinting approach (Sengupta *et al.*, 2019) uses an SHA-512-based hashing module to generate the corresponding hash digest of the application. The fingerprinting algorithm encompasses several key steps:

(a) *Scheduling of input CDFG*: Initially, the input CDFG is scheduled using the IP seller's resource constraints and LIST scheduling algorithm. Post-scheduling, the scheduled data flow graph (SDFG) and the corresponding *"register allocation table (RAT)"* are generated. The RAT indicates the minimum number of registers required for implementing the storage variables. Further, an encoded bit stream is also determined in this step using the generated SDFG and IP seller's chosen encoding mechanism, which serves as input for subsequent hashing block.

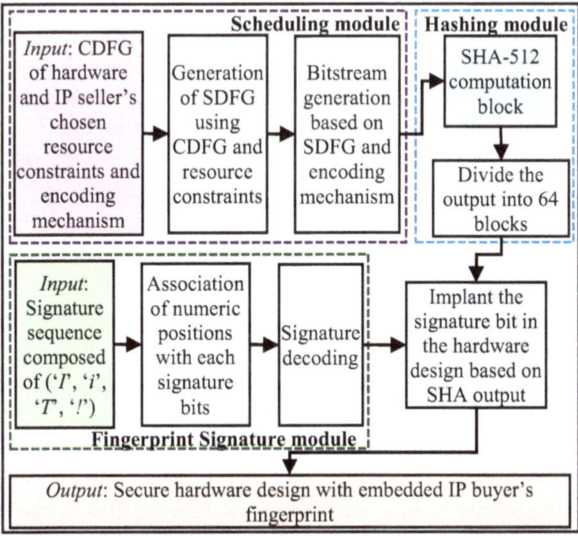

Figure 7.1 Overview of crypto-based multi-variable fingerprint methodology

(b) *Hashing process*: In this phase, the generated bit stream undergoes SHA-512 hash computation, producing a 512-bit hash digest. This obtained hash digest is further divided into 64-bit blocks/chunks. Each chunk is converted into its decimal value. SHA-512 is specifically chosen for its robust security properties, including collision resistance (one-way function), making it computationally infeasible to derive the original text from the hash value or to find two inputs with the same hash value. Moreover, the uniform-length output of SHA-512 facilitates uniform breakdown and association with the input signature.

(c) *Signature embedding*: The final step involves embedding the IP buyer's fingerprint (signature) into the IP core. Firstly, the fingerprint signature digits are associated with unique position numbers. Secondly, the decimal chunk values of SHA output (obtained from the previous step) are matched with the position numbers of the fingerprint signature digits. Subsequently, the positions of the fingerprint signature digits that overlap with the SHA decimal value are obtained as "*selected fingerprint signature digits*." These *selected fingerprint signature digits* are used for embedding into the IP design as buyer fingerprint post conversion (decoding) into fingerprint constraints. The fingerprint constraints comprise storage variable pairs (used in the development of the register allocation table) (Sengupta *et al.*, 2019). The next section discusses these steps with an example.

7.2.2 Demonstration of the technique (Sengupta et al., 2019)

This subsection illustrates the demonstration of crypto multi-variable fingerprinting approach on 8-point DCT. The control data flow graph (CDFG) or the transfer function of the hardware (such as 8-point DCT, commonly employed in audio and image compression tasks) is taken as input. The initial phase involves organizing different operations of 8-point DCT into various control steps using a standardized scheduling algorithm (such as LIST scheduling). Let the IP seller's/designer's resource constraints be one adder and two multipliers (which can also be determined by design space exploration effort). The resulting scheduled data flow graph (SDFG) for 8-point DCT post-scheduling is depicted in Figure 7.2. Subsequently, Table 7.1 represents the register allocation table (RAT) corresponding to the generated SDFG (providing details of the registers associated with various storage variables across different control steps). This table corresponding to 8-point DCT encompasses a total of 31 storage variables distributed over 9 control steps, with a maximum requirement of 8 registers per control step (Sengupta *et al.*, 2019).

Post-scheduling, an encoded bitstream is generated based on the IP seller's encoding mechanism. Each operation within the scheduling is assigned a binary value based on the following criterion: *if an operation number is positioned at an even control step (CS) and is identified as odd, it is encoded as "1"; otherwise, it is encoded as "0"*. For example, in the illustrated DCT example (Figure 7.2), operation 1 is situated in CS 1. Hence, the corresponding encoded output for operation 1 is "0". Similarly, operation 3 falls within CS 2, resulting in an encoded bit value of "1". Following this encoding logic, the resultant encoded bitstream for the scheduling is derived as "001000100010001". The encoded bitstream output is

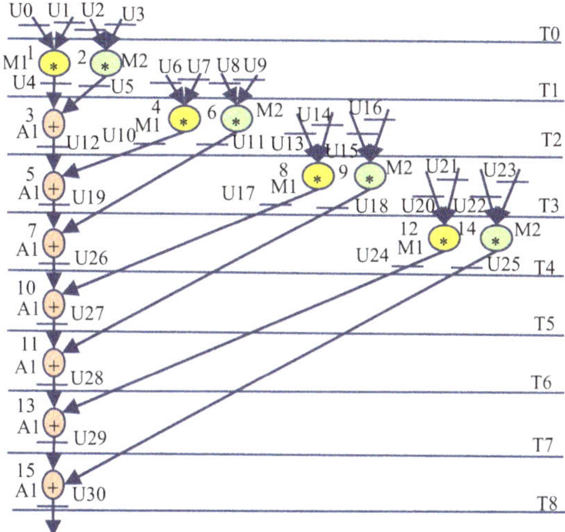

Figure 7.2 *Scheduled data flow graph of 8-point DCT based on IP seller's chosen resource constraint (one adder and two multipliers). Note: Here, U0-U30 are storage variables present in the SDFG.*

Table 7.1 *Register allocation table post embedding IP buyer's fingerprint corresponding to 8-point DCT*

Control steps	**T0**	**T1**	**T2**	**T3**	**T4**	**T5**	**T6**	**T7**	**T8**
R0	U0	U4	U10	U17	U24/U17	U27	U28	U29	U30
R1	U1	U5	U6/U11	U18/U11	U18	—	—	—	—
R2	U2	U6	U12	U12/U18	U25/U18	U25/U18	U25	—	—
R3	U3	U7	U13	U19	U19/U26	U19	—	—	—
R4	—	U8	U14	U20	U26/U24	U26/U24	U26/U24	U26	—
R5	—	U9	U15	U21	—	—	—	—	—
R6	—	—	U16	U22	—/U25	—/U25	—/U25	—/U25	—
R7	—	—	—/U13	U23	—	—	—	—	—

then fed as input to the SHA-512 block. SHA-512 produces a 512-bit message digest (hash value). Subsequently, this 512-bit hash is segmented into 64 blocks or chunks, each consisting of 8 bits. Table 7.2 represents the 64 chunks corresponding to a 512-bit long message digest. Additionally, the decimal values corresponding to each chunk are also shown in Table 7.2. Consequently, there exists a set of 64 integers (lying between 0 and 255).

After this, the buyer's fingerprint or signature is taken as input. This fingerprint comprises four distinct variables denoted as *"i"*, *"I"*, *"T"*, and *"!"*. Here, an 80-digit-long fingerprint signature for embedding purposes is taken as input (*TII!iiTTii!TITiiiIIIiTTiiiIiIiii!!!ii!IT!!IiI!!!ITi!iiT!iiTT!iTiiTT!!!iTIII!Tii!*). Each fingerprint digit is assigned a unique position. Table 7.3 illustrates the fingerprint

Table 7.2 64 blocks of the obtained SHA output

Decimal value of each chunk/block of the SHA output	Binary equivalent of the decimal value	Decimal value of each chunk/block of the SHA output	Binary equivalent of the decimal value
151	10010111	173	10101101
124	01111100	150	10010110
167	10100111	63	00111111
96	01100000	85	01010101
57	00111001	143	10001111
38	00100110	224	11100000
190	10111110	100	01100100
63	00111111	185	10111001
39	00100111	221	11011101
246	11110110	227	11100011
226	11100010	133	10000101
195	11000011	114	01110010
165	10100101	206	11001110
143	10001111	30	00011110
115	01110011	106	01101010
172	10101100	2	00000010
132	10000100	131	10000011
227	11100011	117	01110101
166	10100110	128	10000000
22	00010110	103	01100111
220	11011100	63	00111111
48	00110000	28	00011100
109	01101101	252	11111100
85	01010101	84	01010100
254	11111110	82	01010010
68	01000100	197	11000101
77	01001101	40	00101000
79	01001111	201	11001001
150	10010110	117	01110101
31	00011111	200	11001000
234	11101010	13	00001101
53	00110101	142	10001110

signature digits and their corresponding unique positions. Not all fingerprint digits contribute to the embedding process. As discussed earlier in Section 7.2.1, the decimal chunk values of the SHA output are matched with the position numbers of the fingerprint signature digits. Subsequently, the positions of the fingerprint signature digits that overlap with the SHA decimal value, are obtained as "*selected fingerprint signature digits.*" Each variable holds a specific semantic interpretation, and each digit within the fingerprint corresponds to a particular storage variables pair. The decoding of the selected fingerprint digits into fingerprinting constraints depends on the rules applied to a sorted list of storage variables. The fingerprinting constraints (edge) generation rule corresponding to each variable within the selected fingerprint is as follows (Sengupta *et al.*, 2019):

Table 7.3 Positions of 80-digit fingerprint

Position number	Signature digit	Position number	Signature digit
1	*T*	41	*!*
2	*I*	42	*!*
3	*I*	43	*I*
4	*!*	44	*i*
5	*i*	45	*I*
6	*i*	46	*!*
7	*T*	47	*!*
8	*T*	48	*!*
9	*i*	49	*I*
10	*i*	50	*T*
11	*!*	51	*i*
12	*T*	52	*!*
13	*I*	53	*i*
14	*T*	54	*i*
15	*i*	55	*T*
16	*i*	56	*!*
17	*i*	57	*i*
18	*I*	58	*i*
19	*I*	59	*T*
20	*I*	60	*T*
21	*i*	61	*!*
22	*T*	62	*i*
23	*T*	63	*T*
24	*i*	64	*i*
25	*i*	65	*i*
26	*i*	66	*T*
27	*I*	67	*T*
28	*i*	68	*!*
29	*I*	69	*!*
30	*i*	70	*!*
31	*i*	71	*i*
32	*i*	72	*T*
33	*!*	73	*I*
34	*!*	74	*I*
35	*!*	75	*I*
36	*i*	76	*!*
37	*i*	77	*T*
38	*!*	78	*i*
39	*I*	79	*i*
40	*T*	80	*!*

- "*i*" represents the encoded value of an edge associated with the storage variables pair of prime numbers.
- "*I*" represents the encoded value of an edge associated with the storage variables pair of even numbers.
- "*T*" represents the encoded value of an edge associated with the storage variables pair comprising an odd number and an even number.

- "*!*" represents the encoded value of an edge associated with the storage variables pair where one number is 0, and the other can be any integer.

These selected fingerprint signature digits are used for embedding into the IP design as buyer fingerprint post-conversion (decoding) into fingerprint constraints. The objective of embedding the fingerprint constraints into the design is to embed IP buyer's digital evidence in the design. The selected set of fingerprint digits and their corresponding constraints are as follows:

> ***57***: *(U5, U19),* ***38***: *(U0, U6),* ***63***: *(U1, U26),****39***: *(U2, U20),* ***22***: *(U1, U12),* ***48***: *(U0, U11),* ***68***: *(U0, U15), 77: (U3, U6),* ***79***: *(U7, U23),* ***31***: *(U3, U17),* ***53***: *(U5, U13),* ***30***: *(U3, U13),* ***2***: *(U2, U4),* ***28***: *(U3, U11),* ***40***: *(U1, U16),* ***13***: *(U2, U8).*

If both variables of the derived fingerprint constraint (edge) share the same register, swapping of the register within the same control step is performed to prevent such overlap. If the desired condition cannot be achieved through swapping alone, an additional register is allocated to accommodate the requirement (Sengupta *et al.*, 2019; Sengupta and Rathor, 2020; Sengupta and Rathor, 2021). The register allocation table post embedding of fingerprinting constraints is shown in Table 7.1 (the modified positions of registers due to fingerprint constraint embedding are depicted in red colors) (Sengupta *et al.*, 2019). Finally, a secure hardware IP design (register transfer level (RTL) datapath) is obtained post-datapath synthesis.

7.2.3 Analysis and comparative perspective

The security robustness of the crypto multi-variable fingerprinting approach is assessed through the probability of coincidence metric, representing the likelihood of detecting the same fingerprinting constraints within an unsecured design. Probability of coincidence serves as a metric indicating the strength of evidence for ownership and is also a measure of false positive. It is defined as (Sengupta *et al.*, 2019; Sengupta *et al.*, 2023c; Koushanfar *et al.*, 2005; Anshul and Sengupta, 2023a; Hu *et al.*, 2021; Potkonjak, 2006):

$$\text{Probability of coincidence} = (1 - 1/b)^k \tag{7.1}$$

where "*b*" represents the number of registers in the SDFG before implanting fingerprinting constraints, and "*k*" denotes the total embedded fingerprinting constraints. A lower value of the probability of coincidence is desirable. Figure 7.3 represents the comparison of the probability of coincidence for the crypto multi-variable fingerprinting approach (Sengupta *et al.*, 2019) with a varying number of signature sizes. On increasing the signature size, the value of the probability of coincidence decreases due to the embedding of larger constraints, making the approach more robust.

Further, the design cost for the crypto multi-variable fingerprinting approach is evaluated using equation (7.2) (Sengupta *et al.*, 2019; Sengupta *et al.*, 2023c; Anshul and Sengupta, 2023a).

$$\begin{aligned} Design\ cost = {} & h1 * (((Design\_Area))/A_{\max}) + h2 \\ & * (((Design\_Latency))/L_{\max}) \end{aligned} \tag{7.2}$$

Here, $h1$ and $h2$ are set to 0.5, indicating equal importance for both area and latency parameters during fingerprinting process. *"Design_Area"* and *"Design_Latency"* denote the final fingerprint embedded hardware area and latency (delay). Next, "A_{max}" and "L_{max}" represent the maximum design area and latency, respectively. Figure 7.4 shows the design cost for the crypto multi-variable fingerprinting approach

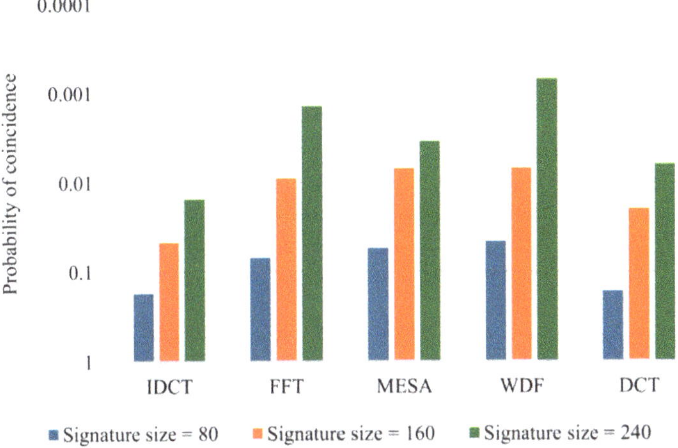

Figure 7.3 Comparison of probability of coincidence with varying signature sizes for cryptographic multi-variable fingerprinting approach (Sengupta et al., 2019)

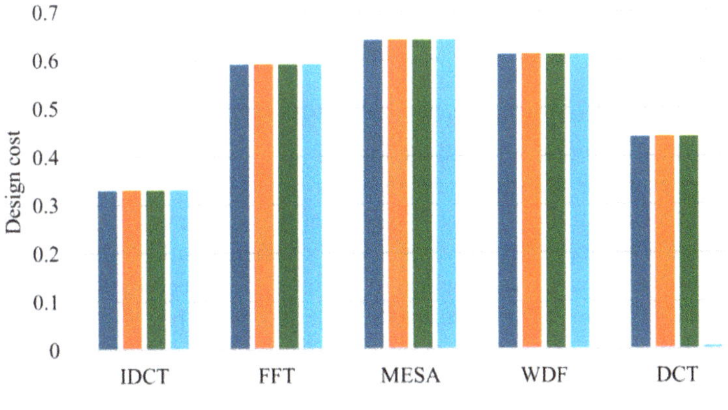

Figure 7.4 Comparison of fingerprint embedded design cost with varying signature sizes for cryptographic multi-variable fingerprinting approach (Sengupta et al., 2019)

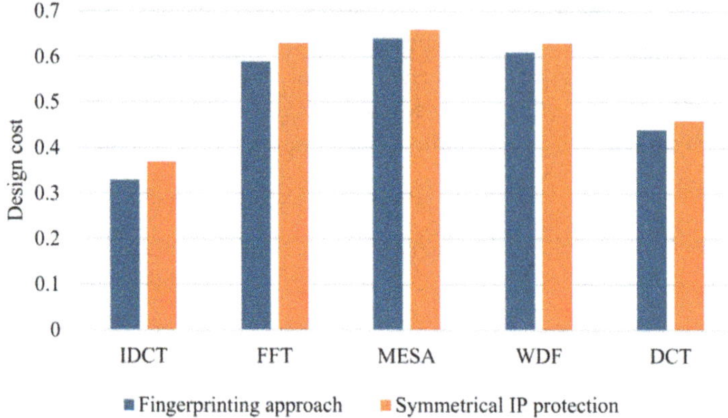

Figure 7.5 Comparison of design cost between cryptographic multi-variable fingerprinting approach (Sengupta et al., 2019) and symmetrical IP protection (Roy and Sengupta, 2017)

(Sengupta *et al.*, 2019) for a varying number of signature sizes. The fingerprinting approach (Sengupta *et al.*, 2019) incurs zero design cost even with an increase in the signature size. Additionally, Figure 7.5 depicts the design cost comparison between fingerprinting approach (Sengupta *et al.*, 2019) and with symmetrical IP protection (Roy and Sengupta, 2017) mechanism. In comparison with Roy and Sengupta (2017), fingerprint approach (Sengupta *et al.*, 2019) demonstrates reduced design cost, primarily due to reduction in register hardware count.

7.3 Symmetrical protection using fingerprinting and watermarking

7.3.1 Details of the technique (Roy and Sengupta, 2017)

The symmetrical IP protection methodology (Roy and Sengupta, 2017) discusses a security approach for securing hardware IPs through the integration of the IP buyer's fingerprinting and IP seller's watermarking constraints by exploiting the HLS process. This approach provides security benefits to both IP buyer and IP seller. In this framework, the IP seller, in addition to embedding their watermark, accommodates the buyer's fingerprint into the design as per the buyer's request during the HLS process. This dual (symmetrical) embedding strategy serves to uphold the rights and interests of both parties, shielding them against various potential threats (as outlined in the introduction section). Figure 7.6 illustrates the overview of the symmetrical IP protection mechanism (Roy and Sengupta, 2017). As shown in Figure 7.6, the symmetrical IP protection approach accepts the CDFG/ transfer function of the application, resource constraints, module library, IP buyer's fingerprint, and IP seller's watermark along with their respective constraints

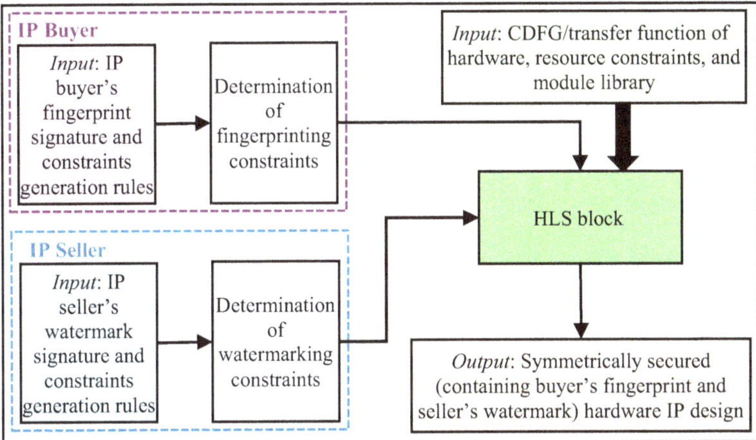

Figure 7.6 Overview of symmetrical IP protection mechanism with fingerprint and watermark (Roy and Sengupta, 2017)

generation rules, as primary inputs. The buyer's fingerprint and seller's watermark are first translated into their corresponding fingerprinting and watermarking constraints based on respective constraint generation/encoding rules. Subsequently, the translated constraints of the seller's watermark signature and the buyer's fingerprint are integrated/embedded into the design across different HLS phases. Initially, during the scheduling phase of the HLS process, the buyer's fingerprinting constraints are integrated. Following this, the seller's watermarking constraints are embedded during the register allocation phase of the HLS process. Based on the final obtained schedule and register allocation information (containing the multiplexing scheme for each resource, detailing interconnectivity specifics, multiplexer type and size, and inputs and outputs), the symmetrically secure (containing buyer's fingerprint and seller's watermark) RTL datapath is generated post-datapath synthesis. Further, the flow diagram providing the details of the symmetrical IP protection framework is depicted in Figure 7.7. There are two primary components of the symmetrical IP protection mechanism (Roy and Sengupta, 2017): (a) fingerprinting block and (b) watermarking block.

Fingerprinting block: As shown in Figure 7.7, the fingerprinting block of the symmetrical IP protection mechanism accepts the IP buyer's fingerprint along with its respective constraint generation rules, as input. Here, fingerprinting constraints are strategically integrated during both the scheduling and register allocation phases of the HLS process, such that it results in low overhead while ensuring robustness. During the scheduling phase, the insertion of buyer fingerprint constraints is achieved through the deliberate enforcement of specific operations within designated control steps during the schedule conflict resolution process. There are three unique digits in the IP buyer's fingerprint. The constraints generation rules or encoding rules for fingerprint signature are outlined as follows (Roy and Sengupta, 2017):

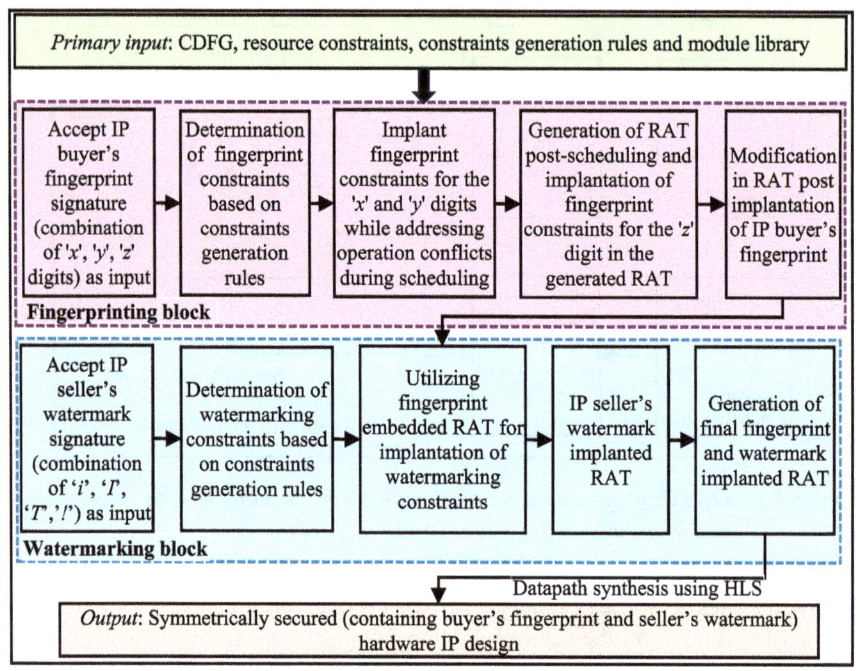

Figure 7.7 Details of symmetrical IP protection technique with fingerprinting and watermarking blocks (Roy and Sengupta, 2017)

- "*x*": Denotes the forced allocation of an even operation within an odd control step while resolving conflicts during scheduling process.
- "*y*": Denotes the forced allocation of an odd operation within an even control step while resolving conflicts during scheduling process.
- "*z*": Represents insertion of an encoded edge (artificial constraint) in RAT (register allocation phase), indicating storage variable pair of odd numbers.

To implement the fingerprint constraints for digits "*x*" and "*y*", firstly operations (in the scheduled data flow graph (SDFG)) are organized/listed in sorted order. Similarly, for the constraints corresponding to the digit "*z*", the storage variables in the scheduled data flow graph must be listed in sorted order. The generation of SDFG and its respective RAT corresponding to an application has already been discussed in Section 7.2.2. The strength/robustness of the buyer's fingerprint is directly proportional to its signature size. Notably, the symmetrical fingerprint signature scheme consists of three distinct variables, each mapped with a unique constraints generation/encoding mechanism that helps in embedding of IP buyer's digital data at minimum design cost. It is recommended that the IP buyer opts for a higher proportion of "*x*" and "*y*" digits and a lesser number of "*z*" digits while selecting the combination of fingerprint signature. This is because both "*x*" and "*y*" fingerprint digit encodings incur minimal hardware area overhead and negligible latency overhead. Conversely, an abundance of "*z*" digits within the fingerprint signature may result in increased hardware overhead. However, it adds an additional layer of security by embedding fingerprint constraints during the register allocation phase of the HLS process (Roy and Sengupta, 2017).

Watermarking block: In the symmetrical IP protection mechanism (Roy and Sengupta, 2017), the IP seller's watermarking constraints are seamlessly integrated into the register allocation step of the HLS process. This integration/embedding is achieved by adding watermarking constraints (artificial edges) within register allocation phase (fingerprint-embedded RAT obtained in the previous step). By introducing these watermarking constraints, the storage variable pair (corresponding to additional constraints) within the design is forcibly allocated to distinct registers (colors) (Roy and Sengupta, 2017). The strength of the watermark directly correlates with the number of these watermarking constraints (commonly referred to as the watermark signature size). While increasing the size of the watermark signature enhances robustness, it may concurrently lead to hardware overhead. Thus, it is imperative to adopt a robust watermarking scheme that provides a balance between security robustness and design overhead.

The symmetrical IP protection mechanism (Roy and Sengupta, 2017) discusses a multi-variable watermarking mechanism, as outlined in Sengupta and Bhadauria (2016). As discussed in Sengupta and Bhadauria (2016), four distinct signature variables are selected, each governed by a unique encoding/constraint generation rule. The IP seller retains the flexibility to select any combination of these four variables to constitute their watermark. The watermarking constraint generation rule corresponding to each variable (within the watermark) is as follows:

- "*i*" represents insertion of an encoded edge (artificial constraint) in RAT, indicated by storage variable pair of prime numbers.
- "*I*" represents insertion of an encoded edge (artificial constraint) in RAT, indicated by storage variable pair of even numbers.
- "*T*" represents insertion of an encoded edge (artificial constraint) in RAT, indicated by storage variable pair comprising an odd number and an even number.
- "*!*" represents insertion of an encoded edge (artificial constraint) in RAT, indicated by storage variable pair, where one number is 0, and the other can be any integer.

Although both the IP seller's watermark and the "*z*" digit of the buyer fingerprint embed constraints during the register allocation phase. However, they result in distinct sets of constraints due to variations in encoding rules. Post embedding of all the generated fingerprinting and watermarking constraints, the symmetrically secure (containing buyer's fingerprint and seller's watermark) RTL datapath corresponding to the hardware IP design is generated (Roy and Sengupta, 2017).

7.3.2 Analysis and comparative perspective

This section discusses the impact of the symmetrical IP protection mechanism on hardware design area, latency, and cost. Figure 7.8 depicts the design cost corresponding to different benchmarks for symmetrical IP protection mechanism

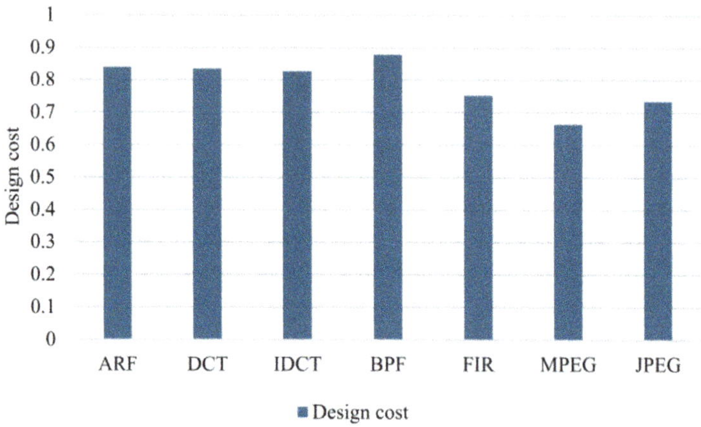

Figure 7.8 Comparison of design cost corresponding to different benchmarks for symmetrical IP protection mechanism (Roy and Sengupta, 2017)

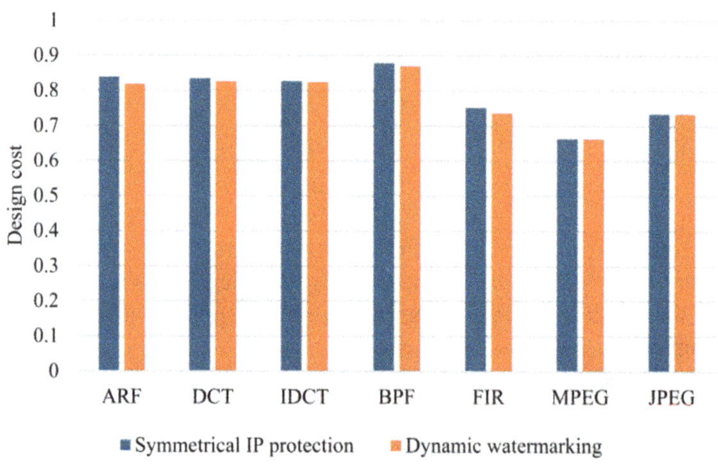

Figure 7.9 Comparison of design cost corresponding to different benchmarks between symmetrical IP protection mechanism (Roy and Sengupta, 2017) and dynamic watermarking (Koushanfar et al., 2005)

(Roy and Sengupta, 2017). The design cost function is adopted from (Roy and Sengupta, 2017). Further, Figures 7.9–7.11 highlight the comparison of design cost, latency, and the area between the symmetrical IP protection mechanism and dynamic watermarking (Koushanfar *et al.*, 2005). As illustrated in Figures 7.9–7.11, symmetrical IP protection (Roy and Sengupta, 2017) offers security benefits to both the seller and the buyer without imposing any additional area overhead. Furthermore, it incurs only minimal latency overhead, of ~1.02%, in comparison to Koushanfar *et al.* (2005). Therefore, symmetrical IP protection provides security with a remarkably low cost overhead, of ~1.93% compared to Koushanfar *et al.* (2005), which only provides protection to IP seller.

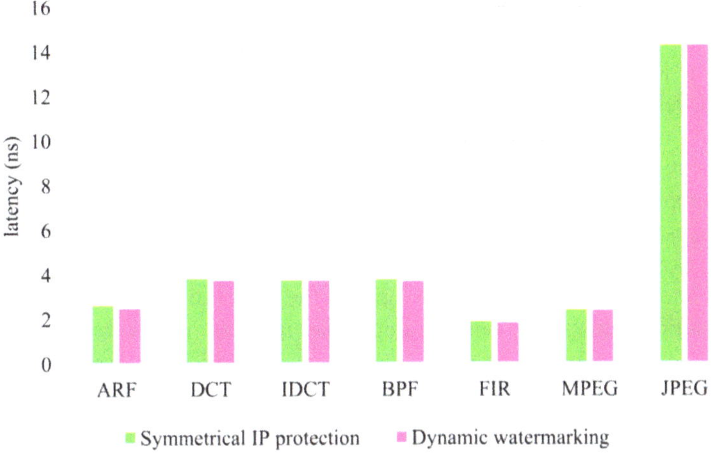

Figure 7.10 Comparison of design latency corresponding to different benchmarks between symmetrical IP protection mechanism (Roy and Sengupta, 2017) and dynamic watermarking (Koushanfar et al., 2005)

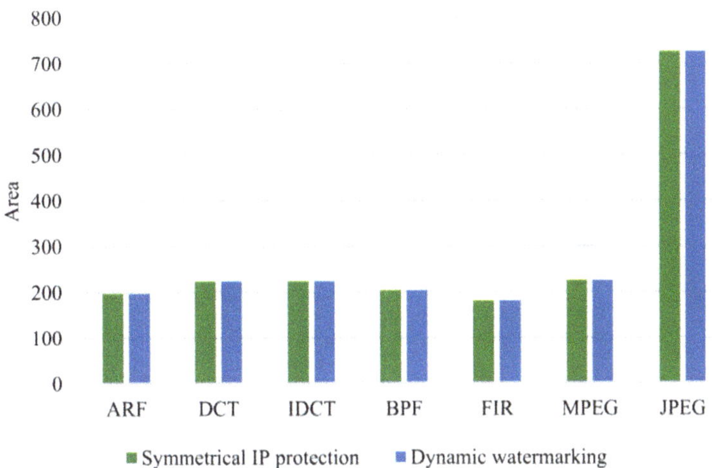

Figure 7.11 Comparison of design area corresponding to different benchmarks between symmetrical IP protection mechanism (Roy and Sengupta, 2017) and dynamic watermarking (Koushanfar et al., 2005)

7.4 Exploration of optimal biometric fingerprint as watermark

Figure 7.12 presents an outline of the low-cost watermarking mechanism to secure loop-based hardware applications. This framework comprises various inputs, such as (a) control data flow graph (CDFG)/C/C++ code or transfer function of the application, (b) the fingerprint of the IP seller, (c) termination criteria and mapping/embedding rules, (d) parameters for design space exploration (DSE) initialization, (e) count of loop iterations, and (f) module library (containing critical information like area, delay, etc., corresponding to utilized functional units, such as adder, multipliers, etc.). This approach generates a low-cost watermarked hardware IP design (or register transfer level (RTL) IP) corresponding to loop-based application. It comprises three primary blocks: (i) design space exploration (DSE) block driven by particle swarm optimization or firefly algorithm or genetic algorithm, (ii) fingerprint biometric block, and (iii) constraints embedding block. The DSE (which is integral to the HLS process), aids in the exploration of low-cost hardware design solutions along with optimal fingerprint minutiae points for embedding as

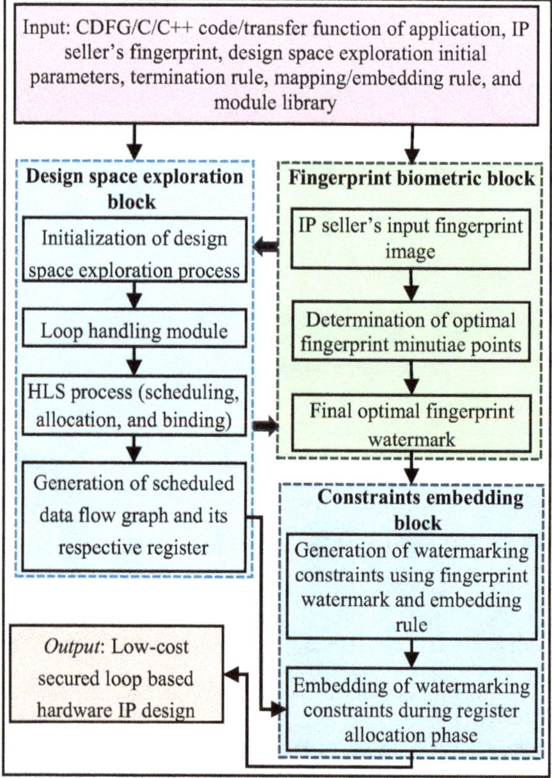

Figure 7.12 Details of low-cost watermarking approach

digital evidence (for securing the design against IP piracy and false IP ownership claim). The DSE is initialized based on provided parameters such as population size, total iterations, random numbers, etc. The particle/chromosomal/firefly encoding plays a crucial role in the initialization of DSE, where each solution is represented across four dimensions: (i) number of adders, (ii) number of multipliers, (iii) number of subtractors, and (iv) total fingerprint minutiae points. After the DSE initialization, the input CDFG and loop iteration count value are fed into the loop handling module, which outputs a loop-flattened CDFG. Subsequently, the HLS processes (scheduling, allocation, and binding) accept the explored resource configuration (such as number of adders, multipliers, and subtractors) along with the loop-flattened CDFG to generate the corresponding loop-pipelined scheduled data flow graph (SDFG). The obtained SDFG is used to create an initial register allocation table (RAT). Concurrently, the fingerprint biometric block leads to generation of maximum fingerprint minutiae points, which are fed as input parameters to DSE block to determine the optimal minutiae point set (fingerprint watermark) for embedding into the design (incurring either zero or minimal design overhead). Finally, the constraints embedding block aids in producing watermarking constraints using the obtained optimal fingerprint watermark and embedding rule. These watermarking constraints are then integrated into the register allocation phase of IP design (in the RAT), yielding a final optimal watermark constraint embedded IP design. Furthermore, the design cost for each initialized solutions is evaluated, enabling the identification of the global best solution. The DSE module iterates until it converges to an optimal secure design solution or meets the termination criterion. Finally, an optimal secure hardware IP design (or register transfer level (RTL) IP) for loop-based applications is produced as the final output.

7.5 Questions and exercise

1. Discuss the importance/role of IP buyer and IP seller in the context of global design supply chain process.
2. List some common hardware security threats faced by IP buyer and IP seller in the design chain process.
3. Why it is important to consider secure the hardware design from the perspective of both IP buyer and IP seller?
4. Discuss the importance of IP fingerprinting approach for securing hardware IP designs.
5. List the primary inputs of crypto multi-variable fingerprinting approach.
6. Briefly explain the crypto multi-variable fingerprinting approach. What are the different phases in the fingerprinting approach?
7. Explain the advantages of hashing block in the context of fingerprinting process. How hash output helps in the generation of robust watermarking constraints.
8. Explain the differences between fingerprinting and watermarking approach for hardware security.

9. What is symmetrical IP protection framework? Why there is a need to provide symmetrical security to hardware designs?
10. Explain the symmetrical IP protection framework with a suitable example.
11. What are the advantages of symmetrical IP protection mechanism?
12. Explain the IP seller's watermark generation process.
13. What are the security metrics used to compare and analyze the security strength and cost of the discussed security approaches?
14. Discuss the importance of exploration of optimal watermarking constraints for security and security of loop-based applications.

References

A. Anshul and A. Sengupta, (2022) "IP Core Protection of Image Processing Filters with Multi-Level Encryption and Covert Steganographic Security Constraints," *2022 IEEE International Symposium on Smart Electronic Systems (iSES)*, Warangal, India, pp. 83–88.

A. Anshul and A. Sengupta, (2023a) "A Survey of High Level Synthesis Based Hardware Security Approaches for Reusable IP Cores [Feature]," *IEEE Circuits and Systems Magazine*, vol. 23, no. 4, pp. 44–62.

A. Anshul and A. Sengupta, (2023b) "Exploration of Optimal Crypto-chain Signature Embedded Secure JPEG-CODEC Hardware IP during High Level Synthesis," *Microprocessors and Microsystems*, vol. 102, 104916.

A. Anshul and A. Sengupta, (2023c) "PSO Based Exploration of Multi-phase Encryption Based Secured Image Processing Filter Hardware IP Core Datapath during High level synthesis, *Expert Systems with Applications*, vol. 223, 119927.

A. Anshul and A. Sengupta, (2023d) "Low-Cost Hardware Security of Laplace Edge Detection and Embossment Filter Using HLS Based Encryption and PSO," *2023 IEEE International Symposium on Smart Electronic Systems (iSES)*, Ahmedabad, India, pp. 135–140.

A. Anshul, K. Bharath and A. Sengupta, (2022) "Designing Low Cost Secured DSP Core Using Steganography and PSO for CE Systems," *2022 IEEE International Symposium on Smart Electronic Systems (iSES)*, Warangal, India, pp. 95–100.

R. Chaurasia, A. Anshul, A. Sengupta and S. Gupta, (2022) "Palmprint Biometric Versus Encrypted Hash Based Digital Signature for Securing DSP Cores Used in CE Systems," *IEEE Consumer Electronics Magazine*, vol. 11, no. 5, pp. 73–80.

B. Colombier and L. Bossuet, (2015) "Survey of Hardware Protection of Design Data for Integrated Circuits and Intellectual Properties," *IET Computers & Digital Techniques*, vol. 8, no. 6, pp. 274–287.

W. Hu, C. Chang, A. Sengupta, S. Bhunia, R. Kastner and H. Li, (2021) "An Overview of Hardware Oriented Security and Trust: Threats, Countermeasures and Design Tools," *IEEE Transactions on Computer-Aided Design of Integrated Circuits and Systems, Invited Paper*, vol. 40, no. 6, pp. 1010–1038.

S. A. Islam, L. K. Sah and S. Katkoori, (2020) "High-Level Synthesis of Key-Obfuscated RTL IP with Design Lockout and Camouflaging," *ACM Transactions on Design Automation of Electronic Systems (TODAES)*, 26, no. 1, pp. 1–35.

F. Koushanfar, I. Hong and M. Potkonjak, (2005) "Behavioral Synthesis Techniques for Intellectual Property Protection," *ACM Transactions on Design Automation of Electronic Systems*, vol. 10, no. 3, pp. 523–545.

B. Le Gal and L. Bossuet, (2012) "Automatic Low-cost IP Watermarking Technique Based on Output Mark Insertions," *ACM Transactions on Design Automation of Electronic Systems (TODAES)*, vol. 16, no. 2, pp. 71–92.

M. Potkonjak, (2006) "Methods and Systems for the Identification of Circuits and Circuit Designs," United State Patent, US7017043B1.

M. Rathor, A. Sengupta, R. Chaurasia and A. Anshul, (2023a) "Exploring Handwritten Signature Image Features for Hardware Security," *IEEE Transactions on Dependable and Secure Computing*, vol. 20, no. 5, pp. 3687–3698.

M. Rathor, A. Anshul, K. Bharath, R. Chaurasia and A. Sengupta, (2023b) "Quadruple Phase Watermarking during High Level Synthesis for Securing Reusable Hardware Intellectual Property Cores," *Computers and Electrical Engineering*, vol. 105, 108476.

M. Rathor, A. Anshul and A. Sengupta, (2024) "Securing Reusable IP Cores Using Voice Biometric Based Watermark," *IEEE Transactions on Dependable and Secure Computing*, vol. 21, no. 4, pp. 2735–2749.

S. Rizzo, F. Bertini, and D. Montesi, (2019) "Fine-Grain Watermarking for Intellectual Property Protection," *EURASIP Journal on Information Security*, vol. 2019, no. 4, Article no. 10.

M. Rostami, F. Koushanfar and R. Karri, (2014) "A Primer on Hardware Security: Models, Methods, and Metrics," *Proceedings of the IEEE*, vol. 102, no. 8, pp. 1283–1295.

D. Roy and A. Sengupta, (2017) "Low Overhead Symmetrical Protection of Reusable IP Core Using Robust Fingerprinting and Watermarking during High Level Synthesis," *Future Generation Computer Systems*, vol. 71, pp. 89–101.

A. Sengupta and S. Bhadauria, (2016) "Exploring Low Cost Optimal Watermark for Reusable IP Cores during High Level Synthesis," *IEEE Access*, vol. 4, pp. 2198–2215.

A. Sengupta and M. Rathor, (2020) "Securing Hardware Accelerators for CE Systems Using Biometric Fingerprinting," *IEEE Transactions on Very Large Scale Integration (VLSI) Systems*, vol. 28, no. 9, pp. 1979–1992.

A. Sengupta and M. Rathor, (2021) "Facial Biometric for Securing Hardware Accelerators," *IEEE Transactions on Very Large Scale Integration (VLSI) Systems*, vol. 29, no. 1, pp. 112–123.

A. Sengupta, U. K. Singh and P. K. Premchand, (2019) "Crypto Based Multi-Variable Fingerprinting for Protecting DSP Cores," *2019 IEEE 9th*

International Conference on Consumer Electronics (ICCE-Berlin), Berlin, Germany, pp. 1–6.

A. Sengupta, R. Chaurasia and A. Anshul, (2023a) "Hardware Security of Digital Image Filter IP Cores against Piracy Using IP Seller's Fingerprint Encrypted Amino Acid Biometric Sample," *2023 Asian Hardware Oriented Security and Trust Symposium (AsianHOST)*, Tianjin, China, pp. 1–6.

A. Sengupta, A. Anshul and R. Chaurasia, (2023b) "Exploration of Optimal Functional Trojan-Resistant Hardware Intellectual Property (IP) Core Designs during High Level Synthesis," *Microprocessors and Microsystems*, vol. 103, 104973.

A. Sengupta, R. Chaurasia and A. Anshul, (2023c) "Robust Security of Hardware Accelerators Using Protein Molecular Biometric Signature and Facial Biometric Encryption Key," *IEEE Transactions on Very Large Scale Integration (VLSI) Systems*, vol. 31, no. 6, pp. 826–839.

M. Yasin, J. J. Rajendran, O. Sinanoglu and R. Karri, (2016) "On Improving the Security of Logic Locking," *IEEE Transactions on Computer-Aided Design of Integrated Circuits and Systems*, vol. 35, no. 9, pp. 1411–1424.

Chapter 8

Hardware obfuscation-high level synthesis-based structural obfuscation for hardware security and trust

Anirban Sengupta[1] *and Rahul Chaurasia*[1]

The chapter describes a high-level synthesis (HLS)-driven methodology for generating secure hardware intellectual property (IP) core design with fault detectability feature using multi-cut-based structural obfuscation and physical biometric (Chaurasia and Sengupta, 2023a). In this methodology, firstly, the IP design is leveraged with transient fault detectability feature and is transformed/obscured using multi-cut-based structural obfuscation. This transformation serves as the first layer of security, ensuring the protection of the design against potential reverse engineering attempts by an adversary. Subsequently, the design is covertly embedded with physical biometric of IP vendor. The embedded naturally unique biometric information of genuine IP vendor serves as the second layer of security, ensuring the protection of the design against IP piracy. Thus, HLS-driven hardware security methodology can concurrently offer security of "fault-detectable IP designs" against the following hardware threats: (a) potential reverse engineering threat by an adversary from the SoC layout stage in an untrustworthy foundry and (b) IP piracy by an adversary that may present in the SoC integration house.

The rest of the chapter is structured as follows: Section 8.1 provides the introduction of the chapter; Section 8.2 delves into the various threat model scenario; Section 8.3 provides background on transient fault; Section 8.4 explores the multi-cut-based structural obfuscation technique and physical biometrics for enhanced hardware IP security; Section 8.5 presents a case study on IIR filter application framework; Section 8.6 presents security features of the methodology; Section 8.7 analyzes the security of IP design achieved through multi-cut based structural obfuscation and physical biometrics; Section 8.8 concludes the chapter by summarizing its findings and implications.

[1]Department of Computer Science and Engineering, Indian Institute of Technology Indore, India

8.1 Introduction

In today's technological landscape, consumer electronics devices such as smartphones, health bands, digital cameras, and Internet of Things (IoT) enabled devices are specifically crafted to handle data-intensive tasks and achieve optimal efficiency. To fulfill these design objectives in the majority of consumer electronics (CE) systems, the foundational hardware often consists of reusable intellectual property (IP) cores (Schneiderman, 2010; Syed and Lourde, 2016; Hroub and Elrabaa, 2022; Salivahanan and Vallavaraj, 2001; Sengupta and Rathor, 2021; Sengupta, 2016). These cores are seamlessly integrated into CE systems to execute functions such as image/video processing and audio processing/filtering. Moreover, in critical systems like autonomous vehicles and medical devices, these reusable IP cores play a crucial role in carrying out computationally and data-intensive tasks with higher efficiency. Such IP cores may be susceptible to single event upsets (SEU, 2022) resulting in transient faults (Kachave and Sengupta, 2016; Park *et al.*, 2021; Qiu *et al.*, 2019; Ramadhan *et al.*, 2017; Sengupta *et al.*, 2018). In mission-critical scenarios, it becomes imperative to ensure the detection and/or tolerance of transient fault in the reusable hardware IP designs. This precaution is essential as designs vulnerable to single event upsets could yield incorrect outputs, posing potential risks.

Moreover, in pursuit of an expedited design cycle and cost-effectiveness, the design of these IP cores involves collaboration with offshore third-party design houses. However, this introduces security risk, as an adversary may deliberately induce threat during the design process. For instance, malicious logic could be implanted through reverse engineering. Further, unauthorized piracy of the IP design may occur, leading to revenue losses to an authentic IP vendor as well as circulation of unreliable/degraded IPs in the market. Pirated IPs typically do not undergo rigorous quality checks and testing before integration into the system-on-chip (SoC) platforms of consumer devices. Therefore, protecting these IP cores is crucial not only from an IP vendor's perspective as well as from a consumer's perspective, as counterfeit cores pose significant risks in terms of compromising confidential information, safety, and integrity (Pilato *et al.*, 2018a; Wang *et al.*, 2015; Koushanfar *et al.*, 2012; Arafin *et al.*, 2017; Castillo *et al.*, 2007; Colombier and Bossuet, 2015; Kean *et al.*, 2008; Chen *et al.*, 2020; Roy *et al.*, 2008; Hu *et al.*, 2021; Chaurasia and Sengupta, 2022d). Therefore, to ensure the reliability and authenticity of fault-secured IP core designs, it is essential to integrate robust security mechanisms (Chaurasia and Sengupta, 2023a).

Now, we discuss the significance of ensuring hardware security and trust in detail. Hardware IP security and trust are crucial aspects in the design and implementation of electronic systems. This is because a hardware IP refers to the pre-designed and pre-verified functional blocks that can be integrated into larger systems, speeding up the design process. Further as technology has advanced, more and more devices have become interconnected, therefore escalating the importance of ensuring the security and trustworthiness of such IP designs. Here are some key significances of hardware IP security and trust to ensure the following:

Protection against piracy/counterfeiting: Trust in hardware ensures that the components used in a system are genuine and have not been tampered with or replaced by pirated/counterfeited parts. Pirated hardware can lead to system failures, compromised system integrity, and financial losses. Implementing security measures during hardware IP design process helps in mitigating the risk of pirated/counterfeit components being integrated into the CE devices (Sengupta and Chaurasia, 2023; Sengupta *et al.*, 2023b; Sengupta *et al.*, 2023c; Chaurasia and Sengupta, 2022d).

Protection of intellectual property against abuse/misuse: Hardware IP represents valuable intellectual property developed by companies or individuals. Ensuring security helps protect this IP from unauthorized access, reverse engineering, or theft, safeguarding the investments made in research and development (Sengupta *et al.*, 2023c).

Mitigation of hardware Trojan insertion: Hardware Trojans are malicious modifications to hardware designs that can be inserted during the design/manufacturing process. Ensuring security in hardware IP is crucial to prevent the introduction of Trojans that may compromise the functionality, security, or privacy of electronic systems (Chaurasia and Sengupta, 2022a).

Long-term reliability: Hardware IP security is crucial for ensuring the long-term reliability of electronic systems. By protecting against malicious activities and unauthorized modifications, trustworthy hardware contributes to the sustained functionality of systems over their operational lifespan.

Mitigation of side-channel attacks: Side-channel attacks exploit unintended information leakage from the physical implementation of a crypto system, such as power consumption or electromagnetic radiation. Security measures on crypto-hardware IP aim to mitigate these types of attacks, enhancing the overall resilience of the system against sophisticated adversaries (Vateva-Gurova and Suri, 2018).

System trustworthiness: Trust in hardware is fundamental to the overall trustworthiness of a system. This is because in all electronics/computing systems hardware is considered as root of trust, which is not always true. Systems that rely on secure hardware components are better equipped to resist various forms of attacks including reverse engineering, brute force attack, IP tampering, side-channel attacks, power analysis, and fault injection attacks. This is particularly important in safety-critical systems, where a compromised hardware component could have life-threatening consequences (Sengupta *et al.*, 2023c).

Supply chain security: Trustworthy hardware IP is vital for maintaining supply chain security. Ensuring that the components used in electronic systems are free from vulnerabilities or malicious modifications, is essential for the overall security of the supply chain (Sengupta *et al.*, 2023c).

Compliance with standards and regulations: Many industries and applications have specific security standards and regulations that must be adhered to. Implementing secure hardware IP is often a requirement for compliance with these standards, ensuring that products meet the necessary security and privacy criteria (Sengupta *et al.*, 2023c).

In summary, the significance of hardware IP security and trust lies in protecting valuable intellectual property, ensuring the integrity of electronic systems, complying with regulations, and safeguarding against various forms of external threats. These measures are essential for building robust, secure, and reliable electronic systems in an increasingly interconnected digital world.

8.1.1 Introduction and fundamentals of hardware obfuscation

Hardware devices, ranging from simple integrated circuits to complex processors, underpin virtually every aspect of modern life, from communication to finance, transportation, and healthcare. However, the increasing prevalence of attacks targeting hardware vulnerabilities poses a significant threat to these systems. In response to these challenges, hardware obfuscation has emerged as a promising approach to enhance the security of hardware designs. Hardware obfuscation refers to the process of intentionally introducing obscurity into a hardware design to thwart reverse engineering. Unlike traditional security measures, hardware obfuscation focuses on safeguarding the underlying hardware itself. By making the hardware design more difficult to understand, obfuscation techniques aim to deter adversaries from extracting sensitive information (design architecture and functionality) or tampering with the device. There are two important classes of obfuscation techniques such as structural obfuscation and functional obfuscation (Hu *et al.*, 2021; Sengupta *et al.*, 2018; Lao and Parhi, 2014 Sengupta and Roy, 2017). Now, we discuss both of the obfuscation techniques in detail.

8.1.1.1 Significance of structural obfuscation

Structural obfuscation refers to the intentional modification of the design and structure by an IP designer (without affecting their actual functionality) to make it more difficult for an attacker (that may present in untrustworthy foundries) to understand and analyze the underlying functionality. It therefore enables hardware security of hardware IP designs from various threats such as IP piracy and reverse engineering (in terms of providing enhanced hindrance level from an adversarial perspective). Structural obfuscation makes reverse engineering more time-consuming and resource-intensive. It aims to raise the bar for attackers by increasing the effort required to understand the underlying circuitry. Here are some key aspects that highlight the significance of structural obfuscation in hardware security for ensuring:

Countering reverse engineering: Structural obfuscation makes it challenging for attackers to analyze the hardware structure, making reverse engineering a time-consuming and complex task. It therefore hinders an adversary attempting to unfold the underlying functionality of the hardware (Sengupta and Rathor, 2021).

Securing intellectual property: Hardware IP design often represents a significant investment in terms of time, resources, and intellectual property value. Structural obfuscation helps in safeguarding these investments by making it harder for an adversary or other malicious entities to replicate or clone the hardware. This is particularly important in industries where proprietary technology of IP designs are key competitive advantages.

Mitigating hardware attacks: Structural obfuscation can be effective in mitigating certain types of hardware attacks, such as side-channel attacks, fault injection attacks, and hardware Trojan insertion. By obscuring the structure of the hardware, attackers find it more difficult and time-consuming to analyze the relationship between the crypto key and the data input of the IP design. This is particularly important in applications where the compromised hardware IP can lead to severe consequences, such as in military or medical infrastructure systems.

In summary, structural obfuscation plays a crucial role in hardware security by adding layers of complexity and ambiguity, making it more challenging for an attacker to compromise the integrity, confidentiality, and availability of hardware systems. It is part of a broader strategy to create resilient and secure hardware architectures in the face of evolving cybersecurity threats. In the literature, several hardware security techniques based on structural obfuscation have been discussed (Chaurasia and Sengupta, 2023a; Hu *et al.*, 2021; Sengupta *et al.*, 2018; Lao and Parhi, 2014; Sengupta and Roy, 2017; Sengupta and Chaurasia, 2022).

8.1.1.2 Significance of functional obfuscation

There is another branch of hardware obfuscation called functional obfuscation/logic locking that aims to secure integrated circuits (ICs) from various attacks, such as reverse engineering, intellectual property theft, and tampering. While both structural and functional obfuscation aims to enhance the security of hardware designs, they differ in their approaches and strategies. Functional obfuscation focuses on obscuring/locking the actual functionality of the hardware design while maintaining its functionality. This includes adding key-gates, incorporating misleading control flow, and using encryption or obfuscated design logic. The primary goal is to prevent an attacker from exploiting the functionality of the hardware IP by insertion of encrypted key-gates (Vijayakumar *et al.*, 2017; Sengupta *et al.*, 2019a; Rathor and Sengupta, 2019). Even if the structural complexity and functionality are known, an attacker still fails to misuse it since the IP design is locked. It should be noted that the effectiveness of obfuscation techniques depends on factors such as the sophistication of potential attackers, the resources they have, and the specific application or use case.

8.1.2 Significance of physical biometrics for hardware security and trust

Physical biometrics play a crucial role in enhancing hardware IP security by providing a layer of authentication/validation based on the embedded unique physical characteristics of an individual/IP vendor. Here are some key significances of physical biometrics for hardware security (Chaurasia and Sengupta, 2022a; Chaurasia and Sengupta, 2022b; Chaurasia and Sengupta, 2022c; Chaurasia and Sengupta, 2021; Sengupta *et al.*, 2023c; Chaurasia *et al.*, 2022; Sengupta and Chaurasia, 2022; Chaurasia *et al.*, 2023; Sengupta and Chaurasia, 2023):

Unique identification of IP owner: Physical biometrics such as handprints, retina/iris patterns, or facial features are unique to each individual. This uniqueness makes it difficult for an adversary to mimic or replicate, providing a robust means of authentication/validation. This therefore enables seamless IP ownership verification for genuine IP vendor during ownership conflict litigation in IP courts.

Increased authentication accuracy: Biometric authentication can significantly increase accuracy compared to traditional methods for hardware security like watermarking and digital signature. This is because biometric-based security methodologies comprise several non-decodable robust security parameters which makes it difficult for an adversary to regenerate the security constraints for evading piracy detection successfully.

Robust resiliency against tampering attack using brute force: Biometric-based approach offers robust security against an adversary aiming to accurately guess the exact signature/security mark by brute force attack. This is because biometric-based hardware security methodologies are capable of generating massive signature space resulting in an extremely low probability of an attacker guessing the exact security mark, as compared to conventional non-biometric-based approaches such as watermarking and digital signature.

8.1.3 Types of hardware IPs

The hardware IPs are generically designed for computational and data intensive applications (Chaurasia and Sengupta, 2023b). These applications may belong to the domain of digital signal processing (DSP), multimedia, and machine learning.

DSP hardware IP cores are designed to accelerate and optimize digital signal processing tasks in electronic systems. These IP cores are often integrated into programmable devices such as Field-Programmable Gate Arrays (FPGAs) or Application-Specific Integrated Circuits (ASICs). Some of the popular DSP IP cores are: Fast Fourier Transform (FFT), Finite Impulse Response (FIR), Infinite Impulse Response (IIR), etc. (Salivahanan and Vallavaraj, 2001; Sengupta and Rathor, 2021). These DSP hardware IP cores cater to a wide range of applications, including audio processing, communication systems, image and video processing, and more. Depending on the requirements of a specific application, designers can choose the appropriate DSP IP core to optimize performance of the target end system and resource utilization. On the other hand, multimedia hardware IP cores are specialized and designed to accelerate multimedia processing tasks in electronic systems. Some examples of multimedia hardware IP cores are Joint Picture Expert Group (JPEG)-codec, used for image compression and decompression in various applications; and Moving Picture Experts Group (MPEG), used for video compression. Further, some of the hardware IP cores for machine learning applications are convolutional neural network (CNN) convolutional IP and linear regression-based IP. Convolutional IP is used for performing image feature extraction and object detection tasks, and on the other hand, linear regression-based IP are used for precise estimation.

8.1.4 Salient features of the chapter

The salient features of the chapter, comprising of the discussion and analysis on the significance of ensuring HLS-based hardware IP security and trust using multi-cut-based structural obfuscation and physical biometric, are the following:

- This chapter discusses the HLS-based design methodology for creating secure hardware IPs. HLS-based design methodology provides an IP designer with the capability to design the IP, starting from the behavioral description of a computationally intensive application framework and progressing to its Register Transfer Level (RTL) design counterpart (Sengupta and Sedaghat, 2011).
- This chapter explains multi-layer security of hardware IPs:
  (a) preventive countermeasure against reverse engineering, in terms of hindering an adversary from interpreting the actual design functionality, by employing multi-cut based structural obfuscation algorithm and,
  (b) detective countermeasure against IP piracy, in terms of robust detection of pirated IP versions, by employing physical biometric-based security algorithm. The embedded security constraints generated through physical biometric also offer seamless verification of IP ownership during ownership conflict litigation in court.
- This chapter discusses different phases of HLS responsible for generating secure hardware IPs, exploiting the DFG generation phase for employing structural obfuscation to obscure the design functionality from an adversary, and the register allocation phase for covertly implanting generated secret hardware security constraints to enable piracy detection.
- This chapter explains the process of extracting unique features from physical biometric information of IP vendor based on handprint and generating secret hardware security constraints by employing signature bit encoding algorithm (Chaurasia and Sengupta, 2023a; Chaurasia and Sengupta, 2021; Sengupta *et al.*, 2023a; Chaurasia *et al.*, 2022; Sengupta *et al.*, 2023b; Chaurasia *et al.*, 2023; Sengupta *et al.*, 2023c).
- This chapter provides a case study on IIR filter application framework (Salivahanan and Vallavaraj, 2001; Sengupta and Mohanty, 2019) and generates its secure IP version using multi-cut based structural obfuscation and physical biometrics of IP vendor (Chaurasia and Sengupta, 2023a; Sengupta and Rathor, 2021).
- The discussed methodology using multi-cut-based structural obfuscation attains higher strength of obfuscation. Additionally, it ensures lower false positive (probability that an unsecured IP design coincidentally contains the same embedded security constraints) and increased resistance to tampering attacks using IP vendor physical biometric information.

The discussed methodology incurs zero design cost overhead leveraging HLS methodology and multi-layer protection for hardware IP security (Chaurasia and Sengupta, 2023a).

8.2 Threat model scenarios

The IP security methodology (Chaurasia and Sengupta, 2023a) counters the following hardware threats that may induced by an adversary.

Scenario-1: In an untrusted foundry, there is a potential risk of reverse engineering launched by an adversary. An adversary may attempt to decipher the true functionality of the IP design from the system-on-chip (SoC) design layout phase to deliberately introduce malicious logic. This threat scenario has been discussed in existing literature (Sengupta *et al.*, 2017; Vijayakumar *et al.*, 2017; Lao and Parhi, 2014).

To enhance security and make reverse engineering more challenging, structural obfuscation can be employed as a hindrance mechanism. This technique aims to conceal the functionality of an IP core from a potential adversary. Typically, an adversary lacks knowledge about the functionality of a specific IP core among various third-party IPs integrated into an SoC design, therefore structural obfuscation on an IP design to conceal the functionality can be effective. Some prior methodologies (Sengupta *et al.*, 2017; Vijayakumar *et al.*, 2017; Lao and Parhi, 2014; Sengupta and Roy, 2017; Pilato *et al.*, 2018b) have presented solutions for countering the same threat scenario. The hardware security methodology (Chaurasia and Sengupta, 2023a) addressed this threat by incorporating multi-cut-based structural obfuscation, implemented by the IP vendor. Multi-cut-based structural obfuscation is employed to obscure both the functionality and interconnections of the design, effectively thwarting attempts to reverse engineer the IP design by an adversary in untrusted foundries.

Scenario-2: In the SoC design house, an adversary may engage in performing unauthorized IP piracy and making false claim of IP ownership. This has arisen due to globalization in design supply chain. A rogue adversary located within the SoC integrator design house may clandestinely replicate or falsely assert ownership of IP. The repercussions of IP piracy may include:

- Damaging original IP vendor's brand value or reputation.
- Pirated IPs, lack rigorous testing and quality checks, may be sold in the open market, which in turn introduces negative consequences for the end customer.
- Loss of financial revenue for the original IP vendor, and
- Pirated IPs may be tampered, which may potentially incorporate malicious logic resulting in degraded performance, unreliable behavior, or the leakage of confidential information.

Such a similar threat scenario has been discussed in Rostami *et al.*, (2014), Koushanfar *et al.* (2005), Rajendran (2017), Sengupta *et al.* (2023d), Hu *et al.* (2021), and Sengupta *et al.* (2019). The hardware security methodology (Chaurasia and Sengupta, 2023a) addresses this threat by incorporating biometric signature-based encoded security constraints, specifically an IP vendor handprint, as a detective countermeasure. The primary aim of covertly embedding biometric information from the genuine IP vendor is to provide strong detection capabilities against potential IP piracy attempts within the SoC integrator house. In the piracy detection and IP ownership resolution process, the covertly embedded security constraints of the biometric signature are extracted from the suspected chip design and compared with the originally embedded biometric security constraints of the IP design. If an exact match is found, it serves as evidence of illegal IP piracy.

8.3 Background on transient fault

Ensuring robust security measures against transient fault is paramount for the reliability of reusable hardware IP cores and the desired IP functionality for mission critical systems. Notably, alpha particles have been identified as a significant factor contributing to single event transients in integrated circuits including IP cores designed as standalone integrated circuits. They are predominantly emitted as a result of uranium and thorium impurities present in the packaging of system-on-chip. This challenge becomes more pronounced as nanometer-scale technology progresses, enabling millions of transistors to be densely packed into a limited area on a single chip, consequently impacting the reliability of the end system. Even with relatively low linear energy transfer, energized alpha particles can cause transient upsets that affect the computational output of an IP design. Additionally, the increasing speed of devices poses challenges for fault detection, as faster devices can produce transients lasting longer than a single cycle, even with modest linear energy transfer from particles. Therefore, evaluating the resilience against multi-cycle transient faults should account for the worst-case delay of transients in fault-detectable IP designs. Further, as mentioned earlier, the evolution of technology in terms of scalability and faster circuits also significantly contributes to the emergence of multi-cycle transient faults, emphasizing the need for fault-detectable designs against multi-cycle transient to ensure reliable system functionality (Sengupta and Kachave, 2017).

Moreover, it is vital to distinguish between single-cycle/multi-cycle transient fault detection, as discussed earlier, and human-induced fault attacks on physical devices. Human-induced fault attacks involve deliberately manipulating the state of a hardware IP to induce errors or drive the device into an unintended state (Benot, 2011; Karaklajić *et al.*, 2013), manifesting as a security lapse. In contrast, transient faults are natural phenomena caused by alpha particles resulting from uranium and thorium impurities in the packaging of system-on-chip hardware IPs. These natural transient faults cause bit-flips in the functional units of the IP design (Kachave and Sengupta, 2016). The impact of transient faults is functional malfunction, not intentional disclosure or exposure of confidential keys/data.

8.4 Multi-cut-based structural obfuscation and physical biometrics for hardware security

In this section, we discuss the process of generating secure hardware IP for a sample application by employing multi-layer security using a combination of multi-cut-based architectural obfuscation and physical biometrics based on handprint signature during HLS, as illustrated in Figure 8.1. The first layer (security layer-1), comprises two sub-modules for ensuring security: security against transient faults and reverse engineering threats. This is achieved by transforming the sample input application into a structurally obfuscated design

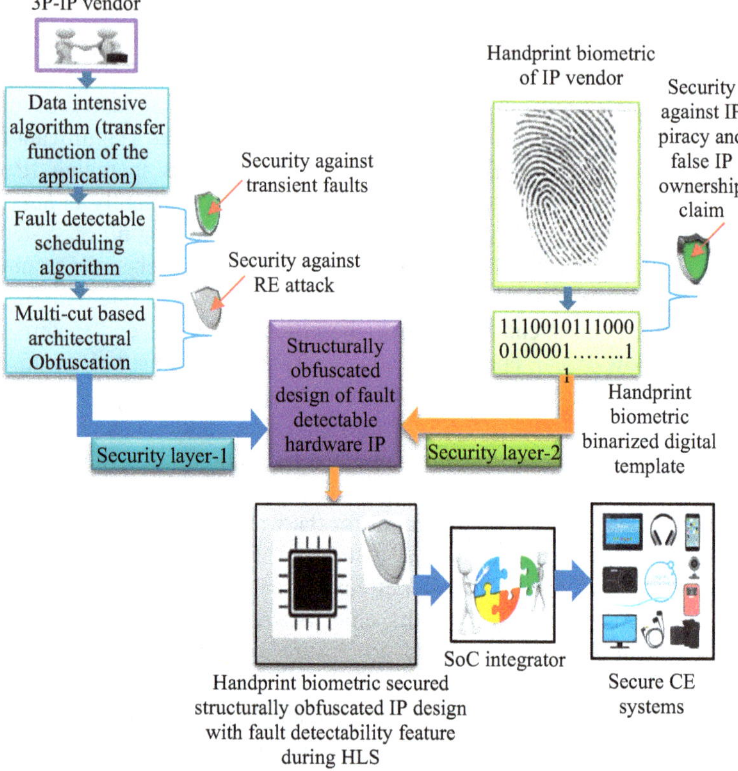

Figure 8.1 Overview of multi-layer hardware security of fault detectable IPs using multi-cut based structural obfuscation and physical biometrics

post enabling fault detectability feature. Additionally, the second layer (security layer-2) is responsible for ensuring security against IP piracy and false IP ownership claim. This is achieved by transforming the IP vendor handprint biometric information into binarized biometric digital template and deriving its corresponding encoded hardware security constraints. Subsequently, the encoded hardware security constraints from genuine IP vendor palmprint are covertly implanted into the design to counter against IP piracy and nullify false IP ownership claim. Thus, the hardware IP leveraged with multi-layer security is generated (during HLS). This is how the discussed security methodology (Chaurasia and Sengupta, 2023a) designs the hardware IP for an application during HLS by concurrently ensuring the hardware security and trust using multi-cut-based architectural obfuscation and physical biometrics of IP vendor. Next, this hardware IP is then forwarded to SoC integrator for integrating into SoCs of CE systems, thereby ensuring the security and trust from an IP vendor's perspective as well as from end user's perspective.

8.4.1 Multi-cut-based structural obfuscation of fault detectable IP cores

Now, we discuss the major steps involved in generating a multi-cut-based structurally obfuscated fault-detectable IP design. The primary inputs include the following: library file (comprising resource information), transfer function of application framework (to be transformed into hardware IP design), and resource constraints (to obtain low design overhead). Moreover, the secondary input is the fault detectability rules to enable transient fault detection. The process involves several sequential steps as depicted in Figure 8.2. Firstly, we build the Data Flow Graph (DFG) that corresponds to the input application or transfer function. Following this, in the subsequent step, the operations of the original processing unit are replicated to form the Dual Modular Redundancy (DMR) schedule. The combination of the original and duplicate units is collectively referred to as the "DMR design." Next, the DMR design is scheduled using IP designer chosen resource constraints (number of addition, multiplication and subtraction units, etc.). This scheduled DMR design is then allocated with hardware resources and their binding is performed. Thus, so far, the input application is transformed into a scheduled DMR design. The scheduled DMR design is then converted into a schedule capable of detecting transient faults by applying the rules for transient fault detectability and its corresponding scheduled design version with transient fault detectability

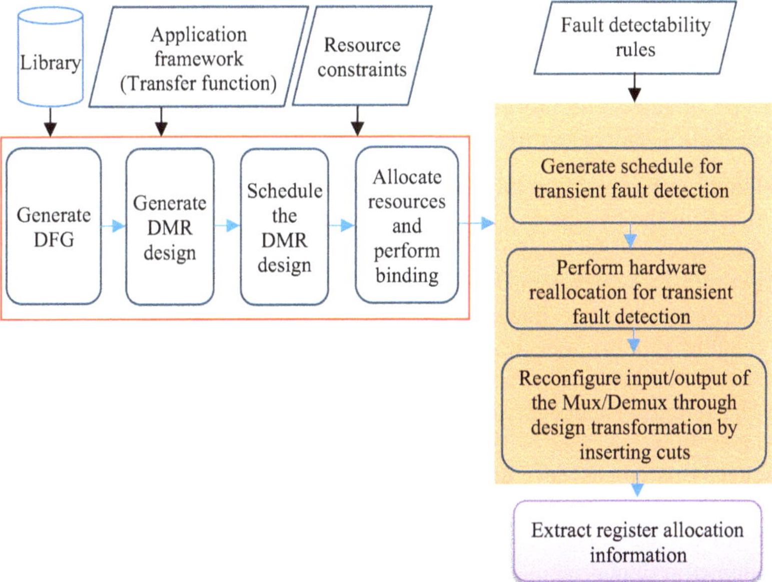

Figure 8.2 Generating register allocation information corresponding to scheduled design with transient fault detectability feature post inserting multi-cuts

feature is generated. Subsequently, the hardware resources are reallocated for transient fault detection, as per detectability rules. Next, the resultant design undergoes structural obfuscation through the implementation of multiple cuts in node operations and the insertion of numerous checkpoints, and subsequently input/output of FUs are reconfigured. This process acts as a robust structural obfuscation technique which effectively obscures the design in terms of making the design uninterpretable/ambiguous to an attacker. High-level architectural transformations, which induce structural changes without affecting functionality, offer a natural approach to achieving security by obscuring or making the functionality less apparent. These transformations fall under the category of structural obfuscation, wherein the IP design becomes obscured in terms of its functionality, posing challenges for a potential attacker (Chaurasia and Sengupta, 2023a; Hu *et al.*, 2021; Sengupta *et al.*, 2018; Lao and Parhi, 2014 Sengupta and Roy, 2017; Sengupta and Chaurasia, 2022). Finally, the register allocation information is extracted from scheduled structurally obfuscated design integrated with transient fault detectability feature. *Note:* This register allocation is exploited for covertly implanting the physical biometric information of IP vendor based on handprint features.

8.4.2 Handprint biometrics

Now, we discuss the major steps involved in generating IP vendor handprint biometric (digital template) and generating a secure RTL datapath of the IP design (structurally obfuscated with transient fault detectability). The primary input includes the captured handprint image of the IP vendor. Moreover, the secondary inputs encompass the specified signature strength and bit encoding rules (provided by the IP vendor). The process involves several sequential steps (Sengupta and Rathor, 2020; Chaurasia *et al.*, 2023), as depicted in Figure 8.3.

(a) Initially, the handprint biometric data of the original IP vendor is obtained by scanning the handprint using an optical scanner.
(b) Subsequently, the handprint image undergoes pre-processing, which involves two sub-processes named binarization and thinning. Binarization, transforms the image into two intensity levels (low “0” and high “255”), whereas the thinning process, reduces the thickness of ridge lines to one-pixel width. This step enhances the accuracy of locating handprint minutiae points precisely.
(c) The thinned handprint image is then processed to extract minutiae points, identifying locations where ridge lines end abruptly or bifurcate into branches. This process leverages the unique features specific to an IP vendor.
(d) Following that, the dimensions of each of the minutiae feature points obtained from the handprint image are determined and its corresponding binarized biometric digital template is generated. Each minutiae feature comprises the following crucial information related to co-ordinates, rotation angle, and crossing number (signifying that a minutia point is a branching or bifurcation). It should be noted that the actual digital template may vary as per the number of IP vendor-chosen minutiae feature points.

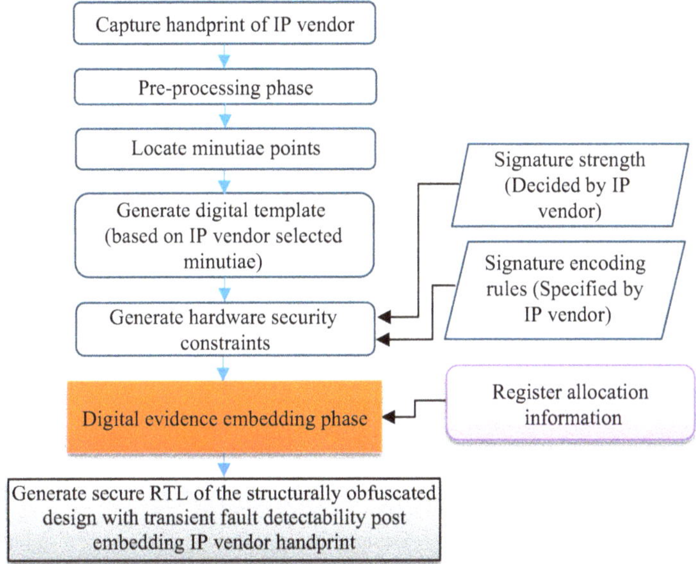

Figure 8.3 Generating secure RTL design for transient fault detectable multi-cut based structural obfuscation with covertly embedded physical biometric of IP vendor

(e) Next, the corresponding hardware security constraints are generated based on IP vendor selected final signature strength and bit encoding rules. This yields secret security constraints aligned with the chosen signature strength and encoding rules.

(f) Subsequently, these generated security constraints are covertly embedded into the design by performing local alteration of storage variables amongst the registers based on "constraints embedding algorithm." These embedded hardware security constraints play a crucial role in implementing detecting/countering IP piracy and nullifying false IP ownership claim.

(g) Finally, the design is initially transformed to enable transient fault detection and is later structurally obfuscated by inserting multiple cuts (to enable hindrance against RE attack). This is additionally implanted with handprint biometric of IP vendor (to detect IP piracy and nullify false IP ownership claim), for generating secure RTL datapath (post-synthesis).

8.4.3 Handprint versus conventional hardware security techniques

This section emphasizes the advantages of handprint biometric over the other conventional hardware security techniques. A comparative analysis based on several crucial features is presented in Table 8.1. As evident handprint biometric-based hardware security technique offers more tampering resiliency, seamless verification of IP ownership from a genuine IP vendor perspective, robustness against key-based attacks, and robust security through unique palmprint features for enabling IP piracy detection with lesser implementation complexity.

Table 8.1 Comparison of handprint and other conventional hardware security techniques

Characteristics/ Attributes	**Handprint biometric (Chaurasia and Sengupta, 2023a)**	**IP watermarking (Koushanfar *et al.*, 2005)**	**IP steganography (Sengupta and Rathor, 2019)**	**Digital signature (Sengupta *et al.*, 2019b)**
Recreation of an identical digital template by an attacker	Highly unlikely	Attainable	Attainable	Attainable
Tampering resiliency	More	Lower than handprint	N/A	Lower than IP water-marking
Verification of IP ownership by the legitimate owner	Seamless	requiring considerable effort	requiring considerable effort	requiring considerable effort
Susceptibility against key based attacks	No	Vulnerable	No	Vulnerable
Security	Relies on inherently distinctive, irreplicable biometric characteristics	Relies on the signature chosen by the author	Relies on stego constraints	Relies on auxiliary security attributes
Complexity in implementation	Less	More	More	More
Detection of pirated IPs	Robust	Less robust	Less robust	Less robust

8.5 Case study on IIR filter application framework

So far, we have discussed the major steps involved in the process of generating secure fault detectable hardware IP design during HLS. Now let us discuss the security methodology (Chaurasia and Sengupta, 2023a) by considering IIR filter application for demonstration purpose. The digital IIR Butterworth filter is amongst the data-intensive applications extensively utilized in audio processing, communication, and control systems, as well as graphic equalizers.

8.5.1 Transfer function of IIR filter

The transfer function of the IIR Butterworth filter representing input and output relationship can be expressed as follows (Salivahanan and Vallavaraj, 2001; Sengupta and Mohanty, 2019):

$$\mathrm{H}(Z) = \frac{Y(Z)}{X(Z)}\left(\frac{16.5171Z^3 + 49.5513.5171Z^2 + 49.5513z + 16.5171}{70.83Z^3 + 31.1205Z^2 + 27.2351z + 2.948}\right)$$
$$= \left(\frac{0.2332 + 0.4664Z^{-1} + 0.4664Z^{-2} + 0.2332Z^{-3}}{1 + 0.4394Z^{-1} + 0.3845Z^{-2} + 0.0416Z^{-3}}\right) \tag{8.1}$$

where *H(Z)* represents system function. Following this, its corresponding transfer function in the time domain can be represented as shown below:

$$T[n] = 0.233a(n) + 0.466a(n-1) + 0.466a(n-2) + 0.233a(n-3) - 0.439b(n-1) - 0.384b(n-2) - 0.041b(n-3) \tag{8.2}$$

Here, $a(n)$, $a(n-1)$, $a(n-2)$ represent the current and past input, and $b(n)$, $b(n-1)$, $b(n-2)$, and $b(n-3)$ denote the current and past output impulses of the filter in the time domain. However, for a more comprehensive understanding, a detailed description of the IIR Butterworth filter can be found in Salivahanan and Vallavaraj (2001).

8.5.2 Algorithm for generating fault detectable IP design schedule

Initially, we created an IIR digital filter IP design that leveraged detectability feature against transient faults emanating from SEU. This design is subsequently structurally obfuscated and exploited for embedding handprint biometric signature to offer multi-layer security of hardware IP design.

Now, let us discuss the algorithm for designing fault detectable schedule that comprises the following steps:

Step-1-DFG construction: initially, a data flow graph is constructed that represents the information flow corresponding to inputs and output of application (as per its transfer function). The data flow graph of IIR digital filter is illustrated in Figure 8.4.
Step-2-DMR construction: Subsequently, the operations of the original unit are duplicated in the next step to form a Dual Modular Redundancy (DMR) design. The combination of the main and duplicate/replica units is collectively referred to as the "DMR design."
Step-3-Schedule DMR design: Following this, both units in the DMR design are scheduled concurrently, adhering to IP designer-specified resource constraints. For illustrative purposes, we have considered an IIR Butterworth digital filter as a case study, and Figure 8.5 depicts the scheduled DMR design corresponding to the IIR Butterworth. The LIST scheduling algorithm is employed to achieve the scheduled DMR design, where each unit comprises 13

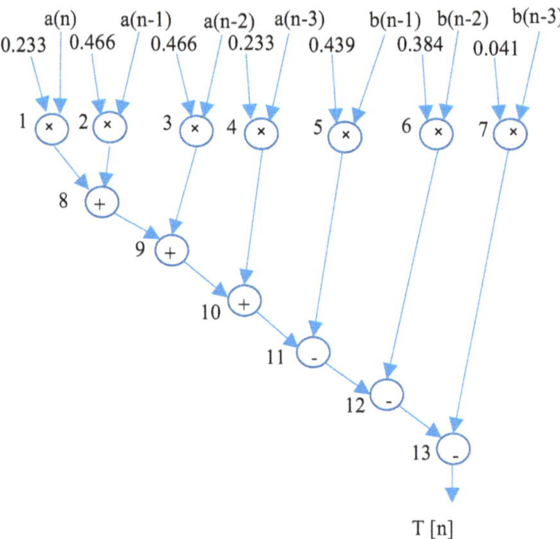

Figure 8.4 DFG of IIR filter

operations (main unit: 1–13, replica unit: 1' to 13'). The respective storage elements required for register assignment are denoted as R1 to R28.

Step-4-Generate fault detectable schedule: Subsequently, the DMR design schedule is transformed into a fault-detectable schedule by applying fault-detectable hardware rescheduling algorithm, which are discussed below:

- Rule-1: If the respective operation timings of the original and duplicate units differ by a Tc-control step (CS), then allocate the same hardware unit such as adder (X_n), multiplier (Y_n), or subtractor (Z_n) in the corresponding duplicate unit. This rule for fault detectability allows the detection of potential transient faults in the design in the sense that the reallocation of the same hardware occurs only after the temporal effects of the transient fault have subsided.
- Rule-2: If the corresponding operations of duplicate unit fail to adhere to the criteria outlined in Rule-1, then reassign the task to an alternate available hardware resource within its duplicate unit to ensure fault detection.
- Rule-3: If any operation continues to violate the aforementioned detection rules, then proceed with rescheduling the operations through incremental downward shifting (one control step at a time) until compliance is satisfied with either Rule-1 or Rule-2 mentioned earlier.

The generated fault detectable IP design schedule post applying transient fault detectability rules, resulted in rescheduling and reallocating of hardware units to operations, as shown in Figure 8.6.

The alterations made through the application of fault-detectability rules have been delineated by a dotted boundary (in brown). Next, the computed output of

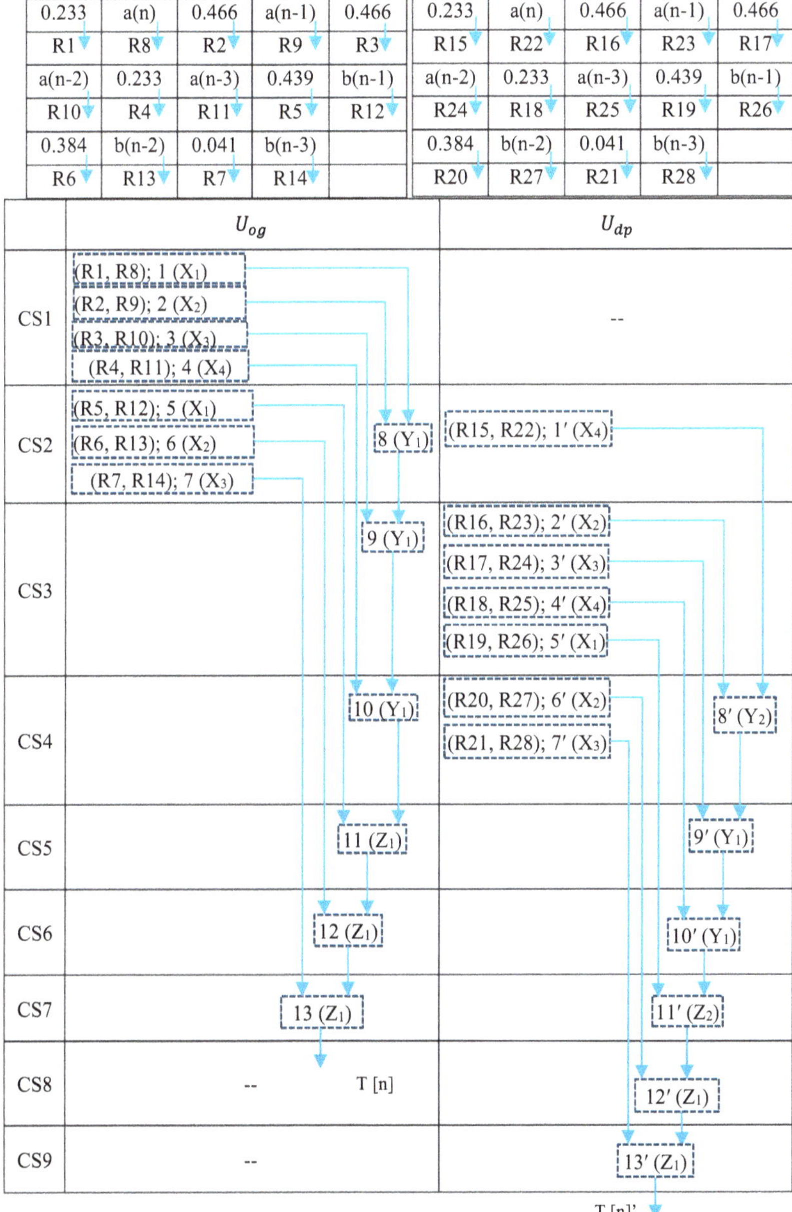

Figure 8.5 Scheduled DMR design of IIR filter involving 4 multipliers, 2 adders, and 2 subtractor resources chosen by IP designer

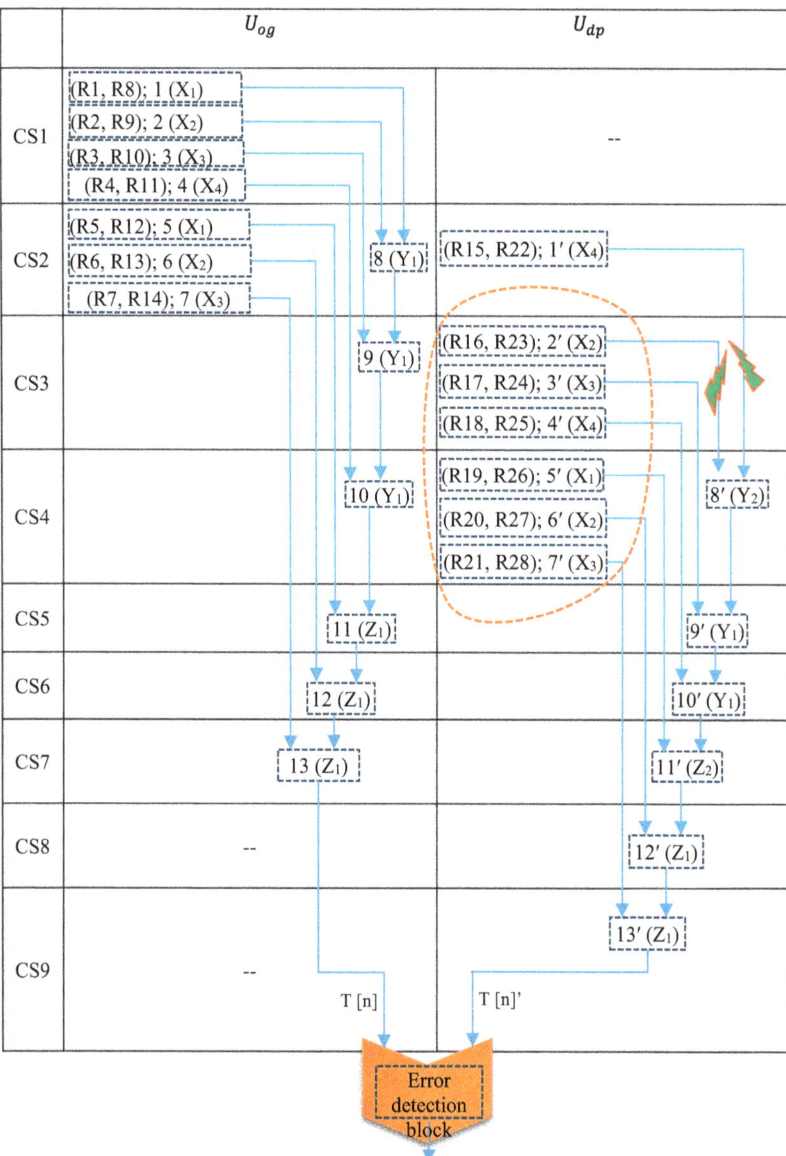

Figure 8.6 Design schedule with transient fault detectability feature

both the main and replica units (T [n] and T [n]' respectively) are fed as inputs to an error detection (ED) block, which consists of multiple comparators (C1, C2, and C3) and a fault-tolerant voter (FTV), as illustrated in Figure 8.7. This multi-stage arrangement comprising comparators and voters is referred to as the error detection block, designed to counter any possible scenario of faulty comparators. In the event

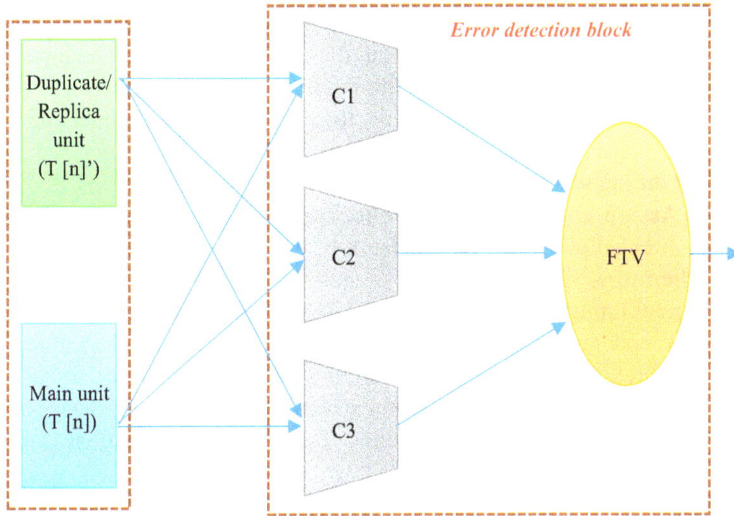

Figure 8.7 Details of error detection framework. Note: C1, C2, and C3 represent the comparator and FTV represents a fault-tolerant voter.

of a fault in one comparator, the defective comparator within the multi-stage configuration of the ED block will yield a complementary output to the remaining two. As a result, the fault-tolerant voter ensures the accurate determination of the majority output, leveraging the two remaining correct comparators. The additional area overhead introduced by the employed ED block is negligible (Rajendran, 2017). Moreover, the approach described in Chaurasia and Sengupta (2023a) assumes the occurrence of a transient fault induced by a SEU, potentially affecting multiple cycles (control steps) in the worst-case scenario. The maximum pulse width of this temporal effect is estimated to be up to 2000 ps, with 1000 ps per control step (Sengupta and Kachave, 2017; Rezgui *et al.*, 2008; Lisboa and Carro, 2007). By configuring *Tc*=2 control steps and reallocating hardware after every 2 steps, the methodology effectively addresses transient faults impact, ensuring their detection.

8.5.3 Algorithm for multi-cut insertion in fault detectable IP design schedule

So far, we have discussed the algorithm for designing fault detectable schedule and its demonstration on the IIR digital filter. Now, we discuss the algorithm for employing structural obfuscation into scheduled fault detectable IP design of IIR digital filter through inserting multi-cuts in some selected regions (Chaurasia and Sengupta, 2023a; Sengupta and Sedaghat, 2015).

Step-1-Applying cuts: Insertion of multiple cuts in the fault-secured scheduled design, results in multi-checkpointing without affecting actual design functionality. The process of multi-cut insertion involves freeing a dependent operation node in replica unit by eliminating its dependency on its parent operation(s).

Step-2-Adjust the operation nodes: Consequently, the operation node, formerly dependent, can be shifted upward in the scheduled design. The inputs to the now independent or free node (from the duplicate unit) are linked and supplied from the main/primary unit. This leads to an altered design interconnectivity without compromising its intended functionality, while also reducing the required control steps during scheduling.

Step-3-Assign comparators for checkpointing: Every inserted cut introduces a fresh checkpoint for fault detection, employing new comparator hardware. Furthermore, the inclusion of each new comparator in the design establishes new input/output connections. This iterative process extends to the remaining dependent operation nodes, leading to altered interconnectivity in the hardware architecture without jeopardizing the final output functionality. Consequently, this approach results in a structurally obfuscated design through adjustments to the corresponding Muxes, De-Muxes, comparators, and other hardware resources.

The resulting scheduled fault-detectable IIR filter design post employing multi-cut insertion algorithm is shown in Figure 8.8. For example, in Figure 8.8, during cut insertion into the design (shown in Figure 8.6), node Y2 (at control step CS4) was initially dependent on X4 (at control step CS2) and X2 (at control step CS3). Cutting the edges (removing dependency) of operation Y2 renders it independent, allowing its upward shift from control step CS4 to CS3, as evident in Figure 8.8. Subsequently, the inputs for Y2 are fed from the respective parent node of the original unit. Similar cut insertion operations are performed in other nodes to achieve structural obfuscation. As depicted, seven resulting comparators corresponding to the seven cuts inserted, have now been added into the IP design that significantly obscures the design architecture. Moreover, as a consequence of incorporating multi-cuts into the design, the latency of the design is also decreased. For instance, the un-obfuscated fault-detectable scheduled design initially necessitated 11 control steps, as illustrated in Figure 8.6. However, following the insertion of cuts, only 10 control steps are now needed, as depicted in Figure 8.8. *Note: Each input and output operation of the scheduled design is assigned to a storage variable for registers storage, where each storage variable is indicated as (d_i); i=0 to 61. The usage of storage variable pairing will be used for embedding the IP vendor handprint as encoded bits (Section 8.6.4).* In the next section, we delve into the details of how a handprint biometric signature is utilized to enhance the security of the IP design against potential threats of IP piracy.

8.5.4 Producing handprint biometric signature for embedding into structurally transformed fault detectable IP design schedule

As discussed earlier in Section 8.5.2, the process for producing handprint biometric signature requires capturing the IP vendor handprint biometric. The captured handprint image is shown in Figure 8.9 (a). Next, the fingerprint image post

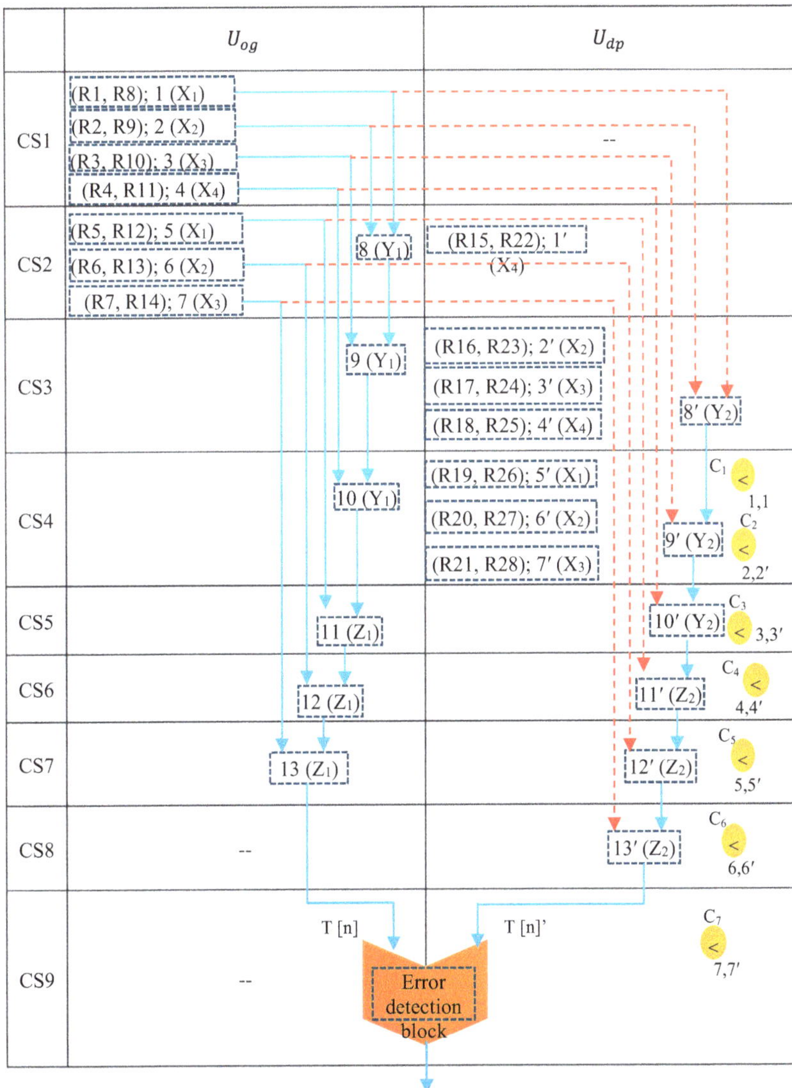

Figure 8.8 Schedule for multi-cut based structurally obfuscated design with fault detectability feature

pre-processing, resulting image after the binarization process is shown in Figure 8.9(b), and the resulting image after the thinning process is shown in Figure 8.9(c). Further, the handprint image with located minutiae points (based on IP vendor selected handprint feature types viz. ridge ending and ridge bifurcation) is shown in Figure 8.9(d).

So far, we obtained the handprint image with located minutiae points using the algorithm discussed in Sengupta and Rathor (2020). The resulting minutiae points with their features comprising the information about its co-ordinates (p1 and p2), crossing number (N_M), and rotation angle (λ) are shown in Table 8.2. As evident,

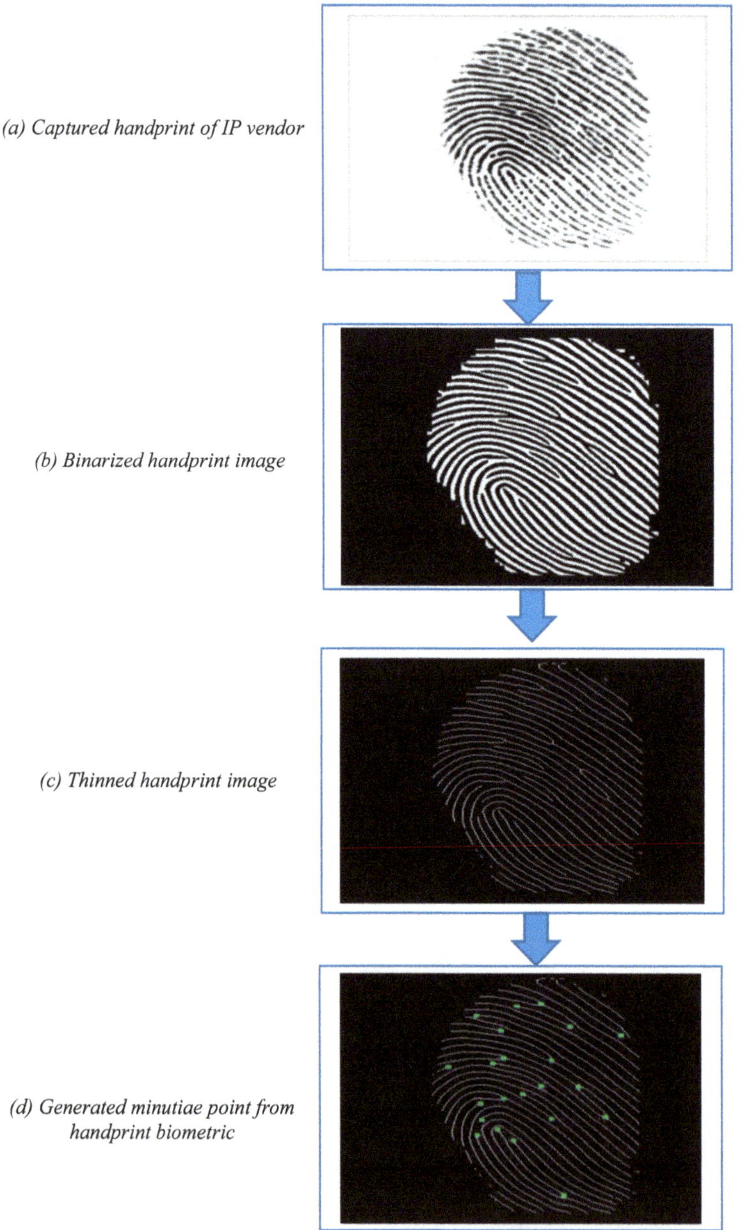

Figure 8.9 Extraction of minutiae points information through the processing of IP vendor handprint biometric

Table 8.2 Determining dimensions corresponding to minutiae feature points

Feature no.	**Coordinate-“p1”**	**Coordinate-“p2”**	N_M	**Ridge angle “λ” (in degrees)**
1	216	46	3→ (ridge bifurcation)	29
2	190	49	1→ (ridge ending)	205
3	146	64	1→ (ridge ending)	187
4	247	80	1→ (ridge ending)	40
5	173	86	1→ (ridge ending)	21
6	302	93	1→ (ridge ending)	48
7	176	127	3→ (ridge bifurcation)	16
8	227	131	3→ (ridge bifurcation)	32
9	164	135	1→ (ridge ending)	190
10	117	140	1→ (ridge ending)	330
11	216	169	1→ (ridge ending)	42
12	256	170	3→ (ridge bifurcation)	223
13	196	181	1→ (ridge ending)	214
14	176	187	3→ (ridge bifurcation)	26
15	151	195	1→ (ridge ending)	328
16	285	215	3→ (ridge bifurcation)	45
17	227	218	1→ (ridge ending)	47
18	152	219	1→ (ridge ending)	131
19	169	233	1→ (ridge ending)	237
20	147	242	1→ (ridge ending)	264
21	186	250	1→ (ridge ending)	239
22	240	332	1→ (ridge ending)	29

there are 22 minutiae points located in the captured handprint image of the IP vendor. Subsequently, the dimensions of each minutiae feature are transformed into its binarized template. The binarized template corresponding to each of the 22 minutiae features is presented in Table 8.3. The binarized template of each minutiae feature is obtained by concatenating the binary dimensions of co-ordinates (p1 and p2), crossing number (N_M), and rotation angle (λ). Thus, the final signature is generated by concatenating the binarized template of minutiae features. However, the final handprint signature may vary as per the number of IP vendor chosen minutiae feature points. The resulting handprint signature by considering all minutiae features and by concatenating them in order is shown below:

“1101100010111011 11011011111011 0 001111001101100 1001010000001 1011101111110111101000011010001010110110101101101011001011101011 10 1111000010110000111111111100001110001110000011111000001010010010 00 0111110111110111010110001100110100101011011000101010011101010100 00 0000101010101111011111110001001011010111101011010110000101110111 11 1010100101111100001111010010001000111011101011111101101 11100011110 1101011011111001100011011011110000011101010011110100111110110110 01001 111110010110000100010111010111101011110111111111000010100110011 1101”.

Table 8.3 Binarized representation of minutiae feature points

Feature no.	Binarized representation (p1- p2- N_M-λ)
1	11011000-101110-11-11101
2	10111110-110001-1-11001101
3	10010010-1000000-1-10111011
4	11110111-1010000-1-101000
5	10101101-1010110-1-10101
6	100101110-1011101-1-110000
7	10110000-1111111-11-10000
8	11100011-10000011-11-100000
9	10100100-10000111-1-10111110
10	1110101-10001100-1-101001010
11	11011000-10101001-1-101010
12	100000000-10101010-11-11011111
13	11000100-10110101-1-11010110
14	10110000-10111011-11-11010
15	10010111-11000011-1-101001000
16	100011101-11010111-11-101101
17	11100011-11011010-1-101111
18	10011000-11011011-1-10000011
19	10101001-11101001-1-11101101
20	10010011-11110010-1-100001000
21	10111010-11111010-1-11101111
22	11110000-101001100-1-11101

The binarized handprint digital template is of size 526 bits, comprising 224 number of 0s and 302 number of 1s.

Next, the generated handprint template is transformed into hardware security constraints using IP vendor chosen signature strength and specified encoding rules. Assuming that an IP vendor has chosen complete signature to generate/form hardware security constraints with the vision of achieving enhanced security. This is because the more the possibility of embedding the hardware security constraints into design, the more the robustness against IP piracy due to enhanced digital evidence. However, sometimes design cost overhead may occur while satisfying all the hardware security constraints. The process of generating the hardware security constraints for embedding using the following bit encoding algorithm is discussed below:

Encoding of signature bit "0": formulate a security constraint by pairing the "<even-even>" storage variable indices in sorted ascending order (starting from 0) and repeat the process for all the remaining signature bits of 0s.

Encoding of signature bit "1": formulate a security constraint by pairing the "<odd-odd>" storage variable indices in sorted ascending order (starting from 1) and repeat the process for all the remaining signature bits of 1s.

Further, it is to be noted that IP vendor may specify any of the possible bit encoding algorithm to generate hardware security constraints differently. The resulting hardware security constraints corresponding to bit 0 and bit 1, by following the above encoding, are listed in Tables 8.4 and 8.5, respectively.

Next, the generated hardware security constraints corresponding to handprint signature are covertly embedded into the scheduled design (during register allocation phase of HLS), obtained after employing fault detectability algorithm and multi-cut based structural obfuscation (as presented in Figure 8.8). In order to perform embedding of security constraints into the design, its register allocation information is extracted, as presented in Table 8.6.

The embedding process of generated security constraints (in Tables 8.4 and 8.5) represents pairs of storage variables serving as covert security information, adhering to the rule that a storage variable pair cannot be simultaneously assigned to one common register. While adhering to the rule, storage variables are locally altered between available registers. If local alteration is not possible to satisfy the rule, then a new register is assigned. Table 8.7 presents the register allocation information after

Table 8.4 Handprint signature-based secret hardware security constrains corresponding to signature bit "0"

<d0, d2>	<–,–>	<–,–>	<–,–>
<d0, d4>	<–,–>	<–,–>	<–,–>
<d0, d6>	<–,–>	<–,–>	<–,–>
<d0, d8>	<d0, d60>	<d2, d60>	<d4, d60>
<d0, d10>	<d2, d4>	<d4, d6>	<–,–>
<d0, d12>	<d2, d6>	<d4, d8>	<–,–>
<d0, d14>	<d2, d8>	<d4, d10>	<–,–>
<d0, d16>	<d2, d10>	<d4, d12>	<d16, d40>
<d0, d18>	<d2, d12>	<d4, d14>	
<d0, d20>	<d2, d14>	<d4, d16>	
<d0, d22>	<d2, d16>	<d4, d18>	
<d0, d24>	<d2, d18>	<d4, d20>	

Table 8.5 Handprint signature-based secret hardware security constrains corresponding to signature bit "1"

<d1, d3>	<–,–>	<–,–>	<–,–>
<d1, d5>	<–,–>	<–,–>	<–,–>
<d1, d7>	<–,–>	<–,–>	<–,–>
<d1, d9>	<d1, d61>	<d3, d61>	<d25, d41>
<d1, d11>	<d3, d5>	<d5, d7>	
<d1, d13>	<d3, d7>	<d5, d9>	
<d1, d15>	<d3, d9>	<d5, d11>	
<d1, d17>	<d3, d11>	<d5, d13>	
<d1, d19>	<d3, d13>	<d5, d15>	
<d1, d21>	<d3, d15>	<d5, d17>	
<d1, d23>	<d3, d17>	<d5, d19>	
<d1, d25>	<d3, d19>	<d5, d21>	

Table 8.6 Extracted register allocation information corresponding to scheduled structurally obfuscated IIR filter design with fault detectability feature

CS	R^1	R^2	R^3	R^4	R^5	R^6	R^7	R^8	R^9	R^{10}	R^{11}	R^{12}	R^{13}	R^{14}	R^{15}	R^{16}	R^{17}	R^{18}	R^{19}	R^{20}	R^{21}	R^{22}	R^{23}	R^{24}	R^{25}	R^{26}	R^{27}	R^{28}
C^0	d0	d1	d2	d3	d4	d5	d6	d7	d8	d9	d10	d11	d12	d13														
C^1	d14	d15	d16	d17					d8	d9	d10	d11	d12	d13	d18							d19						
C^2	d14	d15	d16	d17	d20	d21	d22	d24							d23	d25	d27	d29					d26	d28	d30			
C^3	d34		d16	d17	d20	d21	d22	d35								d31	d32	d33	d36	d38	d40					d37	d39	d41
C^4	d47	d46		d17	d20	d21	d22	d48							d45				d42	d43	d44							
C^5	d50		d49		d20	d21	d22	d51																				
C^6	d53			d52		d21	d22	d54																				
C^7	d56				d55		d22	d57																				
C^8	d59					d58		d57																				
C^9							d60	d61																				

Table 8.7 Locally altered (modified) register allocation information post embedding IP vendor handprint signature into scheduled structurally obfuscated IIR filter design with fault detectability feature

CS	R^1	R^2	R^3	R^4	R^5	R^6	R^7	R^8	R^9	R^{10}	R^{11}	R^{12}	R^{13}	R^{14}	R^{15}	R^{16}	R^{17}	R^{18}	R^{19}	R^{20}	R^{21}	R^{22}	R^{23}	R^{24}	R^{25}	R^{26}	R^{27}	R^{28}
C^0	d0	d1	d2	d3	d4	d5	d6	d7	d8	d9	d10	d11	d12	d13	-	-	-	-	-	-	-	-	-	-	-	-	-	-
C^1	d15	d14	d17	d16	-	-	-	-	d8	d9	d10	d11	d12	d13	d18	-	-	-	-	-	-	d19						
C^2	d15	d14	d17	d16	d21	d20	-	d24	-	d22	-	-	-	-	d23	d25	d27	d29					d26	d28	d30			
C^3	-	-	d17	d16	d21	d20	-	d35	-	d22	-	d34	-	-	-	d31	d32	d33	d36	d38	d40					d37	d39	d41
C^4	-	-	d17	-	d21	d20	d47	d48	-	d22	-	d46	-	-	-	d45			d42	d43	d44	-	-	-	-	-	-	-
C^5	-	-	-	-	d21	d20	d49	-	d51	d22	-	d50	-	-	-	-	-	—	—	-	-	-	-	-	-	-	-	
C^6	-	-	-	d52	d21	-	d53	d54	-	d22	-	-	-	-	-	-	-	-	-	-	-	-	-	-	-	-	-	-
C^7	-	-	-	-	-	-	d55	-	d57	d22	-	d56	-	-	-	-	-	-	-	-	-	-	-	-	-	-	-	-
C^8	-	-	-	-	-	d58	d59	-	d57	-	-	-	-	-	-	-	-	-	-	-	-	-	-	-	-	-	-	-
C^9	-	-	-	-	-	-	-	-	d61	d60	-	-	-	-	-	-	-	-	-	-	-	-	-	-	-	-	-	-

embedding the IP vendor's handprint biometric signature-driven security constraints into the design. These embedded security constraints enable detecting IP piracy and nullifying false IP ownership claim. Finally, post synthesis of the resulting design with embedded handprint biometric information of IP vendor, a secure RTL data path of IP design is generated through HLS.

8.6 Security features/attributes of the methodology

The discussed hardware IP security methodology based on multi-cut-based structural obfuscation and physical biometrics offers several key security features as follows:

- It employs a multi-layer defense mechanism, utilizing multi-cut-based architectural obfuscation to thwart reverse engineering and incorporating handprint biometric to detect IP piracy and nullify false claim of IP ownership.
- The second layer of defense retains its indirect contextual nature unless an adversary successfully de-obfuscates the design, which itself is highly challenging. This is because a higher strength of obfuscation is offered by multi-cut insertion.
- Even if an adversary obtains access to the handprint biometric of an IP vendor, following Kirchhoff's principle, where the attacker possesses knowledge of both the handprint biometric and the embedding process, it remains infeasible to accurately replicate the implanted biometric-based hardware security measures. This difficulty arises because the original handprint biometric of the IP vendor is securely stored offline in an encrypted format, utilizing AES encryption keys. Consequently, even if an attacker illicitly acquires the handprint biometric and its minutia points, they would be unable to decrypt it to recreate its precise security measures due to the robust encryption employed for the handprint biometric data. As a result, attempts by an attacker to falsely claim ownership of the IP and evade detection for IP piracy would prove unsuccessful.
- The embedded handprint signature inherently associates with the unique identity of the original IP vendor. Therefore, it can serve as compelling digital evidence to prove the IP ownership of the original owner and refute fraudulent IP claims. This distinguishes it from watermarking, steganography, and digital signature-based security approaches in terms of more unique security features.

8.7 Analysis of security and design cost of IP design generated through multi-cut-based structural obfuscation and physical biometrics

Analysis of security through multi-cut insertion:
The assessment of security offered against reverse engineering the design or RTL design alteration focuses on the effectiveness/robustness of obfuscation. A higher degree of obfuscation results in increased resilience of the design against reverse

engineering attempts. The robustness of structural obfuscation can be measured using the power of obfuscation metric (Sengupta *et al.*, 2017,2018). The power of Obfuscation (Obf_p) is defined by the following equation, expressing as a normalized value within the range of 0 to 1.

$$Obf_p = \frac{u}{u^Y} \tag{8.3}$$

Here, "u" and "u^Y" represent the number of modified nodes resulting from multi-cut insertion, and the total number of nodes before the application of multi-cut insertion, respectively. A higher value of "Obf_p" signifies stronger security for the design. Further, it should be noted that a node is considered modified if any of the following conditions holds true:

(a) The parent node or primary input of a node in an obfuscated scheduled design (DFG) differs from its original state.
(b) The child of a node in an obfuscated scheduled design (DFG) differs from its original state.
(c) The operation type (e.g., addition, multiplication, and subtraction) of a node in an obfuscated scheduled DFG is not the same.
(d) A node in the original design does not exist in the corresponding obfuscated scheduled DFG.
(e) Generation of new nodes that do not exist in the original design.

The power of obfuscation with respect to different benchmarks is presented in Figure 8.10. The results demonstrate that obfuscation using multi-cut insertion offers a stronger power of obfuscation for various applications and case studies.

Further, the difficulty of a structural obfuscation technique can also be assessed in terms of the effort required by an attacker to de-obfuscate the IP design. In the multi-cut insertion-based technique, an attacker's effort needed for de-obfuscation can be measured by the probability of identifying all the multi-cuts applied to the design for achieving structural obfuscation. Assuming there are "s"-dependent nodes in the scheduled DMR design's duplicate unit, the number of input edges of the dependent nodes of the duplicate unit is 2*s. Thus, from an

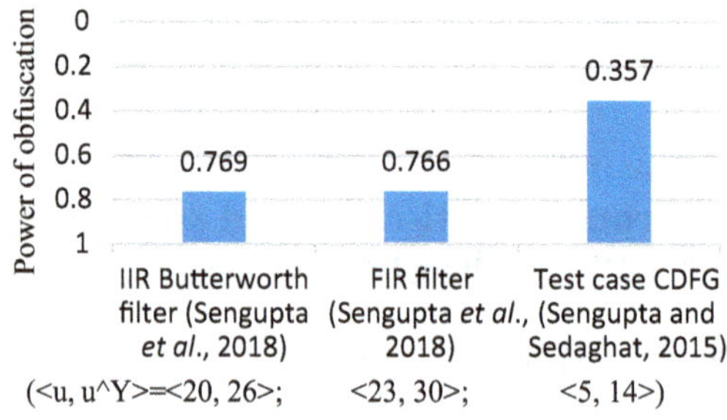

Figure 8.10 Power of obfuscation for different benchmarks

attacker's perspective, the probability of identifying all cut edges (E) due to multi-cut insertion can be computed as follows:

$$I_p = \prod_{i=1}^{E} (\frac{1}{2.s}) \tag{8.4}$$

Moreover, the likelihood of identifying an affected gate (obfuscated logic) due to a single cut can be expressed as:

$$O_p = \frac{1}{|(G^o - G^{uo})|} \tag{8.5}$$

Here, "G^o" represents the number of gates in the obfuscated design, and "G^{uo}" indicates the number of gates in the un-obfuscated design. Thus, from an attacker's perspective, the probability of identifying all obfuscated gates due to all cut edges (E), by performing brute force attack can be computed as follows:

$$I_p^O = \prod_{i=1}^{E} (\frac{1}{|(G^o - G^{uo})|}) \tag{8.6}$$

Here, "I_p^O" reflects the measure of uncertainty that an attacker encounters in terms of his/her effort while launching a brute force attack. Consequently, it serves as an indicator of the entropy of the multi-cut-based obfuscation approach. Entropy is commonly employed as a standard measure to assess the difficulty/hardness of an obfuscation method (Amir *et al.*, 2018). The entropy (probability of identifying all obfuscated gates from the attacker's perspective) of the approach (Chaurasia and Sengupta, 2023a), evaluated using (8.6) is presented in Figure 8.11. As apparent, multi-cut-based structural obfuscation attains robust entropy, providing resilience against adversarial attacks. Further, it also offers a comparative perspective of

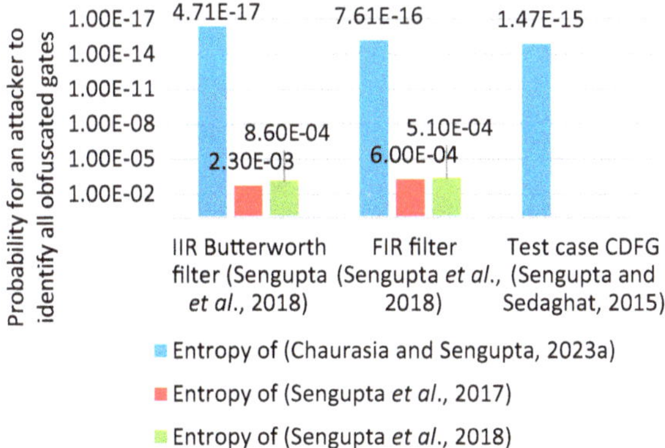

Figure 8.11 Comparison of probability for an attacker to de-obfuscate the IP design

security achieved in terms of entropy for various approaches (Chaurasia and Sengupta, 2023a; Sengupta *et al.*, 2017, 2018). Notably, the entropy values of Sengupta *et al.* (2017) and Sengupta *et al.* (2018) are weaker (non-desirable) than Chaurasia and Sengupta (2023a) due to the reduced effort required by attackers to de-obfuscate the IP design for structural obfuscation. This suggests that the approaches (Sengupta *et al.*, 2017, 2018) exhibit lower security against brute-force attacks during de-obfuscation attempt. Furthermore, Sengupta *et al.* (2017) and Sengupta *et al.* (2018) rely on high-level transformation (HLT) and logic transformation-driven structural obfuscation, which may not be applicable to certain applications like the "Test Case CDFG." In contrast, the multi-cut insertion-based obfuscation (Chaurasia and Sengupta, 2023a) is applicable to all applications, ensuring robust structural obfuscation.

Analysis of security through embedded handprint-based physical biometric: Moreover, the security against IP piracy, achieved by embedding IP vendor handprint biometric signature into the design, is scrutinized through the probability of coincidence (indicating digital evidence as a watermark) and the tamper tolerance metric.

The probability of coincidence (X_p^c) quantifies the likelihood of obtaining, by chance, the security information of a secured design in its unsecured counterpart. Thus, a lower "X_p^c" is desirable, and its calculation is outlined as follows (Koushanfar *et al.*, 2005; Sengupta and Rathor, 2019; Rathor and Sengupta, 2020):

$$X_p^c = \left(1 - \frac{1}{\text{Rn}}\right)^H \tag{8.7}$$

Where "Rn" and "H" represent the utilized registers in the baseline scheduled design and the number of embedded security constraints (hidden watermark strength) in that design, respectively. The comparison of the probability of coincidence (false positive) corresponding to varying handprint features for IIR filter design is presented in Figure 8.12. As apparent, a lower probability of coincidence is reported for a greater number of handprint features. This is

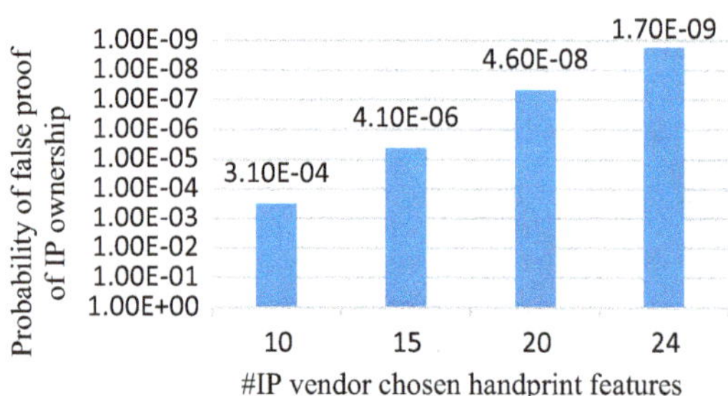

Figure 8.12 Comparison of probability of coincidence (false positive) corresponding to varying handprint features for IIR filter design from an adversarial perspective

because the more the number of features, the more the possibility of embedding larger number of security constraints into IP design. Further, the comparative analysis of probability of coincidence obtained for Chaurasia and Sengupta (2023a) corresponding to different hardware security techniques, has been reported in Figure 8.13. As apparent, handprint biometric offers significantly lower values of X_p^c (desirable). Additionally, the value of probability of coincidence for different IP vendor handprint images has been reported in Table 8.8 (handprint images have been obtained from Sengupta and Rathor, 2020). It can be observed that the handprint with the greater number of security constraints yields a lower value of X_p^c.

Further, the tamper tolerance ability of the design reflects its capability to withstand tampering attacks using brute force. Therefore, a higher tamper tolerance strength indicates greater resilience of the design against tampering attacks.

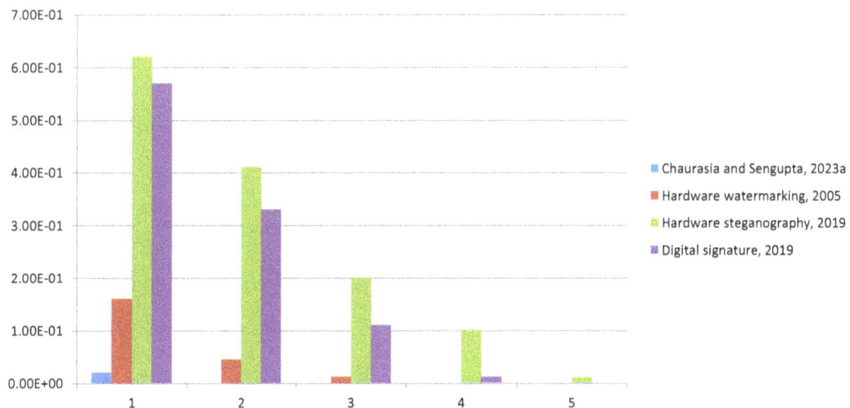

Figure 8.13 Analyzing probability of coincidence (false positive) of handprint embedded secure IIR filter design from an adversarial perspective for varying security constraint sizes corresponding to different hardware security techniques

Table 8.8 Analyzing variation in X_p^c and T_R corresponding to different sizes of handprint constraints

	Handprint sample image-2	**Handprint sample image-3**	**Handprint sample image-4**	**Handprint sample image-5**
# Constraints (*H*)	526	350	538	418
X_p^c	$4.9e^{-9}$	$2.9e^{-6}$	$3.1e^{-9}$	$2.5e^{-7}$
T_R	$2.19e^{158}$	$2.29e^{105}$	$8.99\ e^{161}$	$6.76e^{125}$

A higher tamper tolerance (T_R) ability is desirable and is measured as follows (Koushanfar *et al.*, 2005; Sengupta *et al.*, 2019b):

$$T_R = (S_b)^H \tag{8.8}$$

where "S_b" represents the type of signature bits. The comparison of tamper tolerance or tampering resiliency corresponding to varying handprint features for Chaurasia and Sengupta (2023a) is presented in Figure 8.14. As apparent, higher tamper tolerance (desirable) is reported for a greater number of handprint features. This is because the more the number of features, the more the number of generated security constraints. Further, the comparative analysis of tamper tolerance for Chaurasia and Sengupta (2023a) corresponding to different hardware security techniques has been reported in Figure 8.15. As apparent, handprint biometric offers significantly higher tamper

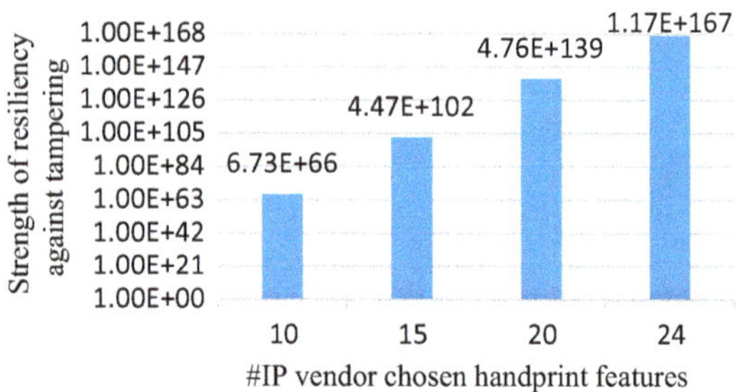

Figure 8.14 Comparison of tamper tolerance corresponding to varying handprint features

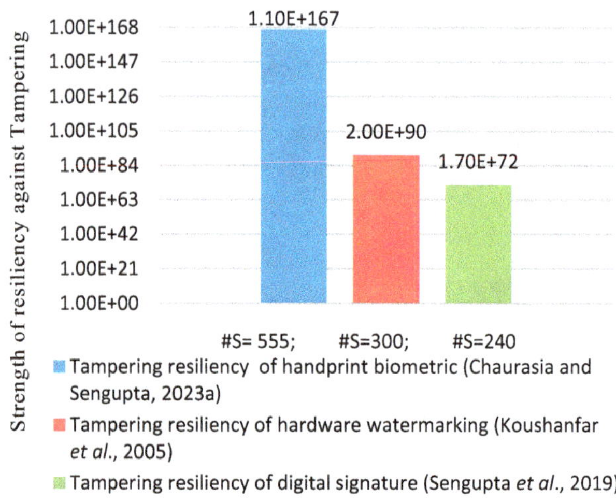

Figure 8.15 Analyzing strength of handprint embedded secure IIR filter design against tampering attack from an adversarial perspective corresponding to different hardware security techniques

tolerance than Koushanfar *et al.* (2005) and Sengupta *et al.* (2019). Additionally, the value of tamper tolerance computed for different IP vendor handprint images has been reported in Table 8.8. It can be observed that the handprint image with the greater number of security constraints yields a higher value of (T_R).

Analysis of entropy:

Entropy is defined as the measure of effort required for an adversary to successfully estimate the exact value of the security mark implanted into the hardware IP design. The effort required from an adversarial perspective, or entropy (E), can be formulated as follows (NIST, 2022; Sengupta *et al.*, 2023):

$$E_x = \frac{1}{2^K} \tag{8.9}$$

Here "K" represents the size of secret signature bits associated with the respective security algorithm. The entropy comparison of the methodology utilizing handprint biometric (Sengupta and Chaurasia, 2023) with other approaches is presented in Figure 8.16. As evident, the security methodology utilizing handprint biometric

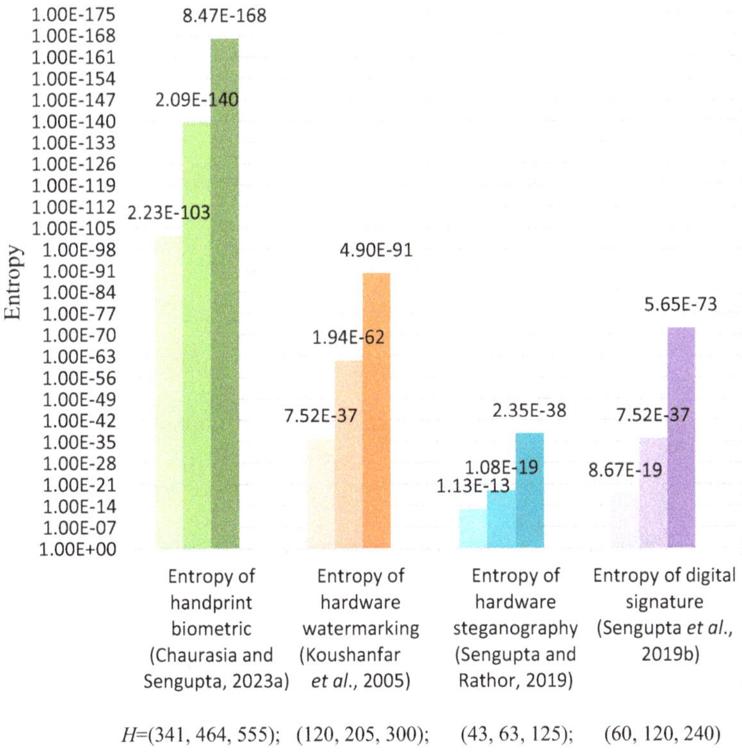

Figure 8.16 Analyzing entropy of handprint biometric and other hardware security techniques for varying signature size

(Sengupta and Chaurasia, 2023) provides more robust security against an adversary attempting to compromise or guess the security signature successfully, rendering stronger entropy.

Analysis of design cost:
The impact in the design cost for enabling multi-layer security of transient fault detectable hardware IP through structural obfuscation and embedded handprint biometric is analyzed using the following normalized function (Koushanfar *et al.*, 2005; Sengupta and Rathor, 2019; Sengupta *et al.*, 2019b):

$$C_f(R_{conf}) = j_1 \frac{T_d}{T_{max}} + j_2 \frac{R_d}{R_{max}} \tag{8.10}$$

Here, "$C_f(R_{conf})$" represents the design cost utilizing the IP vendor-specified hardware resources during the scheduling of the design, T_d and R_d represent the area and latency of the IP design (handprint-embedded obfuscated fault-detectable). Further, T_{max} and R_{max} represent maximum area and latency, respectively, and $j1$ and $j2$ represent the weighting factors to normalize both orthogonal parameters (area and latency). The design cost overhead post embedding IP vendor handprint-driven security constraints into IIR filter design (with fault detectability feature and multi-cut-based obfuscation) is presented in Figure 8.17. Notably, the methodology (Chaurasia and Sengupta, 2023a) incurs a 0.0% design overhead across different strengths of the handprint security constraints. This is attributed to the fact that no additional registers were required during the embedding of generated handprint secret security constraints (associated with handprint signature and encoding method specified by the IP vendor) in the register allocation phase of the HLS. Any conflicts among the secret security constraints are resolved by making local adjustments among the available registers.

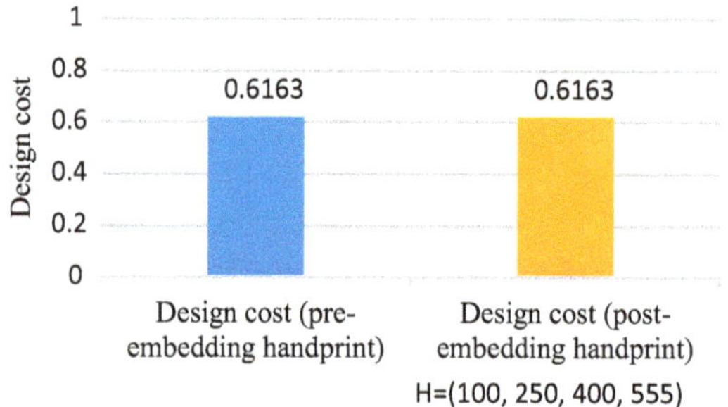

Figure 8.17 Analyzing design cost overhead post embedding handprint-driven security constraints of IP vendor (of varying size) into IIR filter with fault detectability feature and multi-cut-based obfuscation

8.8 Conclusion

This chapter discussed a security-aware HLS-based IP design methodology leveraged with transient fault detectability feature and equipped with multi-layer security against potential reverse engineering and IP piracy/false IP ownership. The discussed approach offered robust security against reverse engineering and RTL alteration through multi-cut-based architectural obfuscation, rendering the design non-obvious and non-interpretable to attackers, thus hindering design analysis and identification. Additionally, it ensured robust piracy detection with covertly embedded handprint biometric information of authentic IP vendor at zero design cost overhead. Further, the embedded unique handprint biometric mark enables an IP vendor to seamlessly prove his/her IP ownership as compared to other security techniques. Thus, it provides an IP design methodology leveraged with multi-layer security for IP protection.

At the end of this chapter, a reader gains the knowledge about the following:

- Importance of hardware IP security and significance of structural obfuscation as well as physical biometric for the same.
- Different hardware threat scenarios and their impact on IP design chain and its stakeholders.
- Crucial basic details of transient fault and their impact on IP functionality.
- End-to-end description and demonstration of multi-cut-based structural obfuscation on IIR digital filter design.
- End-to-end description and demonstration of generating unique security mark by exploiting IP vendor handprint biometric for enhanced security against IP piracy and fraudulent IP ownership claim.
- Significance of physical biometric over the other hardware security techniques such as IP watermarking, hardware steganography, and digital signature insertion.
- A comprehensive understanding of the HLS-based IP design methodology for achieving robust IP protection.

8.9 Questions and exercise

1. Discuss the need to design hardware IPs for computationally intensive applications.
2. Discuss the significance of hardware IP security and trust.
3. Discuss the significance of structural obfuscation for secure IP design.
4. Discuss the significance of physical biometrics for hardware IP security and trust.
5. Discuss different types of hardware IPs and their functionality.
6. Why is the hardware security so crucial? Explain different hardware security threats.
7. Discuss the cause of occurrence of transient faults and their impact on an IP core functionality.

8. Discuss the major steps involved in multi-cut-based structural obfuscation mechanism for transient fault detectable IP designs.
9. Discuss the major steps involved in handprint biometric signature generation.
10. Discuss the comparative analysis of handprint biometric with other conventional techniques for hardware security.
11. Explain the algorithm for generating fault detectable IP design schedule, with an example.
12. Explain the algorithm for multi-cut insertion in fault detectable IP design schedule, with an example.
13. Demonstrate security mark embedding process during register allocation of HLS for IIR filter with a sample handprint biometric signature.
14. Discuss security features of the methodology incorporating multi-cut-based structural obfuscation and physical biometric.
15. Explain the computation of power of obfuscation.
16. Explain the computation of security metrics such as probability of coincidence, tamper tolerance, and entropy.
17. Explain the design cost computation function in terms of orthogonal parameters of area and latency.

References

S. Amir, B. Shakya, X. Xu, *et al.*, (2018) "Development and Evaluation of Hardware Obfuscation Benchmarks," *Journal of Hardware and Systems Security*, vol. 2, pp. 142–161.

M. T. Arafin, A. Stanley and P. Sharma, (2017) "Hardware-based Anti-counterfeiting Techniques for Safeguarding Supply Chain Integrity," *2017 IEEE International Symposium on Circuits and Systems (ISCAS)*, pp. 1–4.

O. Benot, (2011) "Fault Attack," In: van Tilborg, H. C. A., Jajodia, S. (eds) *Encyclopedia of Cryptography and Security*. Boston, MA: Springer. https://doi.org/10.1007/978-1-4419-5906-5_505.

E. Castillo, U. Meyer-Baese, A. Garcia, L. Parrilla and A. Lloris, (2007) "IPP@HDL: Efficient Intellectual Property Protection Scheme for IP Cores," *IEEE Transactions on Very Large Scale Integration (VLSI) Systems*, vol. 15, no. 5, pp. 578–591.

R. Chaurasia and A. Sengupta, (2021) "Securing Reusable Hardware IP Cores Using Palmprint Biometric," *2021 IEEE International Symposium on Smart Electronic Systems (iSES)*, pp. 410–413, doi:10.1109/iSES52644.2021.00099.

R. Chaurasia and A. Sengupta, (2022a) "Protecting Trojan Secured DSP Cores against IP Piracy Using Facial Biometrics," *2022 IEEE 19th India Council International Conference (INDICON)*, India, pp. 1–6, doi:10.1109/INDICON56171.2022.10039864.

R. Chaurasia and A. Sengupta, (2022b) "Security Vs Design Cost of Signature Driven Security Methodologies for Reusable Hardware IP Core," *2022 IEEE*

International Symposium on Smart Electronic Systems (iSES), India, pp. 283–288, doi:10.1109/iSES54909.2022.00064.

R. Chaurasia and A. Sengupta, (2022c) "Symmetrical Protection of Ownership Right's for IP Buyer and IP Vendor Using Facial Biometric Pairing," *2022 IEEE International Symposium on Smart Electronic Systems (iSES)*, India, pp. 272–277, doi:10.1109/iSES54909.2022.00062.

R. Chaurasia and A. Sengupta, (2022d) "Crypto-Genome Signature for Securing Hardware Accelerators," *2022 IEEE 19th India Council International Conference (INDICON)*, India, pp. 1–6, doi:10.1109/INDICON56171.2022.10039955.

R. Chaurasia, A. Anshul, A. Sengupta and S. Gupta, (2022) "Palmprint Biometric Versus Encrypted Hash Based Digital Signature for Securing DSP Cores Used in CE Systems," *IEEE Consumer Electronics Magazine (CEM)*, vol. 11, no. 5, pp. 73–80, doi:10.1109/MCE.2022.3153276.

R. Chaurasia, A. Reddy Asireddy and A. Sengupta, (2023) "Fault Secured JPEG-Codec Hardware Accelerator with Piracy Detective Control Using Secure Fingerprint Template," *2023 IEEE International Symposium on Defect and Fault Tolerance in VLSI and Nanotechnology Systems (DFT)*, Juan-Les-Pins, France, pp. 1–6, doi:10.1109/DFT59622.2023.10313536.

R. Chaurasia and A. Sengupta, (2023a) "Multi-cut Based Architectural Obfuscation and Handprint Biometric Signature for Securing Transient Fault Detectable IP Cores during HLS," *Integration, the VLSI Journal*, vol. 95, 102114, doi:10.1016/j.vlsi.2023.102114.

R. Chaurasia and A. Sengupta, (2023b) "Designing Optimized and Secured Reusable Convolutional Hardware Accelerator against IP Piracy Using Retina Biometrics," *Proceedings of 9th IEEE International Symposium on Smart Electronic Systems (IEEE – iSES)*, India. 2023.

S. Chen, J. Jung, P. Song, K. Chakrabarty and G.-J. Nam, (2020) "BISTLock: Efficient IP Piracy Protection Using BIST," *2020 IEEE International Test Conference (ITC)*, pp. 1–5.

B. Colombier and L. Bossuet, (2015) "Survey of Hardware Protection of Design Data for Integrated Circuits and Intellectual Properties," *IET Computers & Digital Techniques*, vol. 8, no. 6, pp. 274–287, https://doi.org/10.1049/iet-cdt.2014.0028.

A. Hroub and M. E. S. Elrabaa, (2022) "SecSoC: A Secure System on Chip Architecture for IoT Devices," *2022 IEEE International Symposium on Hardware Oriented Security and Trust (HOST)*, pp. 41–44.

W. Hu, C.-H. Chang, A. Sengupta, S. Bhunia, R. Kastner and H. Li, (2021) "An Overview of Hardware Security and Trust: Threats, Countermeasures, and Design Tools," *IEEE Transactions on Computer-Aided Design of Integrated Circuits and Systems*, vol. 40, no. 6, pp. 1010–1038.

D. Kachave and A. Sengupta, (2016) "Integrating Physical Level Design and High Level Synthesis for Simultaneous Multi-Cycle Transient and Multiple Transient Fault Resiliency of Application Specific Datapath Processors," *Elsevier Journal on Microelectronics Reliability*, vol. 60, pp. 141–152.

D. Karaklajić, J.-M. Schmidt and I. Verbauwhede, (2013) "Hardware Designer's Guide to Fault Attacks," *IEEE Transactions on Very Large Scale Integration (VLSI) Systems*, vol. 21, no. 12, pp. 2295–2306.

T. Kean, D. McLaren, and C. Marsh, (2008) "Verifying the Authenticity of Chip Designs with the Designtag System," *IEEE International Workshop on Hardware-Oriented Security and Trust*, pp. 59–64.

F. Koushanfar, S. Fazzari, C. McCants, *et al.*, (2012) "Can EDA Combat the Rise of Electronic Counterfeiting?" *DAC Design Automation Conference 2012*, pp. 133–138.

F. Koushanfar, I. Hong and M. Potkonjak, (2005) "Behavioral Synthesis Techniques for Intellectual Property Protection," *ACM Transactions on Design Automation of Electronic Systems (TODAES)*, vol. 10, no. 3, pp. 523–545.

Y. Lao and Parhi, K., (2014) "Protecting DSP Circuits Through obfuscation," *Proceedings – IEEE International Symposium on Circuits and Systems (ISCAS)*, pp. 798–801.

C.A. Lisboa and L. Carro, (2007) "System Level Approaches for Mitigation of Long Duration Transient Faults in Future Technologies," *12th IEEE European Test Symposium – ETS 2007*, in Freiburg, Germany, pp. 165–172s.

NIST Computer Security Resource Center, Glossary, https://csrc.nist.gov/glossary/term/entropy#:~:text=A%20measure%20of%20the%20amount,is%20usually%20stated%20in%20bits, accessed on Feb 2022.

S. Park, S. Jeon, B. Kim and J. Lee, (2021) "Methods for Improving the Reliability of Intelligent Semiconductor," *2021 IEEE International Conference on Consumer Electronics-Asia (ICCE-Asia)*, pp. 1–4.

C. Pilato, F. Regazzoni, R. Karri and S. Garg, (2018a) "TAO: Techniques for Algorithm-Level Obfuscation during High-Level Synthesis," *2018 55th ACM/ESDA/IEEE Design Automation Conference (DAC)*, San Francisco, CA, USA, pp. 1–6.

C. Pilato, S. Garg, K. Wu, R. Karri and F. Regazzoni, (2018b) "Securing Hardware Accelerators: A New Challenge for High-Level Synthesis," *IEEE Embedded Systems Letters*, vol. 10, no. 3, pp. 77–80.

P. Qiu, D. Wang, Y. Lyu and G. Qu, (2019) "VoltJockey: Breaking SGX by Software-Controlled Voltage-Induced Hardware Faults," *2019 Asian Hardware Oriented Security and Trust Symposium (AsianHOST)*, pp. 1–6.

J. J. V. Rajendran, (2017) "An Overview of Hardware Intellectual Property Protection," *2017 IEEE International Symposium on Circuits and Systems (ISCAS)*, pp. 1–4.

S. Ramadhan, F. I. Hariadi and A. S. Ahmad, (2017) "FPGA Based Hardware Implementation of Fault Detection for Microgrid Applications," *2017 International Symposium on Electronics and Smart Devices (ISESD)*, Yogyakarta, Indonesia, pp. 154–157.

M. Rathor and A. Sengupta, (2019) "Robust Logic Locking for Securing Reusable DSP Cores," *IEEE Access*, vol. 7, pp. 120052–120064, doi:10.1109/ACCESS.2019.2936401.

M. Rathor and A. Sengupta, (2020) "IP Core Steganography Using Switch Based Key-Driven Hash-Chaining and Encoding for Securing DSP Kernels Used in CE Systems," *IEEE Transactions on Consumer Electronics*, vol. 66, no. 3, pp. 251–260.

S. Rezgui, J.J. Wang, Y. Sun, B. Cronquist and J. McCollum, (2008) "Configuration and Routing Effects on the SET Propagation in Flash-based FPGAs," *IEEE Transactions on Nuclear Science*, vol. 55, no. 6, pp. 3328–3335.

M. Rostami, F. Koushanfar and R. Karri, (2014) "A Primer on Hardware Security: Models, Methods, and Metrics," *Proceedings of the IEEE*, vol. 102, no. 8, pp. 1283–1295.

J. A. Roy, F. Koushanfar and I. L. Markov, (2008) "EPIC: Ending Piracy of Integrated Circuits," *2008 Design, Automation and Test in Europe*, pp. 1069–1074.

S. Salivahanan and A. Vallavaraj, (2001) *Digital Signal Processing*. New Delhi: McGraw-Hill Education (India) Pvt Limited.

R. Schneiderman, (2010) "DSPs Evolving in Consumer Electronics Applications," *IEEE Signal Processing Magazine*, vol. 27, no. 3, pp. 6–10.

"Single Event Upsets," Intel [online]. Available: https://www.intel.com/content/www/us/en/support/programmable/support-resources/quality/seu.html, Jan. 2022.

A. Sengupta, (2016) "Intellectual Property Cores: Protection Designs for CE Products," *IEEE Consumer Electronics Magazine*, vol. 5, no. 1, pp. 83–88, doi:10.1109/MCE.2015.2484745.

A. Sengupta and R. Chaurasia, (2022) "Securing IP Cores for DSP Applications Using Structural Obfuscation and Chromosomal DNA Impression," *IEEE Access*, vol. 10, pp. 50903–50913, doi:10.1109/ACCESS.2022.3174349.

A. Sengupta and R. Chaurasia, (2023) "Securing Fault-Detectable CNN Hardware Accelerator against False Claim of IP Ownership Using Embedded Fingerprint as Countermeasure," *Proceedings of 9th IEEE International Symposium on Smart Electronic Systems (IEEE – iSES)*, India, 2023.

A. Sengupta and D. Kachave, (2017) "Low Cost Fault Tolerance against kc-cycle and km-unit Transient for Loop Based Control Data Flow Graphs during Physically Aware High Level Synthesis," *Elsevier Journal on Microelectronics Reliability*, vol. 74, pp. 88–99.

A. Sengupta and M. Rathor, (2020) "Securing Hardware Accelerators for CE Systems Using Biometric Fingerprinting," *IEEE Transactions on Very Large Scale Integration (VLSI) Systems*, doi:10.1109/TVLSI.2020.2999514.

A. Sengupta and M. Rathor (2021) "Structural Transformation and Obfuscation Frameworks for Data-intensive IPs," *"Secured Hardware Accelerators for DSP and Image Processing Applications."* London: IET.

A. Sengupta and M. Rathor, (2019) "IP Core Steganography for Protecting DSP Kernels Used in CE Systems," *IEEE Transactions on Consumer Electronics*, vol. 65, no. 4, pp. 506–515.

A. Sengupta and D. Roy, (2017) "Protecting an Intellectual Property Core during Architectural Synthesis Using High-Level Transformation Based Obfuscation," *IET Electronics Letters*, vol. 53, no. 13, pp. 849–851.

A. Sengupta and R. Sedaghat, (2011) “A High Level Synthesis Design Flow from ESL to RTL with Multi-parametric Optimization Objective,” *IETE Journal of Research*, vol. 57, no. 2, pp. 169–186.

A. Sengupta and R. Sedaghat, (2015) “Swarm Intelligence Driven Design Space Exploration of Optimal k-Cycle Transient Fault Secured Datapath during High Level Synthesis Based on User Power-Delay Budget,” *Elsevier Journal on Microelectronics Reliability*, vol. 55, no. 6, pp. 990–1004.

A. Sengupta, R. Chaurasia, and A. Anshul, (2023a) “Hardware Security of Digital Image Filter IP Cores against Piracy Using IP Seller’s Fingerprint Encrypted Amino Acid Biometric Sample,” *Proceedings of 8th Asian Hardware Oriented Security and Trust Symposium (AsianHOST)*, China, Sept. 2023.

A. Sengupta, R. Chaurasia and K. Bharath, (2023b) “Exploring Unified Biometrics with Encoded Dictionary for Hardware Security of Fault Secured IP Core Designs,” *Elsevier Journal Computers & Electrical Engineering*, vol. 111, Part A, 108928.

A. Sengupta, M. Rathor and R. Chaurasia, (2023c) “Biometrics for Hardware Security and Trust: Discussion and Analysis,” *IT Professional*, vol. 25, no. 4, pp. 36–44, doi:10.1109/MITP.2023.3277594.

A. Sengupta, R. Chaurasia and M. Rathor, (2023d) “HLS based Swarm Intelligence Driven Optimized Hardware IP Core for Linear Regression Based Machine Learning,” *IET Journal of Engineering*, vol. 2023, no. 8, e12299.

A. Sengupta, D. Kachave and D. Roy, (2019a) “Low Cost Functional Obfuscation of Reusable IP Cores Used in CE Hardware Through Robust Locking,” *IEEE Transactions on Computer Aided Design of Integrated Circuits & Systems (TCAD)*, vol. 38, no. 4, pp. 604–616.

A. Sengupta, E. R. Kumar and N. P. Chandra, (2019b) “Embedding Digital Signature Using Encrypted-Hashing for Protection of DSP Cores in CE,” *IEEE Transactions on Consumer Electronics*, vol. 65, no. 3, pp. 398–407.

A. Sengupta, S. P. Mohanty, F. Pescador and P. Corcoran, (2018) “Multi-Phase Obfuscation of Fault Secured DSP Designs with Enhanced Security Feature,” *IEEE Transactions on Consumer Electronics*, vol. 64, no. 3, pp. 356–364.

A. Sengupta, R. Chaurasia and Aditya Anshul, (2023) “Robust Security of Hardware Accelerators Using Protein Molecular Biometric Signature and Facial Biometric Encryption Key,” *IEEE Transactions on Very Large Scale Integration (VLSI) System*, vol. 31, no. 6, pp. 826–839.

A. Sengupta, D. Roy, S. P. Mohanty and P. Corcoran, (2017) “DSP Design Protection in CE Through Algorithmic Transformation Based Structural Obfuscation,” *IEEE Transactions on Consumer Electronics*, vol. 63, no. 4, pp. 467–476.

A. Sengupta and S. P. Mohanty, (2019) *IP Core Protection and Hardware-Assisted Security for Consumer Electronics*, London: The Institute of Engineering and Technology (IET).

A. Syed and R. M. Lourde, (2016) “Hardware Security Threats to DSP Applications in an IoT Network,” *2016 IEEE International Symposium on Nanoelectronic and Information Systems (iNIS)*, pp. 62–66.

T. Vateva-Gurova and N. Suri, (2018) “On the Detection of Side-Channel Attacks,” *2018 IEEE 23rd Pacific Rim International Symposium on Dependable Computing (PRDC)*, Taipei, Taiwan, pp. 185–186, doi:10.1109/PRDC.2018.00031.

A. Vijayakumar, V. C. Patil, D. E. Holcomb, C. Paar and S. Kundu, (2017) “Physical Design Obfuscation of Hardware: A Comprehensive Investigation of Device and Logic-Level Techniques,” *IEEE Transactions on Information Forensics and Security.*, vol. 12, no. 1, pp. 64–77.

X. Wang, Y. Zheng, A. Basak and S. Bhunia, (2015) “IIPS: Infrastructure IP for Secure SoC Design,” *IEEE Transactions on Computers*, vol. 64, no. 8, pp. 2226–2238.

Chapter 9

Hardware obfuscation-algorithmic transformation-based obfuscation for secure floorplan-driven high-level synthesis

Anirban Sengupta[1] *and Rahul Chaurasia*[1]

The chapter describes an integrated methodology that leverages algorithmic transformation-based obfuscation with crypto-watermarking for generating HLS-driven secure floorplan for loop-based IP designs (Sengupta and Kachave, 2017; Sengupta and Rathor, 2020; Sengupta *et al.*, 2017). The significance is that most of the modern electronics systems are integrated with complex loop-based applications to perform important functions such as filtering, convolution, image processing, etc. Loop-based applications can deliver efficient performance if designed using dedicated hardware intellectual property (IP) cores using high-level synthesis (HLS). However, globalization in the design supply chain introduces security vulnerabilities during IP design that need robust countermeasures. More explicitly, the chapter presents a discussion on the following: (a) methodology for generating HLS-driven secure floorplan for loop unrolled hardware IPs; (b) integration of algorithmic transformation-based obfuscation with crypto-watermarking during HLS to provide sturdy detective countermeasure against IP piracy and false IP ownership claim, while simultaneously reducing latency; (c) security-aware physical design-based HLS that is capable to convert a loop-based high-level code (representing a computation-intensive application) into its respective early secure floorplan that considers watermark embedding information of the datapath modules.

The rest of the chapter has been organized as follows: Section 9.1 provides the introduction of the chapter; Section 9.2 discusses prior similar methodologies; Section 9.3 provides details on methodology for algorithmic transformation-based obfuscation for secure floorplan-driven HLS; Section 9.4 provides a detailed analysis and discussion; and Section 9.5 concludes the chapter by summarizing the chapter's findings and implications.

[1]Department of Computer Science and Engineering, Indian Institute of Technology Indore, India

9.1 Introduction

Hardware IP designs are an indispensable part of any modern electronics devices and computing systems, where reliable and efficient functioning matters the most, not only for the end user but also for its stakeholders. Their demand and usage become crucial specifically for applications that require performing computationally intensive operations/tasks. This is because a dedicated reusable hardware IP design for performing specific tasks accelerates the computation of underlying applications and thereby of the system (Pilato *et al.*, 2018; Mohanty and Meher, 2016; Zhang, 2016; Castillo *et al.*, 2007; Sengupta and Chaurasia, 2022; Sengupta *et al.*, 2023). Hardware IP designs enable rapid prototyping, customization, and development of cutting-edge hardware solutions across industries. The significance of hardware IP designs lies not only in their functional capabilities but also in their scalability, portability, and interoperability (Koushanfar *et al.*, 2012; Potluri *et al.*, 2020; Mouris *et al.*, 2022; Chen *et al.*, 2020; Saeed *et al.*, 2019a).

Further, from an IP vendor's perspective, the security of IP design against external hardware threats cannot be overlooked (Saeed *et al.*, 2019b; Colombier and Bossuet, 2015; Chaurasia *et al.*, 2023; Sengupta and Chaurasia, 2023; Sengupta *et al.*, 2023b; Chaurasia and Sengupta, 2021). This is because, in the global integrated circuit (IC) supply chain, due to the involvement of several untrustworthy entities such as IP brokers, untrustworthy design and integration houses, etc. This involvement of untrustworthy entities during IC design cycle may lead the design susceptible to hardware threats (Wang *et al.*, 2015; Arafin *et al.*, 2017; Hroub and Elrabaa, 2022; Chaurasia *et al.*, 2022). An efficient design chain may minimize the involvement of external entities but cannot be completely independent from them. This is because, it is not feasible to design and develop a complete IC under a single house (and in a specific geographic location), due to factors such as design complexity, time to market, and process turnaround time. Thus, it involves untrustworthy design houses such as system-on-chip (SoC) house where integration of several IPs is performed to meet the time-to-market pressure while minimizing the cost overhead. This exposes the incoming IP designs to external hardware threats such as IP piracy (Koushanfar *et al.*, 2005; Sitjongsataporn *et al.*, 2021; Rajendran *et al.*, 2016; Rizo *et al.*, 2022; Wang *et al.*, 2018; Rai *et al.*, 2019; Karmakar and Chattopadhyay, 2020; Le Gal and Bossuet, 2012; Pilato *et al.*, 2019; Castillo *et al.*, 2008; Regazzoni *et al.*, 2021; Ray and Roy, 2020; Rathor and Sengupta, 2020; Chaurasia and Sengupta, 2023; Sengupta *et al.*, 2019; Chaurasia and Sengupta, 2022a; Sengupta *et al.*, 2023). Therefore, it requires that the IP designers develop reusable hardware IPs that are secure against such threats. Further, an IP designer needs to take care of the design functionality and overhead while aiming to enhance the security of the design. A security framework is immensely significant in validating the genuineness of the design. An illegitimate (pirated) design could expose an IP core to various vulnerabilities. Hence, incorporating a resilient secret security mark within the design covertly can serve as a vital tool to identify

IP piracy during the detection process, providing a robust detective countermeasure (Chaurasia and Sengupta 2022a; Chaurasia and Sengupta 2022b; Chaurasia and Sengupta 2022c; Sengupta and Chaurasia, 2022; Sengupta *et al.*, 2021; Sengupta and Chaurasia, 2023; Sengupta *et al.*, 2023; Sengupta and Chaurasia, 2023; Anshul *et al.*, 2023).

The approach in this chapter (Sengupta and Rathor, 2020; Sengupta *et al.*, 2017) presents crypto-watermark and algorithmic transformation-based obfuscation to ensure the generation of high-level synthesis-driven secure floorplan design (generating based on secure datapath modules embedded with secret security constraints and placing them in the form of enveloping rectangle) for loop-based high-level applications. Finite impulse response (FIR) filter and differential equation solver serve as quintessential examples of pivotal application frameworks within digital signal processing and scientific computing, embodying critical algorithms and methodologies. Thus, ensuring the protection of such designs becomes paramount in an environment where external hardware threats could yield substantial adverse repercussions. The approach can be applied to any loop-based hardware IP design.

9.1.1 Threat model

In the IC supply chain, IP cores are designed and supplied by third-party IP vendors. However, a potential adversary in the untrustworthy SoC integrator house may attempt to perform IP piracy and fraudulent claim of IP ownership. The methodology (Sengupta and Rathor, 2020; Sengupta *et al.*, 2017) considers an adversary in the SoC integrator house as a potential attacker, while the IP vendor as the defender. Protection against illegal IP piracy is crucial as it may lead to the following consequences of (a) damaging the brand reputation of the original IP vendor if pirated products, lacking stringent quality checks, are circulated in the market, (b) causing financial losses for the original IP vendor, (c) potentially tampered pirated IPs, containing malicious elements may lead to security hazards. Therefore, it becomes crucial to ensure a detective countermeasure of IP piracy from the IP vendor's perspective.

9.1.2 Salient features of the chapter

The salient features of the chapter, comprising the discussion and analysis on the significance of integrated methodology leveraging algorithmic transformation-based obfuscation with crypto-watermark for secure floorplan-driven HLS, are the following:

- This chapter discusses a robust design methodology for generating secure floorplan design for loop-based hardware IPs using HLS (Sengupta and Rathor, 2020; Sengupta *et al.*, 2017).
- This chapter explains crypto-watermarking methodology using HLS framework for embedding covert security constraints in loop-based IP designs. It enables sturdy detective countermeasures against IP piracy, in the context of loop-based hardware IPs (Chaurasia *et al.*, 2022; Dolev and Yao, 1983; Sengupta *et al.*, 2019).

- The discussed secure hardware IP design methodology attains higher robustness in terms of lower probability of coincidence, higher tamper tolerance, and stronger entropy, at zero floorplan area overhead.

9.2 Discussion on state-of-the-art approaches

In recent years, ensuring the security of hardware IP cores has become a significant concern within the scientific community. The available approaches for securing hardware IP cores are IP watermarking (Wang *et al.*, 2018; Rai *et al.*, 2019; Karmakar and Chattopadhyay, 2020; Le Gal and Bossuet, 2012; Pilato *et al.*, 2019), automatic signature insertion (Castillo *et al.*, 2008), and steganography-based (Rathor and Sengupta, 2020). Watermarking approach (Wang *et al.*, 2018) presented a circuit watermarking scheme aimed at thwarting overbuilding and piracy. It discusses polymorphic gates, which adapt their functionality based on environmental changes, making them suitable for circuit watermarking. It identifies four such gates via an evolutionary method and proposes a watermarking scheme using these gates to replace standard logic gates selectively. However, it results in a significant overhead in the design area, even when embedding a 30-bit watermark. Further, Rai *et al.* (2019) discussed two watermarking methods designed to counter integrated circuit overbuilding and intellectual property piracy. These methods utilize encoding through polymorphic inverter designs, leveraging reconfigurable nanowire technologies. Through a unique nanowire-based fabrication process, nodes within the logic network can be permanently set to either 0 or 1. This strategic manipulation of nodes drives polymorphic inverters in predetermined ways, contributing to the watermark. In summary, this work explores harnessing the distinct qualities of reconfigurable emerging nanotechnologies for hardware security by employing an encoding scheme to embed the designer's signature within an IC. However, it yields area overhead even for a 64-bit watermark signature. Moreover, Karmakar and Chattopadhyay (2020) presented a cellular automata-based finite state machine (FSM) watermarking strategy to detect the potential theft of designers' IPs by an adversary. However, none of these approaches target the security of loop-based hardware IP designs, unlike the methodology discussed in this chapter (Sengupta and Rathor, 2020; Sengupta *et al.*, 2017). Further, the approaches (Wang *et al.*, 2018; Rai *et al.*, 2019; Karmakar and Chattopadhyay, 2020) do not exploit high-level synthesis of IC design in order to integrate watermark for generating a secured design version.

The approach by Le Gal and Bossuet (2012) also attempts to address security issues related to the reuse of IP, such as illegal copying and counterfeiting, by proposing an IP watermarking method integrated into HLS. This approach involves encoding the IP watermark via mathematical sub-marks on design output ports, automatically inserted during synthesis. However, it also incurs design overhead. Further, Pilato *et al.* (2019) introduced a method to incorporate IP watermarking using benevolent Hardware Trojans (HTs). Benevolent HTs, akin to malicious HTs, function as IP watermarks, aiming to safeguard against piracy and counterfeiting throughout the design process or in deployed integrated circuits. These HTs are seamlessly integrated into the IP component during HLS. It employs two triggers: one based on an external pin linked to configuration registers and scan chains, and another based on a predefined input string. However, it also incurs resource overhead for additional Look-Up Tables (LUTs) and Flip-Flops (FFs). Castillo *et al.* (2008) introduced a watermarking method, incorporating digital signature bits within the hardware description language (HDL) design, leveraging system resources. This technique secures the associated watermark signature through the use of message-digest (MD)-5 and the secure hashing algorithm (SHA)-1, yet it leads to increased design area overhead. Indeed, it relies on mathematical relationships between input and output numeric values at specific instances. Mainly, it focuses on automatic IP protection within HLS tools by encoding the watermark through mathematical sub-marks on the output ports of the design. These sub-marks are strategically selected and inserted during synthesis to ensure detection and removal of the watermark. However, the approaches by Le Gal and Bossuet (2012), Pilato *et al.* (2019), and Castillo *et al.* (2008) employ circuit watermarking during HLS but do not target to generate secured design versions for loop-based IP designs.

Furthermore, Regazzoni *et al.* (2021) discuss the significance of safeguarding AI models from piracy, including cloning, unauthorized distribution, and usage. the emphasize the importance of protecting intellectual property (IP) in AI models. Further, Ray and Roy (2020) presented a watermarking for safeguarding multimedia data during interactive digital transmission. Additionally, Rathor and Sengupta (2020) presented a hardware steganography approach to identify pirated versions of digital signal processing (DSP) IPs. This method involves embedding a covert steganographic mark within the DSP design. Moreover, the amount of hidden digital evidence intended for insertion can be fully controlled by the designer using a "thresholding" parameter. The intricate procedure of establishing stego-constraints via a secret stego-key enhances its resilience in comparison to watermarking. However, if an adversary obtains information about the stego-key or the entropy threshold parameter, it could compromise security, enabling evasion during IP piracy detection. In summary, none of the approaches (Wang *et al.*, 2018; Rai *et al.*, 2019; Karmakar and Chattopadhyay, 2020; Le Gal and Bossuet, 2012; Pilato *et al.*, 2019; Castillo *et al.*, 2008; Regazzoni *et al.*, 2021; Ray and Roy, 2020; Rathor and Sengupta, 2020) targeted the security of loop-based IP designs during HLS while addressing reduction in design latency through structural obfuscation and security through covertly embedding the IP vendor crypto-watermark, unlike the methodology discussed in this chapter (Sengupta and Kachave, 2017; Sengupta and Rathor, 2020; Chaurasia *et al.*, 2022). Chaurasia and Sengupta (2023) introduced a security approach for protecting fault-detectable IP designs from piracy in consumer electronics systems. However, none of the prior approaches employ integrated obfuscation and crypto-watermarking for robust security against IP piracy. Further, none of the prior approaches are capable of exploiting the security constraints-embedded datapath modules to generate secure floorplan design corresponding to loop-based high-level applications at zero floorplan area overhead. This is achieved by performing the local alteration of design storage variables during register allocation corresponding to the IP vendor-generated security constraints, during HLS.

9.3 Methodology for algorithmic transformation-based obfuscation for secure floorplan-driven high-level synthesis

So far, we have discussed and analyzed the effectiveness of contemporary hardware security techniques. Now, in this section, we discuss the methodology that leverages algorithmic transformation-based obfuscation with crypto-watermarking for generating HLS-driven secure floorplan for loop-based IP designs (Sengupta and Rathor, 2020; Sengupta *et al.*, 2017).

9.3.1 Overview

The overview of the discussed hardware security methodology that presents algorithmic transformation-based obfuscation and crypto-watermarking for HLS-driven secure floorplan of loop-based IP designs is presented in Figure 9.1. The methodology discussed in this chapter (Sengupta and Rathor, 2020; Sengupta *et al.*, 2017) accepts the following primary inputs: input application in the form of high-level code, IP vendor specified key-set, module library, resource constraints, scheduling algorithm, and encryption key in order to generate secure floorplan (watermark embedded). As shown in Figure 9.1, the methodology comprises three major modules – module-1: algorithmic transformation-based obfuscation, which is responsible for transforming the data flow graph of the design into an optimized design (with lower latency); module-2: crypto-watermarking, which is responsible for generating secret hardware security constraints using encrypted digital signature of the IP vendor; module-3: secure floorplan module using HLS, which accepts transformed design post transformations (from module-1) and encoded hardware

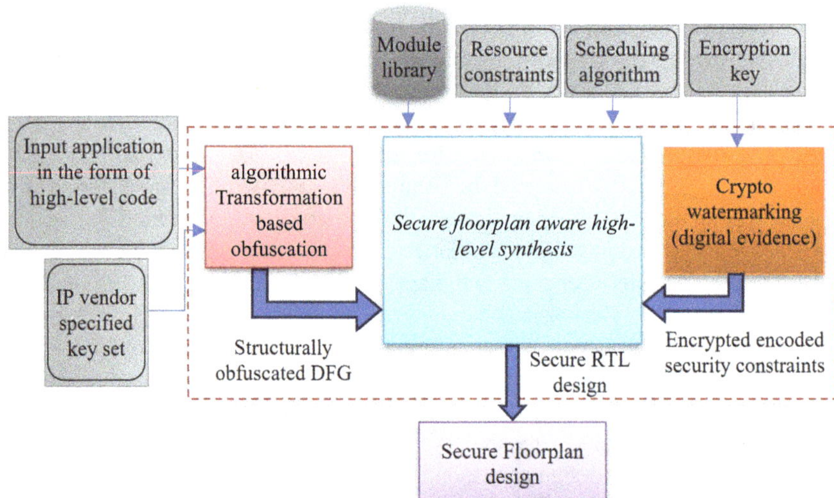

Figure 9.1 Overview of the approach (Sengupta and Rathor, 2020; Sengupta et al., 2017)

security constraints (from module-2) as its inputs and is responsible for generating secure floorplan design corresponding to input loop based high-level application which is obtained using high-level synthesis. The details of each module of the hardware security methodology are discussed in the next section.

9.3.2 Details of methodology

The detailed design flow of the discussed hardware security methodology (Sengupta and Kachave, 2017; Sengupta and Rathor, 2020) for generating high-level synthesis-driven secure floorplan using algorithmic transformation-based obfuscation and crypto-watermarking is depicted in Figure 9.2.

9.3.2.1 Algorithmic transformation-based obfuscation module

The first module structurally transforms the generated data flow graph (DFG) corresponding to high-level code of input application using IP vendor-specified algorithmic transformations for structural obfuscation. The methodology employs three algorithmic transformations-based obfuscation such as: (1) loop unrolling (2) DFG partitioning, and (3) tree-height transformation (THT).

1. ***Loop-unrolling***: Loop unrolling-based transformation, unrolls the loop body of generated DFG design of input application using IP designer/vendor specified value of unrolling factor (*UF*), which acts as the first key K1. For example, if the $UF = 15$, then it is supposed to unroll the loop body 15 times. The size of key-1 is determined by the *ceiling function* ($\{\log_2(UF^{\max})\}$ bit size). "$UF^{\max}$" represents the maximum possible value unrolling factor, equivalent to the maximum iteration count (I) within the loop (e.g., in a 150-tap FIR filter design, $I = 150$). The loop unrolling transforms the design structure (data flow graph) into an optimized latency version that consumes fewer number of control steps to complete the entire iteration count. Loop unrolling is an effective latency reduction strategy for loop-based applications. The employed generic loop-unrolling delay model is shown below:

   Case 1: when $UF = 1$, indicating no unrolling (for instance, pre-algorithmic transformation).

$$\begin{aligned}\text{Total no. of control steps } (CS_T) &= \text{no of CS required to execute loop body one time } (CS_{ot}) \\ &\quad * \text{ no of duplicate loop body iterations } (X) \end{aligned} \tag{9.1}$$

   where no. of duplicate loop body iterations

$$(X) = \left(\frac{\text{loop count}}{\text{unrolling factor}}\right)^{\text{quotient}} = \left(\frac{I}{UF}\right)^{\text{quotient}} \tag{9.2}$$

   Inferring from (9.1) and (9.2)

$$CS_T = (CS_{ot} * X) = \left(CS_{ot} * \left[\frac{I}{UF}\right]^{\text{quotient}}\right)$$

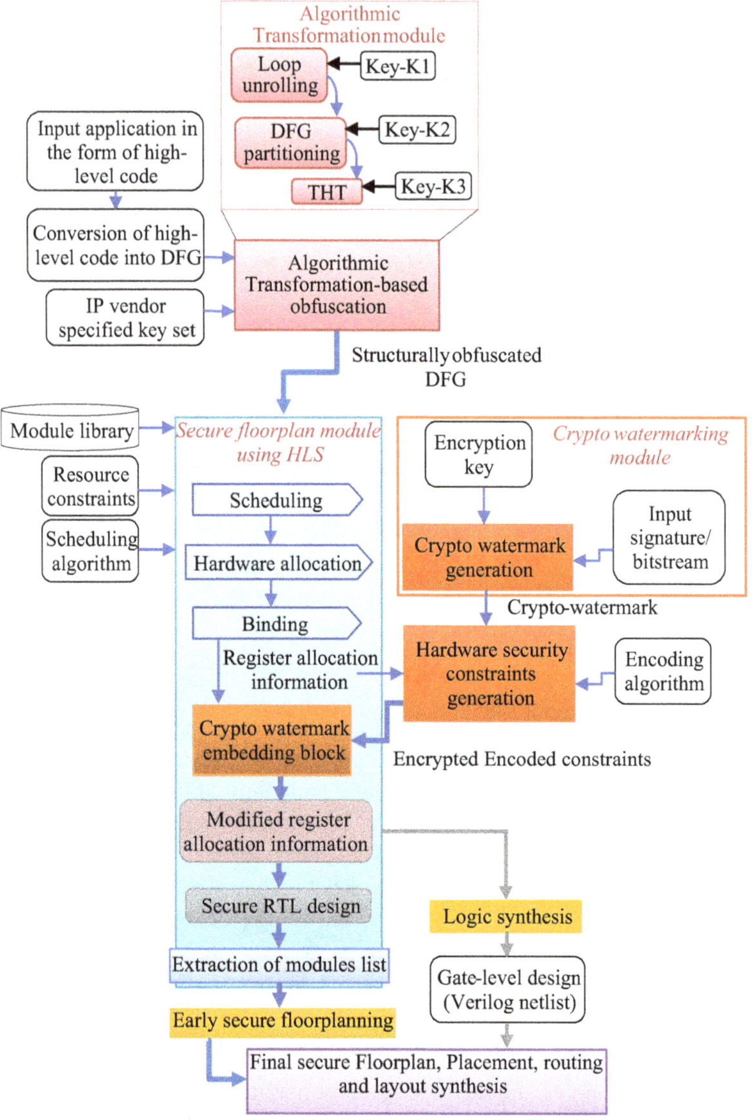

Figure 9.2 Design flow of the algorithmic transformation-based obfuscation with crypto-watermarking for secure floorplan-aware HLS

$$CS_{ot} = CS_f \quad and \quad UF = 1; \quad \text{So}, CS_T = \left(CS_f * I\right)$$

where, CS_f implies the no. of CS required for executing the first loop iteration.

Case 2: when $UF > 1$, indicating loop unrolling (considering UF evenly divides the loop count), e.g., $I \bmod UF = 0$.

No. of CS required in executing loop body one time = no. of *CS* required for executing the first loop iteration (CS_f) + (UF − 1) * no. of *CS* required between initiation of consecutive loop iterations

$$CS_{ot} = \left(CS_f + (UF - 1) * C_{II}\right) \tag{9.3}$$

$$CS_T = C_{ot} * X \tag{9.4}$$

$$CS_T = \left(C_f + (UF - 1) * C_{II}\right) * X \tag{9.5}$$

$$CS_T = \left(CS_f + (UF - 1) * C_{II}\right) * \left(\frac{I}{UF}\right)^{\text{Ceiling}} \tag{9.6}$$

Case 3: when $UF > 1$, indicating loop unrolling (considering UF unevenly divides the loop count), e.g., $I \bmod UF \neq 0$.

$$\text{CS}_T = \text{no. of pipe lined } CS + \text{no. of sequential } CS \tag{9.7}$$

where pipelined control steps are represented by (9.5) and sequential *CS* = no. of iteration executed sequentially * CS_f

$$\text{Sequential CS} = (I \bmod UF) * CS_f \tag{9.8}$$

Now, in order to compute CS_T substitute the values from (9.5) and (9.8) into (9.7)

$$CS_T = \left(CS_f + (UF - 1) * C_{II}\right) * X + (I \bmod UF) * CS_f \tag{9.9}$$

Equation (9.9) represents the total no. of *CS* without performing loop unrolling every time. Unrolling of loop requires only one to find the value of CS_f and C_{II}.

2. ***DFG partitioning***: It accepts the loop-unrolled design obtained using loop-unrolling and serves as the second stage of obfuscation. DFG partitioning further transforms the design by partitioning an unrolled DFG by applying a designer/vendor-specified number of cuts into the design out of the total possible cuts, which acts as the second key K2. For example, applying "n" cuts divides the unrolled DFG into a total of "n + 1" partitions. For example, considering an IP vendor decides to apply four cuts into the DFG design, then it will lead to five partitions. However, cuts are applied in such a way that each resulting partition of DFG design (unrolled) contains at least two nodes/operations with a direct dependency. The size of key-2 is determined using the following *ceiling function* $\{log_2(\text{Max. cut})\}$ bit size, where Max. cut signifies the maximum allowable cuts in the DFG design. Partitions are incorporated into the DFG, such that separate tree-height transformations can be employed in separate partitions.
3. ***Tree-height transformation***: It accepts each of the formed design partitions individually as input and serves as the third stage of obfuscation. THT further transforms the design by modifying the data dependency of specific operations by executing them as parallel sub-computations based on the IP designer-specified key, which acts as the third key K3. In THT, the dependency of

eligible operations is altered in such a way that it does not impact the final output. The size of key-K3 is determined by the *ceiling function* $\{log_2(\text{Max\_THT})\}$ bit size, where Max_THT represents the maximum number of partitions where the THT technique can be applied. For example, if the design comprises five partitions, then the IP designer/vendor can choose to employ THT in any of the partitions up to five. The impact of THT is also the reduction of latency of the final design.

9.3.2.2 Crypto-watermarking module

Crypto-watermarking module accepts the IP vendor-specified encryption key and input watermark signature (bitstream) in order to generate crypto-watermark signature. The details of the crypto-watermark generation module are presented in Figure 9.3. As depicted, firstly the input bitstream is fed as input to the secure hashing algorithm (SHA-512) computation block, where it involves round function computation for 80 iterations in order to generate well distributed (non-linkable) bitstream digest. Next, the generated binarized hashed bitstream undergoes division into blocks, with the vendor selecting their size, typically 128 bits or more, and subsequently each resulting data block is translated into their respective decimal equivalents. Thereafter, in the next block, encryption of each decimal value is performed through Rivest-Samir-Adleman (RSA) encryption using an IP vendor-specified private key (Sengupta *et al.*, 2019; Dolev and Yao, 1983). The RSA encryption block results in encrypted decimal output which is subsequently transformed into the encrypted binary bitstream. This resulting encrypted-hashed binary bitstream is referred to as the crypto-watermark signature which is further exploited

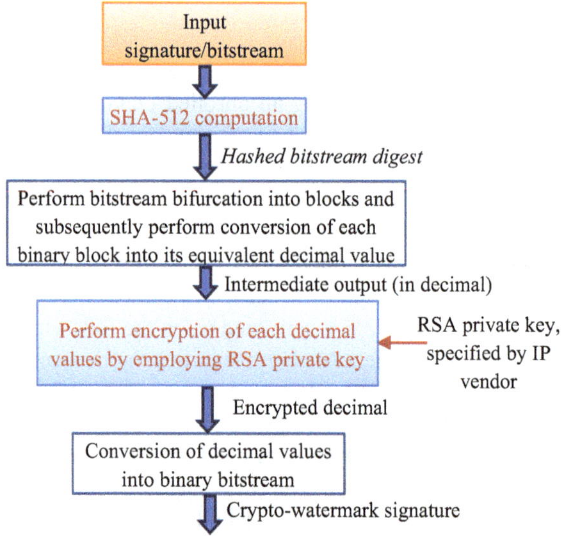

Figure 9.3 Crypto-watermark signature generation

for covertly embedding into the design in the form of encoded hardware security constraints (secret digital evidence) generated using an IP-vendor-specified encoding algorithm. For the sake of brevity, let us consider the generated 128-bit size crypto-watermark signature based on the assumed IP vendor private key "d = 99861536332283195678724683425637018 19," for embedding into the design as shown below:

‘1000001000011111111011010000111110001001101110111000000
0001100111011101000011011110001100100110010001000000001
1010111010111100010’

(9.10)

The details of generating hardware security constraints corresponding to crypto watermark of IP vendor selected signature strength are discussed in the next section.

9.3.2.3 HLS-driven secure floorplan

This module of the discussed methodology generates an HLS-driven secure floorplan design corresponding to the input loop-based high-level application. As shown in Figure 9.2, a *secure floorplan module using HLS* accepts the following inputs such as transformed design post-transformations (from module-1), module library, resource constraints, scheduling algorithm, and encrypted encoded hardware security constraints (extracted from module-2) and generates secure floorplan design which is obtained using high-level synthesis register-transfer level (RTL) datapath output. In this module, firstly the optimized design (using algorithmic transformation) is scheduled using a scheduling algorithm and the designer selected resource constraints/design space exploration. The primary goal of scheduling is to minimize the completion time of a set of tasks by scheduling them onto the available resources efficiently. A LIST scheduling algorithm has been employed for the same. This approach involves scheduling the highest possible number of operations within a single control step, taking into account both resource constraints and data dependencies. Subsequently, the hardware resources are allocated to the scheduled design and their binding is performed. Next, the scheduled design (post resource allocation and binding) is exploited for extracting the register allocation information. The register allocation information comprises the details of the number of registers (storage elements) required for accommodating the design storage variables (holding intermediate design inputs and output) and control steps.

Hardware security constraints generation block: As shown in Figure 9.2, the *hardware security constraints generation* block accepts the following inputs: crypto-watermark signature (shown in 9.10), encoding algorithm specified by IP vendor and register allocation information of the scheduled transformed design.

The details of bitstream encoding algorithm are as follows:

- For bit "0"- encode the storage variables of the design into pairs of even-even storage variables, $<S^i, S^j>$, where i and j indicate even integer value indices in case of sorted storage variable list in ascending order.

- For bit "1"- encode the storage variables of the design into pairs of odd-odd storage variables, <S^i, S^j>, where i and j indicate odd integer value indices in case of sorted storage variable list in ascending order. *Note: IP vendor can perform the encoding in several possible ways.*

Further, the list of storage variables can be sorted in either of the sorted ascending order, sorted descending order, random order, etc. Thus, the encoded hardware security constraints are generated.

Crypto-watermark embedding block: In this block the generated hardware security constraints are fed as input for covertly embedding into the design during the register allocation phase of HLS. Moreover, the generated initial register allocation information is also fed as an input. Using both the inputs, the modified register allocation information (with embedded security constraints of the crypto watermark) is obtained, which is used to generate a crypto-watermark embedded scheduled design corresponding to an input loop-based application. It is further synthesized using HLS flow for generating RTL datapath. Subsequently, the generated datapath information (which yields a collection of RTL modules that encompasses functional units (FUs), interconnecting hardware, and registers used for storing the intermediate input/outputs) is used to produce an early floorplan comprising the security constraints. These RTL modules are organized in descending order based on their size, and in cases of multiple instances of the same module, they are sequenced in ascending order according to their instance number. Utilizing this ordered list, the floorplanning of RTL modules is generated by traversing the list from left to right. Each module is sequentially selected from the list and positioned to expand the floorplan diagonally in the form of enveloping rectangles. Consequently, it results in the creation of an early secure floorplan design tailored for any input loop-based high-level application. Since the early floorplan is created from the RTL modules of the secure IP datapath, therefore the generated early floorplan is a physical design manifestation of the security constraints (crypto-watermark) embedded IP datapath. The early floorplan generated therefore reflects the information of the secured datapath IP at RTL. Henceforth the generated early floorplan is termed as "*secure floorplan*". This is achieved through the employing of algorithmic transformation-based obfuscation and the integrating crypto-watermark signature during HLS. Next, the subsequent physical design phases (final floorplan, placement, routing and layout synthesis, etc.) with the aid of generated early secure floorplan can be obtained.

9.3.3 Demonstration of the methodology

The discussed hardware security methodology for generating high-level synthesis-driven secure floorplan using algorithmic transformation-based obfuscation and crypto-watermarking is demonstrated through FIR filter design and differential equation solver. FIR filters are valued for their stability and linear phase response and are used in a wide range of applications across various domains such as signal processing (for noise reduction, signal equalization, and signal separation), audio processing (for audio equalization, echo cancellation), biomedical signal processing (for filtering out noise from electrocardiogram or electroencephalogram signals), communication and control systems for channel equalization, signal separation and filtering out noise and unwanted disturbances from sensor signals respectively. The equation of 150-tap FIR filter is as follows (Mohanty and Meher, 2016):

$$z[n] = \sum_{i=1}^{150} h[i] * x[n - i] \tag{9.11}$$

where $z[n]$ denotes the output of the FIR filter, $h[i]$ and $x[n - i]$ denotes the constants and input impulses, respectively. The variable "i" ranges from 1 to 150. Figure 9.4 portrays the DFG illustrating the FIR application presented by (9.11).

Further, the function for differential equation solver is provided as follows (Balachandran, 2010):

diffeq {

read (x; y; u; dx);

repeat

ul = u − (3. x. u. dx) − (3.y.dx)

yl = y + u.dx;

u = ul; y = yl;

until(**I**)

} (9.12)

The DFG illustrating the algorithmic representation of differential equation solver, as presented by (9.12), is shown in Figure 9.6.

The process of implementing the methodology for generating high-level synthesis-driven secure floorplan mechanism unfolds in several steps. Steps (1)–(3) elucidate the application of the suggested transformation, steps (4)–(8) demonstrate the embedding of the crypto-watermark into the transformed scheduled design covertly and the generation of secured floorplan:

Step (1) The loop body within the FIR filter application is expanded according to key-K1. For 150 tap FIR filter design, with an iteration count of 150, the maximum $UF^{\max}$ value becomes 150. Consequently, the size of key-K1 is 8 bits computed using *ceiling function* $\{\log_2(UF^{\max})\}$ =*ceiling function* $\{\log_2(150)\}$ bits. Assuming the designer's chosen key is "00001111" denoting UF = 15, the

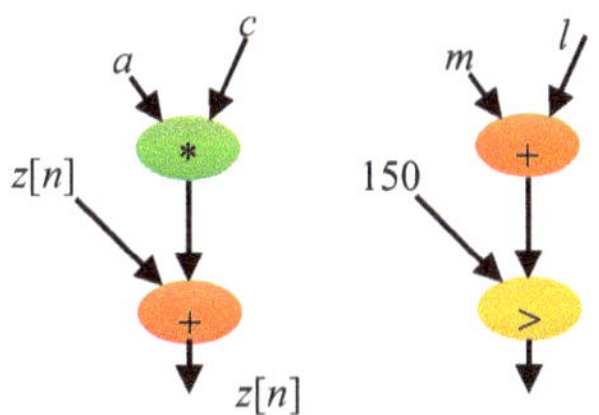

Figure 9.4 DFG corresponding to 150-tap FIR filter design

application's unrolled form using UF = 15 is depicted in Figure 9.5. This renders the initial DFG of FIR filter into loop unrolled design based on key-K1. Similarly, the twice unrolled DFG corresponding to differential equation solver using key K1="0010" is depicted in Figure 9.6.

Step (2) Following the unrolled FIR filter application (as depicted in Figure 9.5), the next step involves DFG partitioning based on key-K2. As the maximum allowable cuts are 14, the size of key-K2 is 4 bits, as computed using *ceiling function* $\{\log_2(\text{Max. cut})\}$= *ceiling function* $\{\log_2(14)\}$ bits. Let's assume the designer's chosen key is "0100" indicating 4 cuts. This division dissects the unrolled FIR filter's DFG into 4 + 1 = 5 partitions (P_1–P_5), depicted in Figure 9.5. This renders the loop unrolled design into a loop unrolled portioned design based on key-K2. Similarly, the DFG design for differential equation solver by applying IP designer chosen K2= "01" is generated as shown in Figure 9.6.

Step (3) The subsequent step involves applying THT within each partition, guided by key-K3. As THT is applicable to a maximum of 5 partitions, the size of

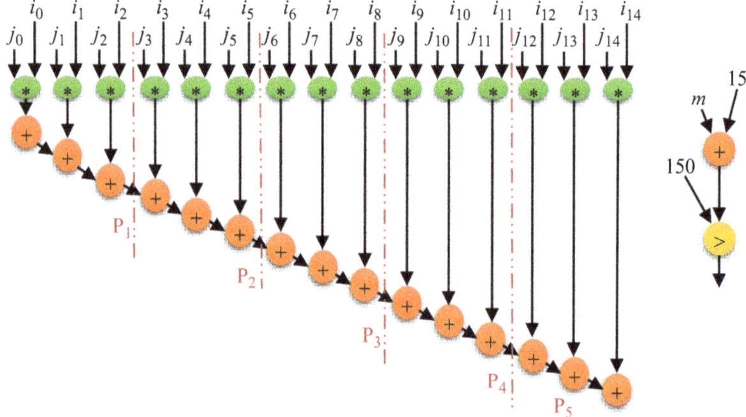

Figure 9.5 Loop unrolled FIR filter design DFG with UF = 15

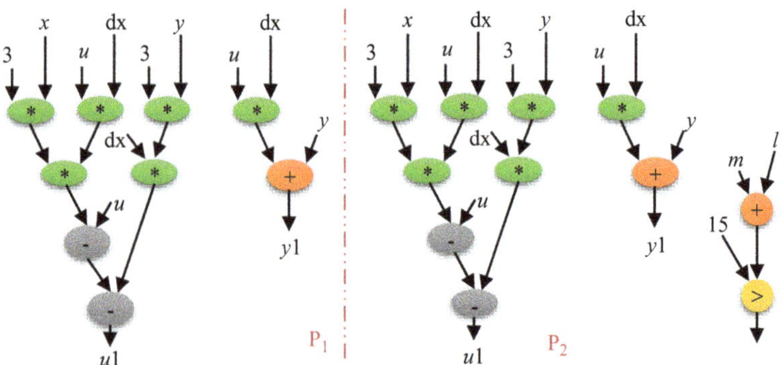

Figure 9.6 DFG corresponding to twice unrolled differential equation solver

key-K3 is 3bits, computed using *ceiling function* {$\log_2$(Max_THT)} = *ceiling function* {$\log_2(5)$} bits. For the sake of demonstration, assume that the designer's chosen key K3 is "101", indicating the application of THT in all five partitions. Similarly, all the transformations are applied to the differential equation solver design.

Step (4) Post performing all the transformations into the DFG design through keys (K1-K3), all DFG partitions are scheduled concurrently using designer-specified resource constraints. For example, the scheduling of the FIR filter design is performed using three multipliers (*) and one adder (+), as IP vendor resource constraints. Subsequently, hardware resources are allocated to the scheduled design and their binding is performed as shown in Figure 9.7. As evident, the scheduled design consumes 17 control steps (C_0 to C_{16}). Further, required storage elements (registers) are indicated using different colors (31 registers required for storing the intermediate input/outputs) of 61 storage variables indicated using (S^0 to S^{60}). Similarly, the obtained scheduled design corresponding to differential equation solver is shown in Figure 9.8. Next, the initial register allocation information corresponding to the scheduled FIR filter design is extracted, which is subsequently exploited for covertly embedding the hardware security constraints. Similarly, the initial register allocation information corresponding to the transformed scheduled differential equation solver design is also obtained.

Step (5) The generated hardware security constraints corresponding to the generated crypto-watermark (as shown in (9.10)) by employing the IP vendor-specified encoding algorithm (discussed earlier) are as follows:

As the IP vendor chose a 128-bit crypto-watermark (for the sake of demonstration) for embedding, composed of 67 0's and 61 1's, therefore, the generated security constraints using the encoding algorithm (explained earlier in Section 9.3.2.3 corresponding to "0" are:

<S^0–S^2>, <S^0–S^4>, <S^0–S^6>, <S^0–S^8>, <S^0–S^{10}>, <S^0–S^{12}>, <S^0–S^{14}>, <S^0–S^{16}>, <S^0–S^{18}>, <S^0–S^{20}>, <S^0–S^{22}>, <S^0–S^{24}>, <S^0–S^{26}>, <S^0–S^{28}>,

………….., <S^0–S^{60}>, <S^2–S^4>, <S^2–S^6>, <S^2–S^8>, ………., <S^2–S^{58}>, <S^2–S^{60}>, <S^4–S^6>, <S^4–S^8>, <S^4–S^{10}>, <S^4–S^{12}>, <S^4–S^{14}>, <S^4–S^{16}>, <S^4–S^{18}>, <S^4–S^{20}>.

Further, the generated security constraints corresponding to "1" are:

<S^1–S^3>, <S^1–S^5>, <S^1–S^7>, ……. <S^1–S^{51}>, <S^1–S^{53}>, <S^1–S^{55}>, <S^1–S^{57}>, <S^1–S^{59}>, <S^3–S^5>, <S^3–S^7>, <S^3–S^9>, ……….., <S^3–S^{59}>, <S^5–S^7>, <S^5–S^9>, <S^5–S^{11}>, <S^5–S^{13}>.

Step (6) Next, these generated hardware security constraints (corresponding to crypto-watermark) are covertly embedded into the initial register allocation table (extracted from the transformed scheduled design). The embedding is done by performing local alteration in such a way that no two storage variable pairs form any generated security constraints pair are accommodated into the same register (color). This is ensured to satisfy the generated security constraints. The security constraints embedded (locally altered) register allocation information corresponding to the

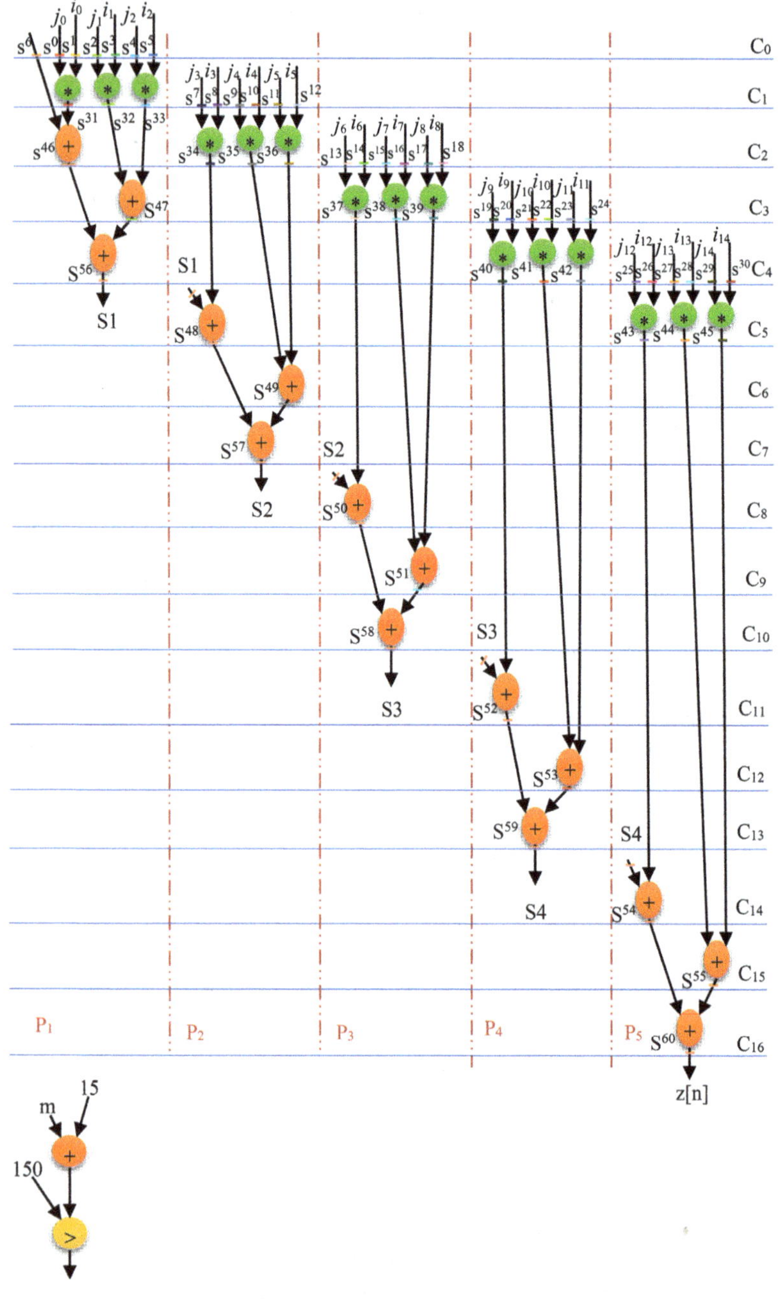

Figure 9.7 Scheduled FIR filter design

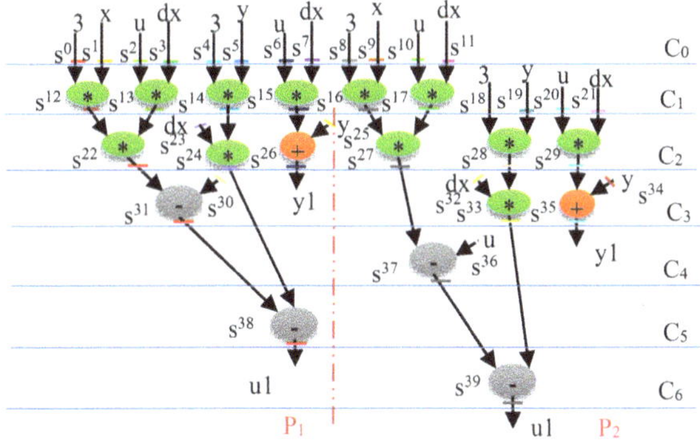

Figure 9.8 Scheduled design for differential equation solver

scheduled FIR filter design is shown in Figure 9.9. In Figure 9.9, the storage variables marked in red represent the altered position of storage variables post-embedding, and the variables marked in blue represent their association pre-embedding the security constraints. As evident from Figure 9.9, no extra registers were required to accommodate the entire list of generated encoded hardware security constraints. However, in case if it is not possible to accommodate any of the security constraints from the list, in the given available registers, then new registers are allocated for accommodating the corresponding storage variable constraint pair. Similarly, the embedding of all the hardware security constraints can be performed. Thus, the locally altered (modified) register allocation information for the FIR filter is obtained post-embedding the crypto-watermark-driven hardware security constraints during the register allocation phase of HLS. Similarly, the register allocation information for differential equation solver application can be obtained.

Step (7) Next, the obtained security constraints (crypto-watermark driven) embedded scheduled design is subjected to HLS for generating RTL datapath IP.

Step (8) Next, the RTL datapath modules that encompass functional units, interconnecting hardware and registers, are extracted and an early secure floorplan (comprising of the security constraints embedded resource list) is generated. Next, the ordered lists of datapath modules are prepared (as per the rules discussed earlier in Section 9.3.2.3). For example, the ordered lists of datapath modules (post-embedding the generated security constraints) corresponding to FIR filter design are as follows:

{d1, x1, (1:32, 32:1), d2, d3, d4, d5, d6, d7, (1:16), x2, x3, x4, x5, x6, x7, x8 (16:1), d8, d9, d10, d11, d12, d13, d14, d15, d16, (1:8), M1, M2, M3, x9, 10, x11, x12, x13, x14, x15, x16, x17, x18, x19, x20,(8:1), C, d17, d18, d19, (1:4), x21, x22, x23, (4:1), A1, R1,R2, R3, R4, R5, R6, R7, R8, R9, R10, R11, R12, R13, R14, R15, R16, R17, R18, R19, R20, R21, R22, R23, R24, R25, R26, R27, R28, R29, R30, R31, d20, d21, (1:2), x24, x25 (2:1)}.

In this context, "d" represents a Demux, "x" represents a Mux, "M" represents a multiplier, "C" represents a comparator, "A" represents an adder, and R

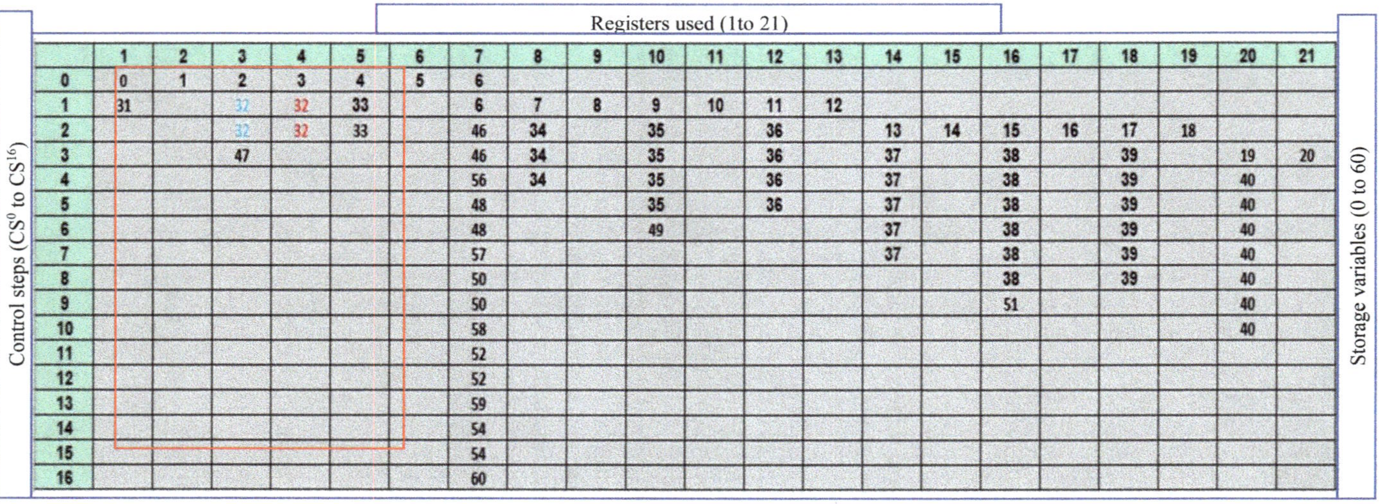

	1	2	3	4	5	6	7	8	9	10	11	12	13	14	15	16	17	18	19	20	21
0	0	1	2	3	4	5	6														
1	31		32	32	33		6	7	8	9	10	11	12								
2			32	32	33		46	34		35		36		13	14	15	16	17	18		
3			47				46	34		35		36		37		38		39		19	20
4							56	34		35		36		37		38		39		40	
5							48			35		36		37		38		39		40	
6							48			49				37		38		39		40	
7							57							37		38		39		40	
8							50									38		39		40	
9							50									51				40	
10							58													40	
11							52														
12							52														
13							59														
14							54														
15							54														
16							60														

Figure 9.9 Locally altered register allocation information of 10-tap FIR filter design, post-embedding hardware security constraints corresponding to crypto-watermark

represents the register. *Note: the resource modules marked in red indicate the resources modified due to the embedding of security constraints.*

Next, an early floorplanning is performed. In order to do so, the above obtained ordered module list (generated post embedding the security constraints) is traversed from left to right and each of its modules is sequentially selected and positioned in such a way that it leads to expand the floorplan diagonally in the form of enveloping rectangle, as shown in Figure 9.10. As evident, the entire floorplan is divided into blocks, where, the size of each block is computed as follows:

$$\begin{aligned}\text{Size of a block} = {} & \text{(largest height of a module among all datapath modules)}\\ & * \text{ largest width of a module among all datapath modules).}\end{aligned}$$

Therefore, the size of each block/rectangle in the floorplan depicted in Figure 9.10 is 38.75*4 units, where 38.75 units is the height of a 1:16 Demux module and 4 units is

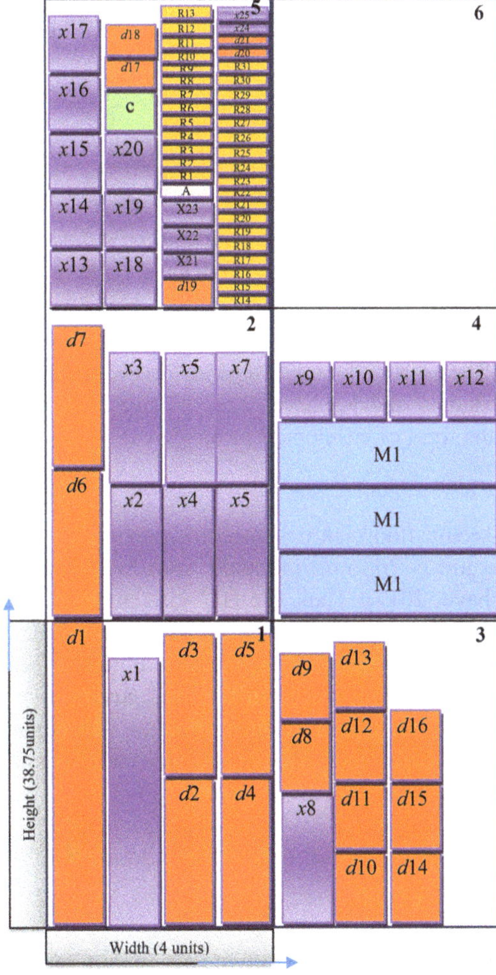

Figure 9.10 HLS-driven secure floorplan design for FIR filter application, in the form of enveloping rectangle

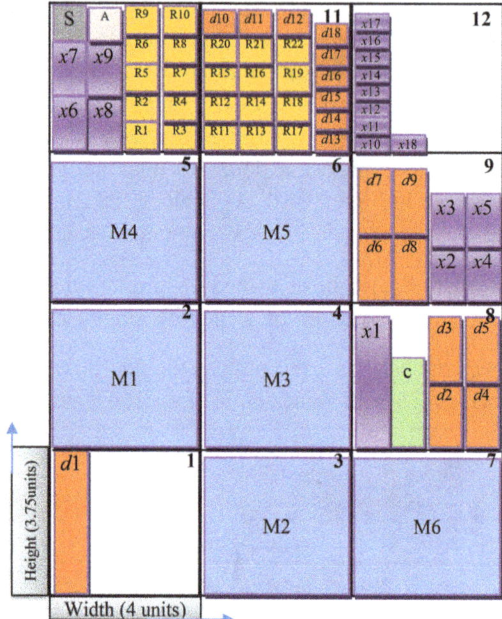

Figure 9.11 HLS-driven secure floorplan design for differential equation solver application, in the form of enveloping rectangle

the width of a multiplier module (largest height and width, respectively, among all datapath modules). The estimated module dimensions are adopted from (Sengupta and Kachave, 2016; Sengupta and Kachave, 2017). Thus, the size of secure floorplan design depicted in the form of enveloping rectangle can be computed as follows:

$$\text{Size of the floorplan design} = \text{size of each block} * \text{total number of blocks.}$$

Thus, the area of secure floorplan design corresponding to FIR filter is 930-unit square (where one unit is 768 nm) (adopted from Sengupta and Kachave, 2016; Sengupta and Kachave, 2017). Thus, the HLS-driven secure floorplan design for FIR filter application using algorithmic transformation-based obfuscation and covertly embedded crypto-watermark signature is obtained. Similarly, the secure floorplan design corresponding to the differential equation solver application is obtained as shown in Figure 9.11. Henceforth, the subsequent physical design phases (final floorplan, placement, routing, layout synthesis, etc.) can be obtained with the aid of generated early secure floorplan.

9.3.4 Detection of crypto-watermark-embedded IP design

This chapter aims to establish a strong countermeasure mechanism using embedded crypto-watermark against potential threats of IP piracy and fraudulent IP ownership claim, enacted by an adversary within the untrustworthy SoC integrator house. It involves a process where, during piracy detection and legal IP ownership resolution, the secret security constraints of crypto-watermark within the suspected chip design (design under test) are extracted and compared against the original security constraints embedded in the IP design. In this case, only if all the extracted security constraints get matched, then IP piracy is detected.

9.3.5 Security properties of the design methodology

The discussed approach presents a formidable challenge to an adversary attempting to regenerate crypto-watermark security constraints. Decrypting the crypto-watermark to regenerate exact security constraints proves exceptionally difficult for an attacker. This is because the strong entropy of the proposed security methodology makes it highly challenging for an adversary to bypass the IP authenticity check process. Entropy gauges the difficulty of an adversary's effort in recreating the exact signature. Further, the proposed methodology during crypto-watermark generation and implantation involves numerous security parameters selected by the IP vendor such as (a) input signature strength, (b) encryption key, (c) encrypted signature combination strength, (d) encoding algorithm, and (e) storage variable storage ordering. Hence, this approach effectively identifies and segregates fake hardware design IPs before their integration into the final system. Moreover, it is important to note that this chapter does not focus on preventing chip over-production typically encountered during the fabrication stage, as that is addressed separately through hardware metering and logic locking.

9.4 Analysis and discussion

This section analyzes the discussed hardware security methodology for generating HLS-driven secure floorplan design for loop-based IPs (Sengupta and Rathor, 2020; Sengupta *et al.*, 2017). The effectiveness of the methodology is analyzed in terms of the attained security level and its impact on associated design parameters such as latency, hardware resources, area, and power.

9.4.1 Analyzing the impact of the methodology on security

The methodology covertly integrates robust crypto-watermark of IP vendor into the design in order to ensure security in terms of the credibility of IP ownership and strength of digital evidence (used for IP piracy detection). The robustness of the achieved security strength offered by the methodology through embedded crypto-watermark is analyzed using the probability of coincidence (X_p), robustness against tampering attack (T_R), and entropy (E_T) metrics.

(a) ***Probability of coincidence (security against ghost signature search attack and false positive)***: It indicates the probability of the authentic crypto-watermark-driven encoded security constraints being detected within an unsecured IP design, coincidently. Therefore, it is desirable for any security methodology to achieve lower X_p value. The lower X_p indicates higher security strength of digital evidence, such that a suspected chip carrying pirated IP can be detected seamlessly. It is also a measure of the false positive of the approach, as well as strength against ghost signature search attack. The X_p of a watermark approach can be measured using the following metric (Koushanfar *et al.*, 2005; Rathor and Sengupta, 2020; Sengupta *et al.*, 2019):

$$X_p = \left(1 - \frac{1}{C}\right)^W \tag{9.13}$$

where "*c*" indicates the number of registers required for accommodating all the storage variables in baseline IP design and "*W*" indicates the strength of watermark-driven hardware security constraints chosen by the IP vendor for integrating into the design. The X_p value of the discussed security methodology in this chapter (Sengupta and Rathor, 2020; Sengupta *et al.*, 2017) has been analyzed corresponding to different loop-based IP designs for varying strength of crypto-watermark constraints based on (9.13). The computed X_p value for different loop-based IP design corresponding to varying sizes of constraints is listed in Table 9.1. As evident, for an IP design, the lower value is achieved with an increase in the number of security constraints. Further, the comparison of the X_p value achieved using the discussed methodology in this chapter (Sengupta and Rathor, 2020; Sengupta *et al.*, 2017) as compared with other contemporary approaches is shown in Table 9.2. As evident, the discussed security methodology using crypto-watermark attains lesser X_p value as compared with other contemporary approaches. Hence, the security methodology presented in this chapter ensures higher security strength in terms of digital evidence for detecting IP piracy. This is because, it is capable of generating a greater number of security constraints than compared to prior approaches (Rai *et al.*, 2019; Castillo *et al.*, 2008; Rathor and Sengupta, 2020; Chaurasia and Sengupta, 2022a).

Table 9.1 Variation in X_P and T_R for different benchmarks

# Benchmarks	**#Storage elements**	**# Constraints (*W*) = 100**		**# Constraints (*W*) = 112**		**# Constraints (*W*) = 128**	
		X_p	T_R	X_p	T_R	X_p	T_R
5-tap FIR filter	11	7.25 E−5	1.26 E+30	2.31 E−5	5.19 E+33	5.03 E−6	3.40 E+38
10-tap FIR filter	21	7.60 E−3	1.26 E+30	4.23 E−3	5.19 E+33	1.93 E−3	3.40 E+38
15-tap FIR filter	31	3.76 E−2	1.26 E+30	2.54 E−2	5.19 E+33	1.50 E−2	3.40 E+38
Differential equation (UF = 2)	22	9.54 E−3	1.26 E+30	5.46 E−3	5.19 E+33	2.59 E−3	3.40 E+38
Differential equation (UF = 4)	44	1.00 E−1	1.26 E+30	7.61 E−2	5.19 E+33	5.27 E−2	3.40 E+38

Table 9.2 Comparison of X_P with respect to other approaches

Benchmarks	**Security methodology in this chapter**	**Hardware watermarking (Rai *et al.*, 2019)**	**Automatic signature (Castillo *et al.*, 2008)**	**Hardware steganography (Rathor and Sengupta, 2020)**	**Facial biometric (Chaurasia and Sengupta, 2022a)**
	X_p	X_p	X_p	X_p	X_p
5-tap FIR filter	5.03E−6	1.06E−4	1.07E−5	1.04E−3	3.66E−4
10-tap FIR filter	1.93E−3	9.24E−3	2.86E−3	2.98E−2	1.74E−2
15-tap FIR filter	1.50E−2	4.29E−2	1.95E−2	9.43E−2	6.57E−2
Differential equation (UF = 2)	2.59E−3	1.14E−2	3.76E−3	3.51E−2	2.10E−2
Differential equation (UF = 4)	5.27E−2	1.10E−1	6.33E−2	1.91E−1	1.48E−1

(b) ***Tamper tolerance (security against brute-force attack)***: It refers to a design's ability to withstand and resist tampering attempts intended to compromise the design's integrity. Hardware IP designs with tamper tolerance are designed to maintain its security measures when subjected to probing, or tampering efforts. Therefore, it is desirable to achieve higher tamper tolerance in order to ensure the integrity of the watermark and the IP design functionality. The discussed methodology in this chapter (Sengupta and Rathor, 2020; Sengupta *et al.*, 2017) employs embedding of robust crypto-watermark into the design. However, the robust tamper tolerance ability of a security approach indicates that adversary cannot guess the exact security mark by brute-force attack for tampering and removal of watermark. The larger the signature space, the lower the probability of guessing the exact watermark signature combination. Therefore, the greater is the tamper tolerance and harder is the attacker effort required. The tamper tolerance or robustness against tampering attack of the discussed security methodology is computed as follows (Koushanfar *et al.*, 2005; Chaurasia and Sengupta, 2022a):

$$T_R = (N)^\wedge W \tag{9.14}$$

where "*N*" refers to the number of different signature bits corresponding to which encoding can be performed (e.g., in the case of crypto-watermark, there are two different types signature bits "0"s and "1"s) and "*(N)^W*" refers to the total size of signature space. Therefore, the larger the signature space, the higher the tamper tolerance (desirable). The tamper tolerance of the discussed methodology for varying numbers of crypto-watermark-driven security constraints is shown in Table 9.1. As evident, higher tamper tolerance is achieved with a greater number of security constraints. Further, the comparison of tamper tolerance of the discussed security methodology as compared to contemporary approaches is shown in Table 9.3. As evident, the discussed methodology with crypto-watermark is capable of achieving higher tamper tolerance. This is because it is capable of enabling the generation of larger signature space than other approaches (Rai *et al.*, 2019; Castillo *et al.*, 2008; Rathor and Sengupta, 2020; Chaurasia and Sengupta, 2022a).

(c) ***Entropy***: The entropy of a security methodology represents the level of effort an adversary needs to successfully estimate the exact value of a security mark implanted within the design. In other words, the entropy of a methodology refers to the randomness or uncertainty associated with the adversarial effort in deducing the exact watermark security constraints (Sengupta *et al.*, 2023; NIST, 2023). Higher entropy implies greater complexity or randomness, often enhancing the attacker's effort needed to decipher the security constraints. It can be expressed through the following formulation (Chaurasia and Sengupta, 2023; Sengupta *et al.*, 2023):

$$\mathrm{E_T} = \frac{1}{2^S} \times \frac{1}{2^K} \tag{9.15}$$

where "S" indicates the total number of signature bits and "K" indicates the encryption key strength, corresponding to the respective security algorithm. The comparison of the discussed security methodology with (Rai *et al.*, 2019; Castillo *et al.*, 2008; Rathor and Sengupta, 2020; Chaurasia and Sengupta, 2022a) is shown in Table 9.4. As evident, the discussed approach in this chapter (Sengupta and Rathor, 2020; Sengupta *et al.*, 2017) offers stronger entropy (higher randomness or uncertainty faced by an attacker), making it more resilient against adversarial attacks.

Table 9.3 Comparison of T_R with respect to other methodologies

Security methodology in this chapter		**Hardware watermarking (Rai *et al.*, 2019)**		**Automatic signature (Castillo *et al.*, 2008)**		**Hardware steganography (Rathor and Sengupta, 2020)**		**Facial biometric (Chaurasia and Sengupta, 2022a)**	
#SC	T_R	#SC	T_R	#SC	T_R	#SC	T_R	#SC	T_R
64	1.84E + 19	30	1.07E + 9	60	1.15E + 18	21	N/A	56	7.20E + 16
100	1.26E + 30	64	1.84E + 19	80	1.20E + 24	59	N/A	64	1.84E + 19
112	5.19E + 33	80	1.20E + 24	100	1.26E + 30	63	N/A	81	2.41E + 24
128	3.40E + 38	96	7.92E + 28	120	1.32E + 36	72	N/A	83	9.67E + 24

Table 9.4 Entropy comparison (E_T)

Security methodology in this chapter	**Hardware watermarking (Rai *et al.*, 2019)**	**Automatic signature (Castillo *et al.*, 2008)**	**Hardware steganography (Rathor and Sengupta, 2020)**	**Facial biometric (Chaurasia and Sengupta, 2022a)**
Entropy	Entropy (E_T)	Entropy (E_T)	Entropy (E_T)	Entropy
1.59E−58	9.31E−10	8.67E−19	4.76E−7	1.38E−17
2.31E−69	5.42E−20	8.27E−25	1.73E−18	5.42E−20
5.65E−73	8.27E−25	7.88E−31	1.08E−19	4.13E−25
8.63E−78	1.26E−29	7.52E−37	2.11E−22	1.03E−25

9.4.2 Analyzing the impact of discussed security methodology on design latency, hardware resources, area, and power

a) ***Impact of algorithmic transformations on design latency***: in the discussed methodology, the employed transformations such as loop unrolling (transforms the design structure into an optimized latency version) and THT (transforms the design by executing them as parallel sub-computations) are performed to minimize the design latency. These, transformations on the loop-based IP designs yield a reduction in latency of the final design without impacting its actual functionality. The impact of the algorithmic transformations based on IP vendor-decided keys is depicted in Table 9.5. As evident, the discussed methodology in this chapter (Sengupta and Rathor, 2020; Sengupta *et al.*, 2017) renders a significant reduction in design latency post performing algorithmic transformations as compared to pre-algorithmic design versions corresponding to different loop-based high-level applications.
b) ***Impact of crypto-watermarking on hardware resources***: In the discussed design methodology, the generated crypto-watermark signature in the form of encoded hardware security constraints is covertly embedded into the IP design during HLS. These embedded security constraints locally alter the assignment of storage variables among the design registers, in order to satisfy the security constraints. This local alteration results in change/modification in interconnectivity among the interconnecting units (Mux and Demux) and their sizes. The comparative analysis of the approach discussed in this chapter (Sengupta and Rathor, 2020; Sengupta *et al.*, 2017), in terms of the impact of crypto-watermark on hardware resources (pre and post-watermarking) is shown in Table 9.6. The design resources marked in red indicate modifications due to the embedding of security constraints.
c) ***Impact of crypto-watermarking on floorplan area***: As the embedding of crypto-watermarks results in changes to datapath resource connectivity, it also affects their sizes. In the discussed methodology, an early floorplan design is

Table 9.5 Impact of key-based algorithmic transformations on design latency

Benchmarks	Pre-algorithmic transformation total latency (for I = 150) (in ns)	IP designer selected key values (K1, K2, K3)	Post-algorithmic transformations (total latency)				
				Latency based on unrolling factor (UF) (in ns)		Total latency for I = 150 (in ns)	% Reduction in latency
FIR filter	49.68	K1 = 00000101, K2 = 0101, K3 = 101	150-tap FIR filter	UF = 5	0.794	23.84	52.0%
		K1 = 00001010, K2 = 0101, K3 = 101	150-tap FIR filter	UF = 10	1.523	22.85	54.0%
		K1 = 00001111, K2 = 0101, K3 = 101	150-tap FIR filter	UF = 15	2.053	20.53	58.66%
Differential equation solver (for I = 150)	103.30	K1 = 0010, K2 = 01, K3 = 010	Differential equation solver	UF = 2	1.033	77.48	25.0%
		K1 = 0100, K2 = 11, K3 = 100	Differential equation solver	UF = 4	1.721	65.08	37.0%

Table 9.6 Impact of crypto watermarking on hardware resources

Benchmarks	Pre-watermark resources			Post-watermark resources		
	FUs	Muxes	Demuxes	FUs	Muxes	Demuxes
5-tap FIR filter	Registers (R1to R11) Three multipliers (M1, M2, M3), one adder (A1), one comparator (C)	x1, (32:1) x2, x3 (16:1) x4 to x10, (8:1), x11 to x13, (4:1) x14 (2:1)	d1 (1:32) d2 (1:16) d3 to d6 (1:8) d7 to d9, (1:4) d10, (1:2)	Registers (R1to R11) Three multipliers (M1, M2, M3), one adder (A1), one comparator (C)	x1, (32:1) x2, x3 (16:1) x4 to x10, (8:1), x11 to x13, (4:1) x14, x15 (2:1)	d1 (1:32) d2 (1:16) d3 to d6 (1:8) d7 to d9, (1:4) d10, d11 (1:2)
10-tap FIR filter	Registers (R1to R21) Three multipliers (M1, M2, M3), one adder (A1), one comparator (C)	x1, (32:1) x2, x3 (16:1) x4 to x15, (8:1), x16 to x18, (4:1) x19 (2:1)	d1 (1:32) d2 (1:16) d3 to d11 (1:8) d12 to d14, (1:4) d15, (1:2)	Registers (R1 to R21) Three multipliers (M1, M2, M3), one adder (A1), one comparator (C)	x1, (32:1) x2, x3 (16:1) x4 to x15, (8:1), x16 to x18, (4:1) x19, x20 (2:1)	d1 (1:32) d2 (1:16) d3 to d11 (1:8) d12 to d14, (1:4) d15, d16 (1:2)
15-tap FIR filter	Registers (R1 to R31) Three multipliers (M1, M2, M3), one adder (A1), one comparator (C)	x1 (32:1) x2 to x8 (16:1) x9 to x10 (8:1)	d1(1:32) d2 to d7 (1:16) d8 to d16 (1:8)	Registers (R1to R31) Three multipliers (M1, M2, M3), one adder (A1), one comparator (C)	x1 (32:1) x2 to x8 (16:1) x9 to x10 (8:1)	d1(1:32) d2 to d7 (1:16) d8 to d16 (1:8)

(Continues)

Table 9.6 (Continued)

	Pre-watermark resources			Post-watermark resources		
Benchmarks	**FUs**	**Muxes**	**Demuxes**	**FUs**	**Muxes**	**Demuxes**
		x21 to x23 (4:1) x24 (2:1)	d17 to d19 (1:4) d20 (1:2)		x21 to x23 (4:1) x24, x25 (2:1)	d17 to d19 (1:4) d20, d21 (1:2)
Differential equation (UF = 2)	Registers (R1to R22) Three multipliers (M1 to M6), one adder (A1), one subtractor (S1), one comparator (C)	x1, x2 (8:1), x3 to x10 (4:1) x11 to x26, (2:1)	d1, d2 (1:8) d3 to d8 (1:4) d9 to d18 (1:2)	Registers (R1to R22) Three multipliers (M1 to M6), one adder (A1), one subtractor (S1), one comparator (C)	x1 (8:1), x2 to x11 (4:1) x12 to x27 (2:1)	d1 (1:8) d2 to d9 (1:4) d10 to d19 (1:2)
Differential equation (UF = 4)	Registers (R1to R44) Three multipliers (M1 to M6), one adder (A1), one subtractor (S1), one comparator (C)	x1, x2, x3, x4 (8:1), x5 to x16 (4:1) x17 to x36, (2:1),	d1, d2, d3, d4 (1:8) d5 to d14 (1:4) d15to d28 (1:2)	Registers (R1to R44) Three multipliers (M1 to M6), one adder (A1), one subtractor (S1), one comparator (C)	x1, x2, x3, (8:1), x4 to x16 (4:1) x17 to x37, (2:1),	d1, d2, d3, (1:8) d4 to d14 (1:4) d15to d29 (1:2)

Table 9.7 Impact of crypto-watermark on floorplan area in terms of enveloping rectangle

Benchmarks	Pre-embedding crypto-watermark	Post-embedding crypto-watermark	%Area overhead
5-tap FIR filter	620-unit sq. = 476,160 nm^2	620-unit sq. = 476,160 nm^2	0.00%
10-tap FIR filter	620-unit sq. = 476,160 nm^2	620-unit sq. = 476,160 nm^2	0.00%
15-tap FIR filter	930-unit sq. = 714,240 nm^2	930-unit sq. = 714,240 nm^2	0.00%
Differential equation (UF = 2)	420-unit sq. = 322,560 nm^2	420-unit sq. = 322,560 nm^2	0.00%
Differential equation (UF = 4)	420-unit sq. = 322,560 nm^2	420-unit sq. = 322,560 nm^2	0.00%

Table 9.8 Impact of crypto-watermark on power consumption

Benchmarks	Pre-embedding crypto-watermark (w)	Post-embedding crypto-watermark (w)	% Power overhead
5-tap FIR filter	8.073	8.139	0.81%
10-tap FIR filter	10.370	10.436	0.63%
15-tap FIR filter	13.396	13.462	0.48%
Differential equation (UF = 2)	6.065	6.000	0.00%
Differential equation (UF = 4)	7.865	7.763	0.00%

demonstrated in the form of an enveloping rectangle. The impact of generating IP vendor crypto-watermark floorplan area is shown in Table 9.7. As evident, the discussed security methodology does not yield any floorplan design area overhead while integrating all the generated crypto-watermark embedded data-path modules (containing secret security constraints).

d) ***Impact of crypto-watermarking on power***: the embedding of generated security constraints may lead to a change in the number of hardware resources. The comparison of power consumption pre and post-embedding crypto-watermark is shown in Table 9.8. As evident from Table 9.8, the discussed approach in this chapter incurs negligible power overhead due to crypto-watermark embedding.

9.5 Conclusion

This chapter discussed a design methodology for generating high-level synthesis-driven secure floorplan design from the behavioral description of loop-based high-level application, using algorithmic transformation-based obfuscation and crypto-watermarking. The effectiveness of the approach is discussed and analyzed in terms of optimized latency and secured floorplan design corresponding to loop-based IPs. The approach ensures enhanced security while occurring no floorplan area overhead.

9.6 Questions and exercise

1. Discuss the need for secure hardware IP designs.
2. Discuss the overview of the methodology for algorithmic transformations-based obfuscation for secure floorplan design.
3. Discuss the algorithmic transformation-based obfuscation technique.
4. Discuss the significance of the tree height transformation-based obfuscation technique.
5. Discuss crypto watermarking for binarized signature generation.
6. Discuss the details of the methodology for HLS-driven secure floor plan.
7. Discuss the algorithmic transformation-based obfuscation and demonstrate it for FIR filter application.
8. Discuss the algorithmic transformation-based obfuscation and demonstrate it for differential equation solver.
9. Explain the steps to generate a secure floorplan for the FIR filter.
10. Discuss the security properties of the integrated security approach using algorithmic transformations-based obfuscation and crypto-watermarking.
11. Discuss the metrics to analyze the security of the methodology.
12. Discuss the impact of crypto-watermark on hardware resources.
13. Discuss the differential equation solver and generate its scheduled graph based on appropriate hardware resources.
14. Discuss the FIR filter and generate its scheduled graph based on appropriate hardware resources.
15. Explain entropy.

References

A. Anshul, R. Chaurasia and A. Sengupta, (2023) “Securing Hardware Coprocessors against Piracy Using Biometrics for Secured IoT systems,” *Artificial Intelligence for Biometrics and Cybersecurity*, Chap. 8, pp. 175–193. Stevenage: The Institution of Engineering and Technology.

M. T. Arafin, A. Stanley and P. Sharma, (2017) “Hardware-Based Anti-counterfeiting Techniques for Safeguarding Supply Chain Integrity,” *IEEE International Symposium on Circuits and Systems (ISCAS)*, Baltimore, MD, USA, pp. 1–4, doi: 10.1109/ISCAS.2017.8050605.

S. Balachandran, (2010) “High Level Synthesis,” *Indian Institute of Technology Madras*, Available at: https://www.ee.iitm.ac.in/vlsi/_media/iep2010/hls.pdf, last accessed on Dec 2023.

E. Castillo, U. Meyer-Baese, A. Garcia, L. Parrilla and A. Lloris, (2007) “IPP@HDL: Efficient Intellectual Property Protection Scheme for IP Cores,”

Transactions on Very Large Scale Integration (VLSI) Systems, vol. 15, no. 5, pp. 578–591, doi: 10.1109/TVLSI.2007.896914.

E. Castillo, L. Parrilla, A. Garcia, U. Meyer-Baese, G. Botella and A. Lloris, (2008) "Automated Signature Insertion in Combinational Logic Patterns for HDL IP Core Protection," *2008 4th Southern Conference on Programmable Logic*, Bariloche, Argentina, pp. 183–186.

R. Chaurasia and A. Sengupta, (2021) "Securing Reusable Hardware IP Cores Using Palmprint Biometric," *2021 IEEE International Symposium on Smart Electronic Systems (iSES)*, pp. 410–413.

R. Chaurasia and A. Sengupta, (2022a) "Protecting Trojan Secured DSP Cores against IP Piracy Using Facial Biometrics," *2022 IEEE 19th India Council International Conference (INDICON)*, India, 2022, pp. 1–6.

R. Chaurasia and A. Sengupta, (2022b) "Crypto-Genome Signature for Securing Hardware Accelerators," *2022 IEEE 19th India Council International Conference (INDICON)*, India, pp. 1–6.

R. Chaurasia and A. Sengupta, (2022b) "Security vs Design Cost of Signature Driven Security Methodologies for Reusable Hardware IP Core," *2022 IEEE International Symposium on Smart Electronic Systems (iSES)*, India, pp. 283–288.

R. Chaurasia and A. Sengupta, (2022c) "Symmetrical Protection of Ownership Right's for IP Buyer and IP Vendor Using Facial Biometric Pairing," *2022 IEEE International Symposium on Smart Electronic Systems (iSES)*, India, pp. 272–277.

R. Chaurasia and A. Sengupta, (2023a) "Designing Optimized and Secured Reusable Convolutional Hardware Accelerator against IP Piracy Using Retina Biometrics," *Proceedings of 9th IEEE International Symposium on Smart Electronic Systems (IEEE – iSES)*, India, Accepted Dec 2023.

R. Chaurasia and A. Sengupta, (2023b) "Multi-cut Based Architectural Obfuscation and Handprint Biometric Signature for Securing Transient Fault Detectable IP Cores during HLS," *Integration, the VLSI Journal*, vol. 95, 102114.

R. Chaurasia, A. Anshul, A. Sengupta and S. Gupta, (2022) "Palmprint Biometric versus Encrypted Hash Based Digital Signature for Securing DSP Cores Used in CE Systems," *IEEE Consumer Electronics Magazine (CEM)*, vol. 11, no. 5, pp. 73–80.

R. Chaurasia, A. Sengupta and P. P. Kanhegaonkar, (2022) "Secured Integrated Circuit (IC/IP) Design Flow," *Nanoelectronics for Next-generation Integrated Circuits*, Chap. 14, pp. 257–274. Boca Ration FL: CRC Press.

R. Chaurasia, A. R. Asireddy and A. Sengupta, (2023) "Fault Secured JPEG-Codec Hardware Accelerator with Piracy Detective Control Using Secure Fingerprint Template," *2023 IEEE International Symposium on Defect and Fault Tolerance in VLSI and Nanotechnology Systems (DFT)*, Juan-Les-Pins, France, pp. 1–6.

S. Chen, J. Jung, P. Song, K. Chakrabarty and G.-J. Nam, (2020) "BISTLock: Efficient IP Piracy Protection Using BIST," *2020 IEEE International Test Conference (ITC)*, Washington, DC, USA.

B. Colombier and L. Bossuet, (2015) "Survey of Hardware Protection of Design Data for Integrated Circuits and Intellectual Properties," *IET Computers and Digital Techniques*, vol. 8, no. 6, pp. 274–287, 2015, https://doi.org/10.1049/iet-cdt.2014.0028.

D. Dolev and A. Yao, (1983) "On the Security of Public Key Protocols," *IEEE Transactions on Information Theory*, vol. 29, no. 2, pp. 198–208.

A. Hroub and M. E. S. Elrabaa, (2022) "SecSoC: A Secure System on Chip Architecture for IoT Devices," *2022 IEEE International Symposium on Hardware Oriented Security and Trust (HOST)*, McLean, VA, USA, pp. 41–44.

R. Karmakar and S. Chattopadhyay, (2020) "Hardware IP Protection Using Logic Encryption and Watermarking," *2020 IEEE International Test Conference (ITC)*, Washington, DC, USA, pp. 1–10.

F. Koushanfar, S. Fazzari, C. McCants, *et al.*, (2012) "Can EDA Combat the Rise of Electronic Counterfeiting?," *DAC Design Automation Conference 2012*, San Francisco, CA, USA, pp. 133–138, doi:10.1145/2228360.2228386.

F. Koushanfar, I. Hong and M. Potkonjak, (2005) "Behavioral Synthesis Techniques for Intellectual Property Protection," *ACM Transactions on Design Automation of Electronic Systems (TODAES)*, vol. 10, no. 3, pp. 523–545.

B. Le Gal and L. Bossuet, (2012) "Automatic Low-cost IP Watermarking Technique Based on Output Mark Insertions," *Design Automation for Embedded Systems*, vol. 16, no. 2, pp. 71–92, https://doi.org/10.1007/s10617-012-9085-y.

B. K. Mohanty and P. K. Meher, (2016) "A High-Performance FIR Filter Architecture for Fixed and Reconfigurable Applications," *IEEE Transactions on Very Large Scale Integration (VLSI) Systems*, vol. 24, no. 2, pp. 444–452.

D. Mouris, C. Gouert and N. G. Tsoutsos, (2022) "Privacy-Preserving IP Verification," *IEEE Transactions on Computer-Aided Design of Integrated Circuits and Systems*, vol. 41, no. 7, pp. 2010–2023.

NIST Computer Security Resource Center, Glossary, https://csrc.nist.gov/glossary/term/entropy#:~:text=A%20measure%20of%20the%20amount,is%20usually%20stated%20in%20bits, last accessed on October 2023.

C. Pilato, K. Basu, M. Shayan, F. Regazzoni and R. Karri, (2019) "High-Level Synthesis of Benevolent Trojans," *2019 Design, Automation & Test in Europe Conference & Exhibition (DATE)*, Florence, Italy, pp. 1124–1129, doi:10.23919/DATE.2019.8715199.

C. Pilato, S. Garg, K. Wu, R. Karri and F. Regazzoni, (2018) "Securing Hardware Accelerators: A New Challenge for High-Level Synthesis," *IEEE Embedded Systems Letters*, vol. 10, no. 3, pp. 77–80.

S. Potluri, A. Aysu and A. Kumar, (2020) "SeqL: Secure Scan-Locking for IP Protection," *2020 21st International Symposium on Quality Electronic Design (ISQED)*, Santa Clara, CA, USA, doi: 10.1109/ISQED48828.2020.9136991.

S. Rai, A. Rupani, P. Nath and A. Kumar, (2019) "Hardware Watermarking Using Polymorphic Inverter Designs Based on Reconfigurable Nanotechnologies," *2019 IEEE Computer Society Annual Symposium on VLSI (ISVLSI)*, Miami, FL, USA, pp. 663–669, doi: 10.1109/ISVLSI.2019.00123.

J. J. Rajendran, O. Sinanoglu and R. Karri, (2016) "Building Trustworthy Systems Using Untrusted Components: A High-Level Synthesis Approach," *IEEE Transactions on Very Large Scale Integration (VLSI) Systems*, vol. 24, no. 9, pp. 2946–2959.

M. Rathor and A. Sengupta, (2020) "IP Core Steganography Using Switch Based Key-driven Hash-chaining and Encoding for Securing DSP Kernels Used in CE Systems," *Transactions on Consumer Electronics*, vol. 66, no. 3, pp. 251–260, doi:10.1109/TCE.2020.3006050.

A. Ray and S. Roy, (2020) "Recent Trends in Image Watermarking Techniques for Copyright Protection: A Survey," *International Journal of Multimedia Information Retrieval,* vol. 9, pp. 249–270.

F. Regazzoni, P. Palmieri, F. Smailbegovic, R. Cammarota, and I. Polian, (2021) "Protecting Artificial Intelligence IPs: A Survey of Watermarking and Fingerprinting for Machine Learning," *CAAI Transactions on Intelligence Technology*, vol. 6, pp. 180–191.

A. R. D. Rizo, J. Leonhard, H. Aboushady and H.-G. Stratigopoulos, (2022) "RF Transceiver Security against Piracy Attacks," *IEEE Transactions on Circuits and Systems II: Express Briefs*, vol. 69, no. 7, pp. 3169–3173.

S. M. Saeed, N. Mahendran, A. Zulehner, R. Wille and R. Karri, (2019a) "Identification of Synthesis Approaches for IP/IC Piracy of Reversible Circuits," *Journal of Emerging Technologies in Computing Systems*, vol. 15, no. 3, pp. 23:1–23:17, https://doi.org/10.1145/3289392.

S. M. Saeed, A. Zulehner, R. Wille, R. Drechsler and R. Karri, (2019b) "Reversible Circuits: IC/IP Piracy Attacks and Countermeasures," *IEEE Transactions on Very Large Scale Integration (VLSI) Systems*, vol. 27, no. 11, pp. 2523–2535.

A. Sengupta and R. Chaurasia, (2022a) "Hardware IP Cores for Image Processing Functions," *Advances in Image and Data Processing Using VLSI Design*, Chap. 7, pp. 7.1–7.14. Bristol: IOP Publishing Ltd.

A. Sengupta and R. Chaurasia, (2022b) "Securing IP Cores for DSP Applications Using Structural Obfuscation and Chromosomal DNA Impression," *IEEE Access*, vol. 10, pp. 50903–50913.

A. Sengupta and R. Chaurasia, (2023a) "Integrated Defense Using Structural Obfuscation and Encrypted DNA Based Biometric for Hardware Security," *Physical Biometrics for Hardware Security of DSP and Machine Learning Coprocessors*, Chap. 2, pp. 25–56. Stevenage: The Institution of Engineering and Technology.

A. Sengupta and R. Chaurasia, (2023b) "Securing Fault-Detectable CNN Hardware Accelerator Against False Claim of IP Ownership Using Embedded Fingerprint as Countermeasure," *Proceedings of 9th IEEE International Symposium on Smart Electronic Systems (IEEE – iSES)*, India, Accepted Dec 2023.

A. Sengupta and D. Kachave, (2016) "Generating Multi-Cycle and Multiple Transient Fault Resilient Design during Physically Aware High Level Synthesis," *Proc. 15th IEEE Computer Society Annual Symposium on VLSI (ISVLSI)*, Pittsburgh, pp. 75–80.

A. Sengupta and D. Kachave, (2017) “Low Cost Fault Tolerance against kc-Cycle and km-Unit Transient for Loop Based Control Data Flow Graphs during Physically Aware High Level Synthesis,” *Elsevier Journal on Microelectronics Reliability*, vol. 74, pp. 88–99.

A. Sengupta and M. Rathor, (2020) “Enhanced Security of DSP Circuits Using Multi-key Based Structural Obfuscation and Physical-Level Watermarking for Consumer Electronics Systems,” *IEEE Transactions on Consumer Electronics (TCE)*, vol. 66, no. 2, pp. 163–172.

A. Sengupta, R. Chaurasia and A. Anshul, (2023a) “Hardware Security of Digital Image Filter IP Cores against Piracy Using IP Seller’s Fingerprint Encrypted Amino Acid Biometric Sample,” *2023 Asian Hardware Oriented Security and Trust Symposium (AsianHOST)*, Tianjin, China, pp. 1–6.

A. Sengupta, R. Chaurasia and A. Anshul, (2023b) “Robust Security of Hardware Accelerators Using Protein Molecular Biometric Signature and Facial Biometric Encryption Key,” *IEEE Transactions on Very Large Scale Integration (VLSI) Systems*, vol. 31, no. 6, pp. 826–839, June 2023.

A. Sengupta, R. Chaurasia and K. Bharath, (2023c) “Exploring Unified Biometrics with Encoded Dictionary for Hardware Security of Fault Secured IP Core Designs,” *Elsevier Journal Computers & Electrical Engineering*, vol. 111, Part A, 108928.

A. Sengupta, R. Chaurasia and M. Rathor, (2023d) “HLS Based Swarm Intelligence Driven Optimized Hardware IP Core for Linear Regression Based Machine Learning,” *IET Journal of Engineering*, vol. 2023, no. 8, e12299.

A. Sengupta, R. Chaurasia and T. Reddy, (2021) “Contact-Less Palmprint Biometric for Securing DSP Coprocessors Used in CE Systems,” *IEEE Transactions on Consumer Electronics*, vol. 67, no. 3, pp. 202–213.

A. Sengupta, E. R. Kumar and N. P. Chandra, (2019) “Embedding Digital Signature Using Encrypted-Hashing for Protection of DSP Cores in CE,” *IEEE Transactions on Consumer Electronics*, vol. 65, no. 3, pp. 398–407.

A. Sengupta, D. Roy, S. Mohanty and P. Corcoran, (2017) “DSP Design Protection in CE through Algorithmic Transformation Based Structural Obfuscation,” *IEEE Transactions on Consumer Electronics*, vol. 63, no. 4, pp. 467–476.

S. Sitjongsataporn, A. Thitinaruemit and S. Prongnuch, (2021) “Implementation of High Level Synthesis for Adaptive FIR Filtering on Embedded System,” *2021 7th International Conference on Engineering, Applied Sciences and Technology (ICEAST)*, Pattaya, Thailand, pp. 257–260.

T. Wang, X. Cui, D. Yu, *et al.*, (2018) “Polymorphic Gate Based IC Watermarking Techniques,” *2018 23rd Asia and South Pacific Design Automation Conference (ASP-DAC)*, Jeju, Korea (South), pp. 90–96.

X. Wang, Y. Zheng, A. Basak and S. Bhunia, (2015) “IIPS: Infrastructure IP for Secure SoC Design,” *IEEE Transactions on Computers*, vol. 64, no. 8, pp. 2226–2238, doi:10.1109/TC.2014.2360535.

J. Zhang, (2016) “A Practical Logic Obfuscation Technique for Hardware Security,” *IEEE Transactions on Very Large Scale Integration (VLSI) Systems*, vol. 24, no. 3, pp. 1193–1197.

Chapter 10

Fundamentals on HLS-based hardware Trojan

Anirban Sengupta[1] *and Mahendra Rathor*[2]

The chapter describes the fundamentals of high-level synthesis (HLS)-based hardware Trojan and its various types. This chapter provides an overview of hardware Trojan and discusses the motivation for an adversary to perform an HLS-based hardware Trojan attack. Further, it highlights the potential sites for hardware Trojan insertion in the semiconductor design supply chain. It discusses in detail the three different types of HLS-based hardware Trojan attacks followed by hardware Trojan in 3PIP modules.

The organization of the chapter is as follows: Section 10.1 discusses a brief overview of hardware Trojan, different ways of the hardware Trojan insertion followed by the Trojan insertion during HLS and its potential effects; Section 10.2 provides an overview of three different types of HLS-based hardware Trojan attacks, viz. battery exhaustion, degradation, and downgrade, and discusses the threat scenario. Section 10.3 presents insights on battery exhaustion attacks with a motivational example. Section 10.4 presents insights on degradation attacks with a motivational example while discussing its impact on security and overhead. Section 10.4 presents insights on downgrade attack and explains with the help of compromising secure hash algorithm; Section 10.5 presents insights on hardware Trojan in 3PIP modules and explains using an example; Section 10.6 concludes the chapter.

10.1 Introduction

Different computer-aided design (CAD) tools are employed by a fabless design company for its system-on-chip (SoC) design. These CAD tools generally belong to third parties. Further, the intended SoC design is shipped to an offshore (third party) foundry for manufacturing. However, this design supply chain has crucial security flaws. For example, an adversary (rogue employee) or a malicious foundry can insert hardware Trojans secretly into the design (Polian *et al.*, 2016; Hurtarte *et al.*, 2007; Xiao *et al.*, 2016; Sengupta *et al.*, 2017).

[1]Department of Computer Science and Engineering, Indian Institute of Technology Indore, India
[2]School of Instrumentation, Devi Ahilya Vishwavidyalaya, India

A brief overview of hardware Trojan: A hardware Trojan can be defined as a backdoor malicious alteration to the design that can be exploited to extract sensitive information such as encryption keys or maliciously impact the functioning or design parameters of a system. A hardware Trojan can offer manifold advantages for the attacker in terms of the following: (i) It can sabotage the economics of rival design houses or companies and (ii) it can impose cyber-threats to national defense/security (Polian *et al.*, 2016; Xiao *et al.*, 2016).

The above discussion highlights the reasons why hardware Trojans have become a major concern for digital integrated circuits (ICs). The hardware Trojan threat can have an impact on all the various stages of design in the electronics supply chain such as register transfer level (RTL) design, gate-level netlist, design layout, mask, and finally the manufactured ICs.

Hardware Trojan insertion: There can be different ways of inserting hardware Trojans. (i) A rogue employee or a dishonest design house can directly place a hardware Trojan in the design at any abstraction level. (ii) Compromised computer-aided design (CAD) tools can insert the Trojan during the design process at different steps of the design cycle. Thus, inserted hardware Trojan leads to malicious modification of the design files in the early stages in the fabless design houses. Moreover, it can also lead to modification during the fabrication process in a third-party malicious foundry. Figure 10.1 shows a typical scenario of hardware Trojan insertion that can affect an SoC design (Pilato *et al.*, 2019).

Hardware Trojan insertion using HLS: Let us discuss the prospect of hardware Trojan insertion during HLS. Automatic HLS (McFarland *et al.*, 1988) is increasingly becoming a popular tool to facilitate the trend of designing modern SoCs by reusing existing (pre-designed and verified/validated) components (reusable

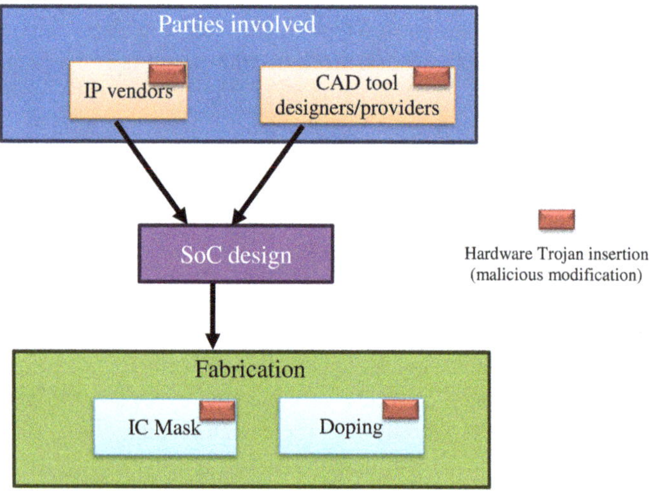

Figure 10.1 SoC designs affected by hardware Trojan insertion due to various parties involved

blocks or intellectual property cores). This has largely addressed the issue of escalated complexity, cost, and error-proneness of hardware designs. By using automatic HLS, the designers can produce an optimized hardware (RTL) in the form of Verilog/VHDL description from a high-level specification (in terms of C or C++ or mathematical function). The generated RTL description from HLS is orders of magnitude larger compared to the high-level specification. This often makes it cumbersome for an engineer/designer to understand and correlate the two descriptions (i.e., the RTL code and the high-level specification). Although the formal methods can be leveraged to correlate the high-level description with the automatically (using HLS tool) generated hardware description. However, the HLS tools may apply some optimizations and re-structuring, therefore the correlation may become laborious. This encourages the simulation-based methods to validate designs generated from HLS. However, this technique cannot cover all possible input patterns. Hence, this provides an opportunity for an adversary to insert hardware Trojan at the HLS level without being detected (Pilato *et al.*, 2019; Basu *et al.*, 2019; Rathor and Sengupta, 2023).

Potential effects of Trojan insertion during HLS: Compromised HLS tools (Trojan inserted) can lead to the generation of such components that can fail repeatedly (due to Trojan activation) in periodic intervals. This makes such components less competitive in the market. Further, the HLS tool designer can insert hidden features into the design to compromise the security properties of the resulting IP. The hardware Trojans are fixed or permanent (as they cannot be patched like software Trojans) and can considerably harm the revenue of the IP (compromised) users. Since HLS can place a Trojan during the initial stages of the design flow, therefore it can affect both the functional and non-functional (such as delay or power consumption) aspects of the resulting design (Pilato *et al.*, 2019; Basu *et al.*, 2019; Sengupta *et al.*, 2015).

The next section emphasizes on the study of the trustworthiness of HLS tools and the potential threat vector imposed by a compromised HLS tool.

10.2 Different types of HLS-based hardware Trojan

Intellectual property (IP) cores are designed to execute specific or specialized functions. This specialized functionality of an IP design can be augmented with extra functionality by modifying the HLS behavior by a potential adversary. This leads to a compromised IP component at the output of HLS. Unlike software Trojans, such malicious functionality (hardware Trojan) is not patchable or filterable post IC fabrication. The malicious functionality introduced during HLS can impact the design in the following different ways and hence can lead to different kinds of hardware Trojan attacks (Pilato *et al.*, 2019):

- It can expedite aging or battery consumption: ***battery exhaustion attack***.
- It can influence the performance of the component: ***degradation attack***.
- It can alleviate the security level of the component: ***downgrade attack***.

Figure 10.2 shows the above-mentioned HLS-based hardware Trojan attacks.

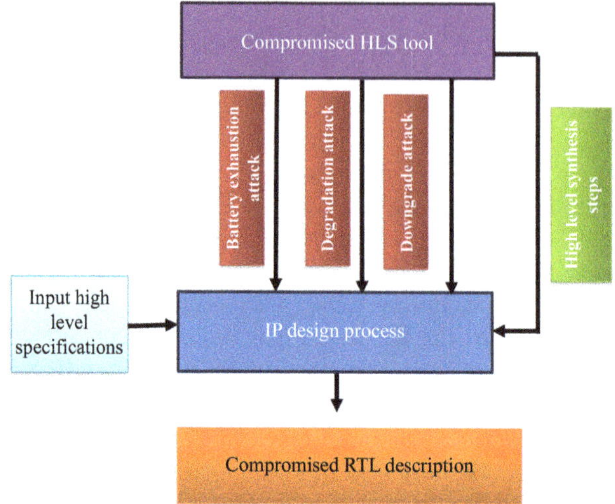

Figure 10.2 HLS-based hardware Trojan attack

10.2.1 Threat scenario considered for HLS-based hardware Trojan

The rogue developer of an HLS tool is assumed to be the adversary. Further, it is assumed that the malicious modifications can be installed into the HLS framework by modifying the regular updates for particular tool users. Moreover, the threats can also be imposed by such small companies that design specific IP blocks for the semiconductor design houses. These IP providers or vendors may not be trustworthy or honest. Such third-party IP (3PIP) vendors or design companies may deliver maliciously modified IP blocks, which can be generated using internal HLS tools. They can do so to target some specific IP users owing to some personal or state interests. This leads to employing the malicious CAD (HLS) tools is an emerging design threat for the SoC designs (Pilato *et al.*, 2019).

In the subsequent subsections, we will discuss three different types of HLS-based hardware Trojan attacks, viz. battery exhaustion, degradation, and downgrade attack, along with the hardware Trojan in 3PIP modules.

10.3 Battery exhaustion Trojan attack

In this section, we discuss battery exhaustion (Pilato *et al.*, 2019), its impact, and how it can be performed as an attack during the HLS design process under the following sections.

10.3.1 What is battery exhaustion and why it is important to study?

Let us first discuss why studying the battery exhaustion attack is so important for the modern consumer devices, internet of things (IoT), and cyber-physical systems (CPS) perspectives. One of the key design principles of modern CPS or consumer/IoT devices is efficient energy management. This has led to major of the devices used in SoCs battery powered. This can be understood with an example of a simple practical need of a consumer electronics device user. The most important criterion often looked at during purchasing a new phone is its standby time, which is the time duration of a single battery charge. The modern CPS are designed and developed such that they can offer energy-efficient operations of devices and may become non-operable when the battery drains. Now let us define the battery exhaustion attack. Enabling or realizing unwanted/undesired/unexpected functionality execution that drains current to a significant extent from the battery to expedite its exhaustion. This undesired functional execution results in additional power consumption upon switching it on. Hence, the battery time shortens without affecting the functionality. This may eventually result in different types of impacts such as (i) declining product competitiveness, (ii) system failure, and so on.

10.3.2 Battery exhaustion attack during HLS

The high-level synthesis design process can be exploited to perform the battery exhaustion. The objective is to modify the datapath in order to consume a considerable amount of current upon Trojan activation. The underlying concept of battery exhaustion is to reuse the idle functional units in the datapath. The impact of executing the undesired functionality by reusing the idle functional units can be amplified by choosing those units that result in a larger dynamic power dissipation such as multipliers. Such alterations in the datapath have negligible area overhead and negligible additional static power. In this case, only multiplexers are required to select between the correct or fake operations. If there are no states with unutilized functional units, then the battery exhaustion attack cannot be realized. In this situation, a malicious HLS tool can construct new states to implement the attack. However, it introduces performance overhead due to additional states. In a nutshell, the idle functional units can be reused in two different ways to perform the battery exhaustion attack.

(i) Reusing unused functional units only in the existing states (or control steps). In this case, no performance overhead (or additional states) incurs.
(ii) Reusing unused functional units by incorporating additional states (If there are no states with unutilized functional units). In this case, performance overhead incurs.

Let us understand with an example how battery exhaustion can be performed by reusing existing functional units during the HLS process.

A high-level description of a sample code is as follows:

```
u = a + b;
v = c * d;
x = c * e;
z = x + u;
```

The corresponding scheduled dataflow graph generated during an HLS step is shown in Figure 10.3.

The above-mentioned four operations are computed to generate u, v, x, and z. These operations are scheduled in three control steps (cs1 to cs3). In their corresponding datapath, the two multiplications can be assigned to the same multiplier unit as they are executed in distinct control steps. Similarly, the two additions can be assigned to the same adder unit.

The fundamental idea of reusing the idle functional units is based on the fact that a functional unit (multiplier or adder etc.) may not be used in all control steps (there can be some control steps where the functional units are idle). Therefore, a malicious HLS framework can leverage the opportunity of executing the unused functional units in the particular control steps during the operation. This opportunity is created by inserting the fake operations in some particular control steps during the scheduling process such that the resource constraints are not violated. During the operation, an adversary can trigger the additional execution of these functional units in such control steps where they were initially unused. Thus, the dynamic power consumption of the component can be increased. The computational errors due to the additional execution of the functional units can be avoided by not storing the result of this spurious operation in any register. This makes the Trojan detection harder. Figure 10.4 shows the insertion of a fake multiplication operation in control step cs3 during the scheduling process (shown using the red-colored node). This insertion only reuses the existing multiplier unit in the datapath using multiplexer hence also complies with the available resource constraints. This alters the datapath in terms of multiplexing of inputs for the execution of additional multiplication operations using the same (existing) multiplier unit. This altered datapath is shown in Figure 10.5, where the additional inputs to the multiplexer are shown using the red arrows. It is noteworthy that the output registers are used to store the result of valid computation only. In case of fake (additionally inserted) operations, register-write operations are disabled.

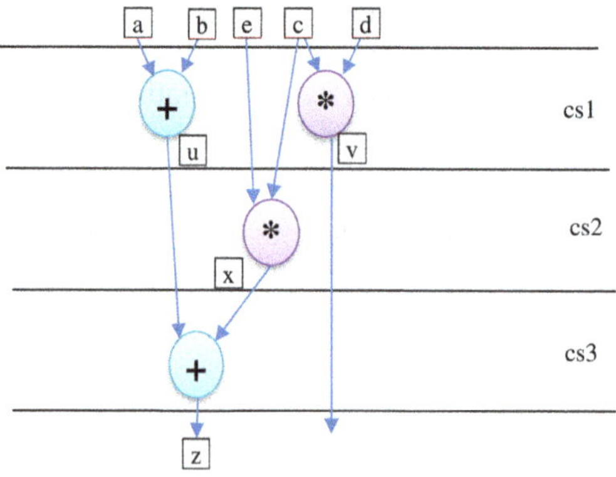

Figure 10.3 Scheduled data flow graph (Pilato et al., 2019)

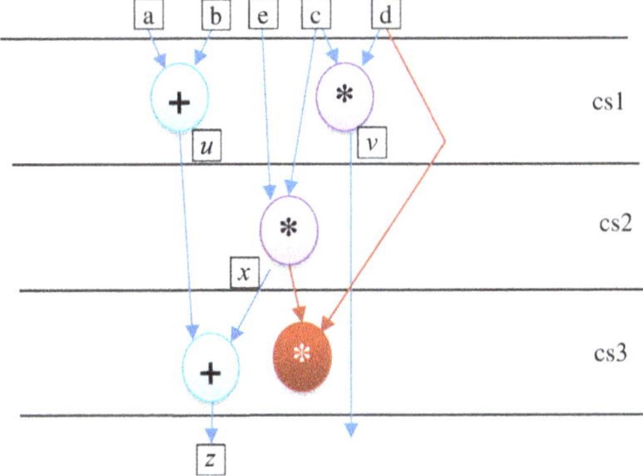

Figure 10.4 Fake operation insertion to enable the reuse of functional units

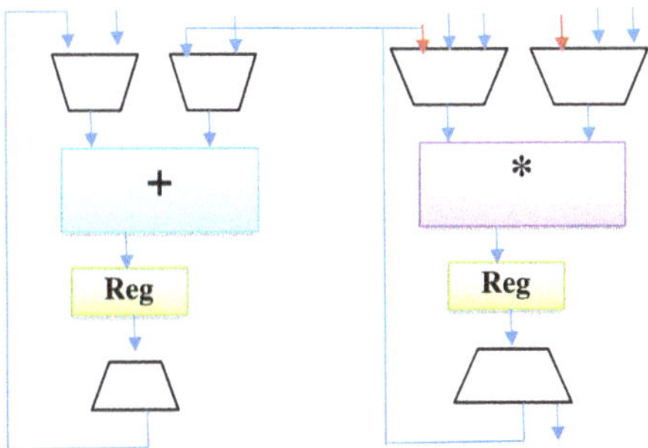

Figure 10.5 RTL datapath (generated post HLS) affected with battery exhaustion attack

10.4 Degradation Trojan attack

In this section, we discuss the motivation for the degradation attack (Pilato *et al.*, 2019) and how it can be applied to the SoC-based products through compromised HLS tools.

10.4.1 Why studying degradation attack is important?

For achieving hardware acceleration, reusable computation-intensive IPs or kernels are suitable candidates. As IP reuse has now become an inevitable part of the SoC design process, most of the fabless companies (SoC design houses) either employ 3PIPs or generate reusable IPs from high-level descriptions from the HLS tools. The success of the reusable hardware accelerator IPs not only determines the quality of the intended SoCs but also the business economics. However, the vendor of the 3PIPs may deliberately supply such IPs that may fail prematurely. Such failures are difficult to identify because they may be realized after a significant amount of time (after the IP is in production). Hence, this encourages an adversary to provide weak/compromised HLS tools in order to hinder the long-term quality of the products of competitive companies.

10.4.2 Threat model of degradation attack

A compromised HLS tool can be developed by a rouge tool vendor or designer to place such as hardware Trojan into the design in order to degrade the performance of an intended IP upon activation. When a certain time period passes on, the added hardware Trojan into the IP design can enable aging degradation. This can sabotage the desired quality and endurance of the IPs.

10.4.3 Description of HLS-aided degradation attack

Let us understand the degradation attack with the help of an example. A C-language specification of a finite-impulse response (FIR) filter application is as follows (Pilato *et al.*, 2019):

```
int FIR (int N, int S){
   int j;
   for (j=0; j<N; j++){
      S += x[i]*h[i];
   }
   return S;
}
```

where N represents the number of taps and S denotes the resulting summation. The designer is intended to design a hardware accelerator for the above N-tap FIR filter. Here, the body of the loop is repeated for the N number of times (number of taps value). The corresponding finite state machine (FSM) of the above FIR filter application is depicted in Figure 10.6. Now let's see how an attacker can modify this FSM during the HLS process in order to enable the degradation attack.

An adversary designs the malicious HLS framework to place alternative paths in some particular points of the computation. These alternative paths do not contain any useful operations but are formed to execute a small number of empty FSM states. The empty FSM states are also termed bubbles. Figure 10.7 depicts this scenario. The malicious HLS framework places an empty FSM state after the two memory-read operations. Transition edges of the new (additionally) added path are highlighted in red color. Upon Trojan activation, this additionally added path is executed. This slows down the computation. It does not affect the desired functionality. However, it increases the number of cycles required for the execution after the Trojan is activated. This is to be noted that when the bubbles are placed inside the body of the loop, the latency overhead is enhanced by the number of iterations of the loop (the more the number of times the loop iterated, the higher the

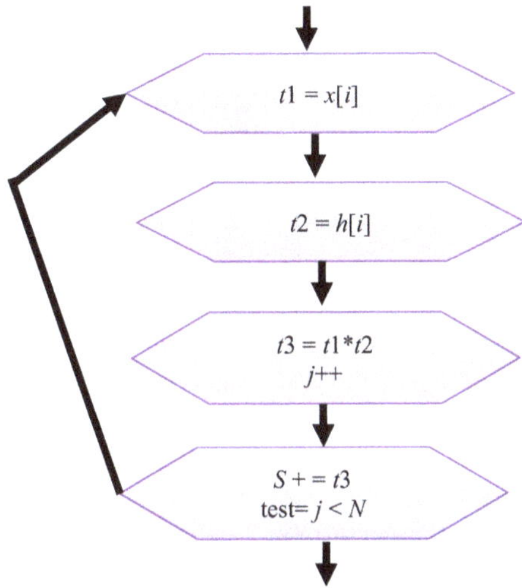

Figure 10.6 FSM before degradation attack, where t1, t2, and t2 denote the temporary variables (Pilato et al., 2019)

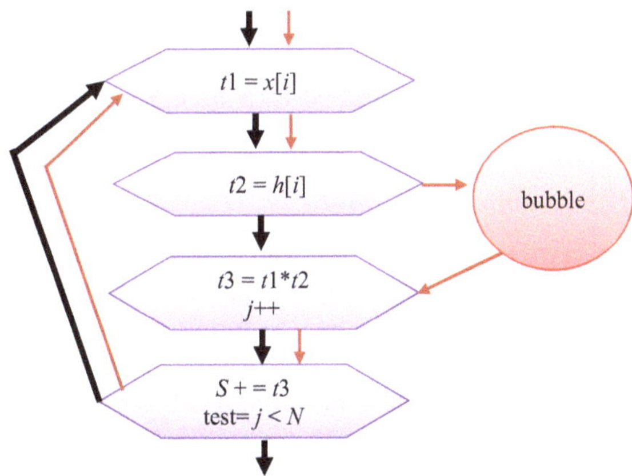

Figure 10.7 FSM after degradation attack (Pilato et al., 2019)

latency overhead will be). In this attack, it is generally difficult to understand if an FSM transition is valid. This is because any prior information about the FSM before HLS is not known to the designer. This makes traditional logic synthesis-based FSM anomaly detection methods unusual.

A few considerations with bubbles insertion into FSM:

(i) Only after a specific period of time, the degradation effects must be perceivable.
(ii) Empty FSM states must be designed in a fashion that they should not incur computational errors (ensure a correct computation) when a transition

between the valid FSM states and the augmented extra paths with the bubbles takes place.

(iii) The FSM bubbles are required to be inserted such that the generated compromised FSM escapes even C-to-HDL formal verification.

10.4.4 Impact on datapath and controller

The degradation attack performed by inserting bubbles does not modify the datapath. However, there is still some area overhead in the overall design. This is because of the additional resources required to implement the bubble states in the FSM or controller. Encoding of the FSM states requires some extra flip-flops. In addition, encoding of the extra elements of the output and next state functions needs extra logic ports. For the example discussed in the previous section, the area overhead due to the insertion of one empty FSM state is estimated to be less than 1%. Empty FSM states do not perform any valid computation. Only the latency of the computation is affected in these bubble states. In the case of an 8-tap FIR filter (i.e., N = 8), overhead in the performance with one bubble insertion is observed to be 3% and with two bubbles insertion is observed to be 6% (Pilato *et al.*, 2019).

10.5 Downgrade attack

In this section, we discuss what a downgrade attack (Pilato *et al.*, 2019) is and how it can be performed using a compromised HLS tool. Further, we also discuss its impact on design parameters.

10.5.1 What is downgrade attack?

Security features of cryptographic algorithms such as Advanced Encryption Standard (AES) encryption algorithm and Secure Hash Algorithm (SHA) can be compromised by an adversary. Typically, these algorithms perform some bit manipulations, which are repeated in a fixed number of rounds. The AES is a symmetric key cryptographic algorithm that converts a fixed length plain text (i.e., 128-bit) into a fixed length cipher text (128-bit) using a specified number of rounds (e.g., 180) and a fixed length key (128-bit). Whereas SHA generates a fixed length hash after performing a fixed number of rounds (e.g., SHA-512 generates a 512-bit hash digest after 80 rounds). The security of such cryptographic algorithms can be undermined by recovering the key. Reducing the number of rounds helps recover the key. This type of attack is termed a downgrade attack as it downgrades the functionality.

10.5.2 Downgrade attack using HLS

A malicious HLS tool can be designed to compromise or weaken the security features of cryptographic algorithms such as the AES encryption algorithm (Sengupta and Mohanty, 2019) and the Secure Hash Algorithm (Sengupta and Rathor, 2019). A rogue IP designer who possesses his/her own HLS tool can launch this attack. In order to realize this type of attack, the HLS designer and the IP designer need to have collusion. They together decide to identify the number of rounds (of cryptographic algorithms) to be performed and the places of insertion of modification. For example, the places of malicious modification (downgrade Trojan) insertion can be the counter variable and the constant defining the number of rounds. It needs to be highlighted that this type of attack requires insight into the algorithm, therefore, cannot be fully automated. In order to perform this attack during HLS, a malicious HLS developer can provide such hidden directives that can be exploited while developing IPs for accessing the undocumented features and produce compromised designs. The attacker can extract sensitive information by accessing the compromised designs.

Now let us discuss this type of attack with an example. A sample C code of a round based security algorithm, i.e., SHA-256 with a main loop having 64 rounds is shown below (Pilato *et al.*, 2019).

```
#define R(x,y) (((p)>>(q)) | ((p) << (32-(q))))

#define CH (a,b,c) (((a) & (b)) ^ (~(a) & (c)))

#define MA (a,b,c) (((a) & (b)) ^ ((a) & (c)) ^ ((b)&(c)))

#define EP0(a) (R(a,2) ^ R(a,13) ^ R(a,22))

#define EP1(a) (R(a,6) ^ R(a,11) ^ R(a,25))

#define SHA_rounds 64

-- array of constants

static const unit32_t m[64] = {/* omitted values */};

-- hash function

void sha256_transform (void *state, const char in[]){

-- internal state variables

unit32_t p,q,r,s,t,u,v,w,j,t1,t2, n[64];

-- execution of 64 rounds

for (j=0; j<SHA_rounds; ++j){

t1= w+EP1(t) + CH(t,u,v) +m[j] + n[j];

t1= EP0(p) + MA(p,q,r);

w=v; v=u; u=t; t=s+t1;

s=r; r=q; q=p; p=t1+t2;

}}
```

An attacker (malicious HLS tool) can reduce the rounds from 64 to 18 rounds and generate message pairs with collision for SHA-256 (Derbez *et al.*, 2013; Sanadhya and Sarkar, 2008). Similarly, the key recovery for AES-128 can simply be performed by reducing the number of rounds from 10 to 7 (Derbez *et al.*, 2013). In order to accomplish this, the attacker can modify the loop constant (i.e., SHA_rounds) or preload a number into the counter variable j greater than 0. This is how a reduced number of rounds are executed by the compromised components or IP cores. Hence, it can be concluded that the round-based security algorithms can be attacked or compromised by downgrading them.

10.5.3 Impact of downgrade attack on design parameters

The design area is minimally affected by the Downgrading. This is because the modification in the design is only applied to the round counter logic. This also put limited effect on overall power consumption. The limited power consumption hinders the power analysis mechanisms to identify anomalies when the Trojan is inactive. Triggering of Trojan leads to executing a reduced number of rounds by the component. Therefore, the security of the IP core is compromised. For example, a reduced number of rounds in the AES encryption algorithm helps in recovering the key.

10.6 Hardware Trojan in 3PIP module

In third-party modules/IPs, a hardware Trojan (malicious logic) can deliberately be placed by an untrustworthy 3PIP vendor (Sengupta *et al.*, 2017; Sengupta and Bhadauria, 2015). The library of an HLS tool may contain these Trojan-infected IPs as the basic modules. Since the Trojans are secretly placed and remain inactive during the usual operating conditions, they may not be detected through functional checking and testing. The hidden hardware Trojan becomes active when specific circuit conditions are formed to trigger the malicious logic. Figure 10.8 shows an

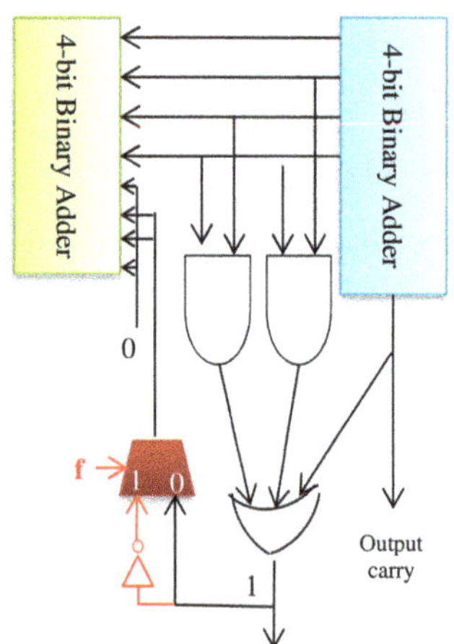

Figure 10.8 A hardware Trojan in a 3PIP module (BCD counter) (Sengupta et al., 2017)

instance of 3PIPs (BCD adder circuit) which is compromised with hardware Trojan that changes computational output value when it is activated. The hidden Trojan is triggered when select line input, i.e., "f" of the 2:1 multiplexer is set to logic 1. In normal conditions, "f" remains at logic 0 and therefore the hardware Trojan does not function in the circuit. The hidden malicious logic in the compromised IP design is denoted using Red wires/portions. Triggering of Trojan can be realized using various possibilities such as activation through an FSM counter (which is also termed as Trojan time bomb), sensing a specific signal in the design and external antenna, etc. When the hardware Trojan is activated, the target system experiences corruption in the computational output value due to the infected component. However, security against such types of Trojans can be achieved during HLS by designing the DMR schedule of the targeted IP core (Sengupta *et al.*, 2017; Sengupta and Bhadauria, 2015).

10.7 Conclusion

Hardware Trojans have emerged as a serious threat to modern SoC designs as they are increasingly being employed in critical infrastructure. There are different stages in the design supply chain where the hardware Trojan can be inserted into the design. A 3PIP vendor, a CAD tool vendor, and an offshore foundry are the critical sites for Trojan insertion. We specifically focused on hardware Trojans which can be placed into the designs during the HLS design process and the hardware Trojan in 3PIP modules which may be included in HLS tool library. In this chapter, we discussed particularly the following three HLS-based high-level hardware Trojans that can be implanted by a compromised HLS tool: (i) battery exhaustion attack, (ii) degradation attack, and (iii) downgrade attack. Further, we also put some light on hardware Trojan in 3PIP.

At the end of this chapter, a reader is expected to gain the following insights:

1. Overview hardware Trojan and the motivation for adversary to perform HLS-based hardware Trojan attack
2. Potential sites for hardware Trojan insertion in the design supply chain
3. Battery exhaustion attack and its realization during HLS with an example
4. Impact of battery exhaustion attack on design parameters
5. Degradation attack and its realization during HLS with an example
6. Impact of degradation attack on design parameters
7. Downgrade attack and its realization during HLS with an example
8. Impact of downgrade attack on design parameters
9. Overview of hardware Trojan in 3PIP modules with an example

10.8 Questions and exercise

1. What is hardware Trojan insertion?
2. What are the different stages in the design supply chain where the hardware Trojan can be inserted?

3. What is a compromised CAD tool?
4. How can a compromised CAD tool impact the system functionality and market economics?
5. What do you mean by a malicious HLS tool?
6. What are the different types of HLS-based hardware Trojan attacks?
7. What is a battery exhaustion attack?
8. How does the battery exhaustion attack impact the area and performance of the design?
9. What is a degradation attack?
10. How does the degradation attack impact the area and performance of the design?
11. How can the security of cryptographic algorithms be compromised during HLS?
12. How can a Trojan-infected 2:4 decoder circuit be designed to produce a compromised 3PIP module?

References

K. Basu, S. Mohamed Saeed, C. Pilato, *et al.*, (2019) "CAD-Base: An Attack Vector into the Electronics Supply Chain," *ACM Transactions on Design Automation of Electronic Systems (TODAES)*. 24, 4, Article 38, 30 pp.

P. Derbez, P.-A. Fouque and J. Jean, (2013) "Improved Key Recovery Attacks on Reduced-Round AES in the Single-Key Setting," in *Advances in Cryptology—EUROCRYPT*, Berlin: Springer, pp. 371–387.

J. S. Hurtarte, E. A. Wolsheimer and L. M. Tafoya, (2007) *Understanding Fabless IC Technology*, Amsterdam, The Netherlands: Elsevier.

M. C. McFarland, A. C. Parker and R. Camposano, (1988) "Tutorial on High-Level Synthesis," *DAC '88 Proceedings of the 25th ACM/IEEE Design Automation*, vol. 27, no. 1, pp. 330–336.

C. Pilato, K. Basu, F. Regazzoni and R. Karri, (2019) "Black-Hat High-Level Synthesis: Myth or Reality?, " *IEEE Transactions on Very Large Scale Integration (VLSI) Systems*, vol. 27, no. 4, pp. 913–926.

I. Polian, G. T. Becker and F. Regazzoni, (2016) "Trojans in Early Design Steps—An Emerging Threat," in *Proceedings of TRUDEVICE*, pp. 55:1–55:6.

M. Rathor and A. Sengupta, (2023) "Revisiting Black-Hat HLS: A Lightweight Countermeasure to HLS-Aided Trojan Attack," *IEEE Embedded Systems Letters*, vol. 16, no. 2, pp. 170–173, doi: 10.1109/LES.2023.3327793.

S. K. Sanadhya and P. Sarkar, (2008) "Attacking Reduced Round SHA-256," in *Applied Cryptography and Network Security*, S. M. Bellovin, R. Gennaro, A. Keromytis, and M. Yung, Eds, Berlin: Springer, pp. 130–143.

A. Sengupta and S. Bhadauria, (2015) "Untrusted Third Party Digital IP Cores: Power-Delay Trade-off Driven Exploration of Hardware Trojan Secured Datapath during High Level Synthesis," in *Proceedings of the 25th Edition on*

Great Lakes Symposium on VLSI (GLSVLSI '15), New York, NY: ACM, pp. 167–172.

A. Sengupta and S. P. Mohanty, (2019) "Advanced Encryption Standard (AES) and Its Hardware Watermarking for Ownership Protection," *IP Core Protection and Hardware-Assisted Security for Consumer Electronics*, pp. 317–335.

A. Sengupta and M. Rathor, (2019) "Security of Functionally Obfuscated DSP Core against Removal Attack Using SHA-512 Based Key Encryption Hardware", *IEEE Access Journal*, vol. 7, pp. 4598–4610.

A. Sengupta, S. Bhadauria and S. P. Mohanty, (2017) "TL-HLS: Methodology for Low Cost Hardware Trojan Security Aware Scheduling with Optimal Loop Unrolling Factor during High Level Synthesis," *IEEE Transactions on CAD of Integrated Circuits and Systems*, vol. 36, no. 4, pp. 655–668.

A. Sengupta, (2017) "Hardware Vulnerabilities and Its Effect on CE Devices: Design-for-Security against Trojan", *IEEE Consumer Electronics*, vol. 6, no. 3, pp. 126–133.

K. Xiao, D. Forte, Y. Jin, R. Karri, S. Bhunia and M. Tehranipoor, (2016) "Hardware Trojans: Lessons Learned after One Decade of Research," *ACM Transactions on Design Automation of Electronic Systems (TODAES)*, vol. 22, no. 1, Art. no. 6.

Chapter 11

Hardware Trojans—detective countermeasure against HLS-based hardware Trojan attack

Anirban Sengupta[1] and Mahendra Rathor[2]

The focus of this chapter is to discuss a lightweight and efficient countermeasure for detecting high-level synthesis (HLS)-based hardware Trojan attacks. This chapter first describes the HLS-based Trojan attack and the contemporary solutions to counter them. Further, a lightweight countermeasure for high-level Trojan detection (LC-HLTD) is discussed in detail along with the case studies on some applications.

The organization of the chapter is as follows: Section 11.1 discusses the trustworthiness of computer-aided-design (CAD) tools; Section 11.2 introduces the black-hat HLS; Section 11.3 discusses contemporary solutions to battery exhaustion attack; Section 11.4 presents the lightweight countermeasure for high-level Trojan detection, i.e., LC-HLTD while discussing salient features of LC-HLTD, threat scenario of LC-HLTD, high-level transformation (HLT) tool for the detective countermeasure, different phases of LC-HLTD approach, and applying LC-HLTD for detecting the battery exhaustion attack (due to pseudo operations insertion); Section 11.5 analyzes some case studies on digital signal processing (DSP) applications; and Section 11.6 concludes the chapter.

11.1 Introduction

A practical semiconductor design flow cannot be realized without computer-aided design (CAD) tools. This is because the CAD tools not only alleviate the design efforts and time but also reduce the cost. The CAD synthesis tools have gained popularity for creating designs at different phases of the very large-scale integration (VLSI) design process. For translating high-level description into register transfer level (RTL) designs, HLS tools are increasingly gaining popularity. However, as discussed in Chapter 10, third-party HLS tools cannot be fully trusted as they can pose a serious hardware threat of Trojan insertion. This exposes the dark side of the horizontal semiconductor business model. In the next section, we will discuss the

[1]Department of Computer Science and Engineering, Indian Institute of Technology Indore, India
[2]School of Instrumentation, Devi Ahilya Vishwavidyalaya, India

hardware Trojan threat due to an HLS tool and formally define what a black-hat HLS is (Polian *et al.*, 2016; Hurtarte *et al.*, 2007).

11.2 What is black-hat HLS?

A black-hat HLS is a compromised (maliciously and intentionally modified) HLS process that aims at secretly placing a hardware Trojan into the intended design during its some particular steps (Pilato *et al.*, 2019; Basu *et al.*, 2019; Rathor and Sengupta, 2023). The targeted steps of a black-hat HLS could be scheduling, allocation, binding, datapath synthesis, and controller synthesis. Figure 11.1 also

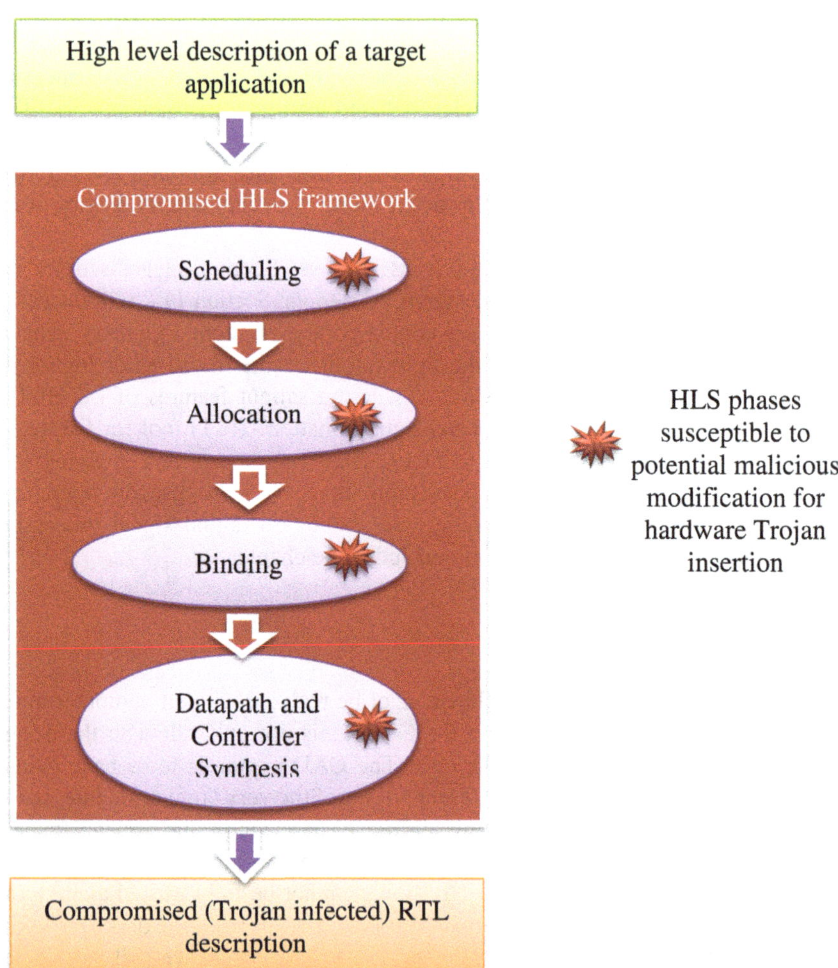

Figure 11.1 Compromised RTL generation through a malicious HLS framework

depicts the different HLS design phases for potential hardware Trojan insertion by a compromised HLS framework. The major impacts of placing hardware Trojan during HLS can be in terms of the corrupted functional behavior of the design and degraded design metrics such as power and delay. As discussed in the previous chapter, the compromised HLS behavior (black-hat HLS) can result in different types of high-level Trojan attacks such as battery exhaustion attacks, degradation attacks, and downgrade attacks. The battery exhaustion attack is primarily based on the insertion of fake operations during the scheduling process performed by a compromised HLS tool (Pilato *et al.*, 2019; Basu *et al.*, 2019). This chapter discusses a countermeasure (Rathor and Sengupta, 2023) specifically for the battery exhaustion attack.

11.3 Contemporary solutions to counter battery exhaustion attack

Researchers have come up with some solutions to address the battery exhaustion (driven by fake operations insertion) Trojan attack. The solutions can broadly be categorized as preventive and detective countermeasures. A preventive countermeasure was proposed by Potkonjak (2010). This solution introduces a model of fully specified design that aims to use all functional units in all the control steps by augmenting the schedule and assignment of operations. This thwarts an adversary from placing fake or pseudo operations during the scheduling phase of a compromised HLS framework. Practically, this preventive approach has limitations such as an excessively high design overhead. Further, high-level Trojan detection-based approach was introduced by Abderehman and its fellows in 2022 (Abderehman *et al.*, 2022). This detective countermeasure extracts the finite-state machine with datapath (FSMD) in order to enable Trojan detection. In order to capture any modification in the datapath, this approach analyzes the FSMD extracted from RTL. However, this detection approach may become cumbersome for highly complex or large RTLs. In addition, this approach leads to a larger detection time as it is based on extracting FSMD from RTL and analyzing for Trojan presence (which is not so easy). Considering the limitations of the above-mentioned mechanisms, a lightweight detective control-based countermeasure for high-level Trojan detection (LC-HLTD) has recently been introduced by Rathor and Sengupta (2023). In the next section, the LC-HLTD is discussed in detail.

11.4 Detective countermeasure techniques against hardware Trojan

In recent years, Rathor and Sengupta (2023) have introduced a lightweight countermeasure LC-HLTD against the HLS-aided Trojan attack by enabling the detection of fake operations (that may potentially be added by a malicious HLS framework during the scheduling phase of the HLS process). This LC-HLTD is a simpler and efficient approach for the detection of potential HLS-aided Trojan attacks. It is based on the following major fact. For an input application, scheduling information along with the final RTL description is provided as output by the existing HLS tools. For example, commercial tools such as Vivado HLS and academic tools such as Bambu (Pilato and Ferrandi, 2013) show the scheduling information for a given application. This enables to use of the scheduling information of the operations of a particular application to identify whether the fake operations are present or not. This approach does not rely on inspecting the RTL description for detecting fake operations (Rathor and Sengupta, 2023).

The salient features of the HLS-aided high-level Trojan detection approach (Rathor and Sengupta, 2023) are presented in the following section.

11.4.1 Salient features of the lightweight detective control-based countermeasure for high-level Trojan detection (LC-HLTD) approach

- This approach (Rathor and Sengupta, 2023) uses a lightweight high-level transformation (HLT) tool which relies on operation-count modification-related transformations only.
- The Trojan detection in a potentially compromised HLS tool is enabled by matching the number of operations obtained from the output of the HLT tool and the output from the HLS (high-level Trojan-infected) tool (Rathor and Sengupta, 2023).
- The Trojan detection is applicable for the following two cases (Rathor and Sengupta, 2023):

  (i) For such applications where the operation count modification-related transformations are not applicable.
  (ii) For such applications wherein the operation count modification-related transformations can be applied before scheduling.

- This approach (Rathor and Sengupta, 2023) is capable of offering very low detection time while simultaneously maintaining a very good accuracy of detection (to be 100%). Achieving very good detection accuracy does not depend on the size or complexity of the designs.

11.4.2 Threat scenario of LC-HLTD (Rathor and Sengupta, 2023)

The first and foremost consideration is a realizable black-hat HLS process that can place hardware Trojan into the components using the maliciously modified HLS steps. The placement of the high-level hardware Trojan for realizing the battery exhaustion attack is encouraged by the fact that the available functional units, which have not been used in the existing control steps, can be reused for the Trojan execution. The reusing of unused functional units in some specific control steps is realized by placing the fake operations during the scheduling phase of the malicious HLS framework. The unnecessary execution of functional units during the operating conditions leads to battery exhaustion/ rapid drainage.

The detection of high-level Trojans using the LC-HTD approach depends on the following files/tools (Rathor and Sengupta, 2023):

- An algorithmic description of the target IP
- The HLS (that can potentially be a compromised one) tool
- A lightweight high-level transformation (HLT) tool

11.4.3 High-level transformation tool for detective countermeasure

A lightweight HLT framework is defined as the HLT process (Sengupta *et al.*, 2017; Sengupta and Roy, 2017) which performs only such high-level transformations (obviously before scheduling) that can affect the total operation count of the target application. The lightweight HLT framework is first fed with the input application being considered. The following major steps are performed: (a) high-level description or transfer function of the target application is taken as input, (ii) only operation count modification-related transformations are applied such as LU, ROE, LICM, and LT, (iii) an intermediate form or representation of the target application is produced and the operation count is identified from it, and (iv) the scheduling, functional units allocation and binding information along with the area and delay estimations can also be reported by the lightweight HLT tool. The operation count modification-related transformations are briefly discussed as follows (complete details are presented in Sengupta *et al.*, 2017):

- **Loop unrolling (LU):** This high-level transformation technique unrolls the body of the loop (in a loop-based application) based on an unrolling factor value. Unrolling the loop certainly increases the number of operations in the application as the operations inside the loop body are repeated as many times as the loop body is unrolled.
- **Redundant operation elimination (ROE):** This high-level transformation technique eliminates the redundant operations (such operations whose operation type and the parent operations are the same as another operation in the application) from the application. Applying the ROE technique certainly reduces the number of operations in an intended application.
- **Loop invariant code motion (LICM):** This high-level transformation technique moves such operations from inside the loop body to outside the loop which does not depend on the loop iteration variable. Therefore, the LICM transformation ensures that the loop independent or invariant operations are not unnecessarily executed multiple times inside the loop, but rather executed once outside the loop. This affects the operation count if LU transformation is applied after the LICM. Please note that performing LU without applying LICM affects the operation count differently.
- **Logic transformation (LT):** This high-level transformation technique transforms the logic of an operation. For example, an multiplication operation that multiplies a variable by 3 can simply be replaced by two addition operations that add the variable to itself. As evident, it affects the overall operation count.

11.4.4 Different phases of LC-HLTD (Rathor and Sengupta, 2023)

As discussed earlier, the LC-HLTD approach has been developed to discern the fake operations (hardware Trojans) placed during the scheduling process by a potential black-hat HLS. Figure 11.2 depicts the detection process which is accomplished in the below-mentioned different phases:

1. **Deducing the operation count of the target application before scheduling using a lightweight HLT framework**:

   When the HLT is applied, the overall count of the application is modified (if the operation count modification-related transformations such as LU, ROE,

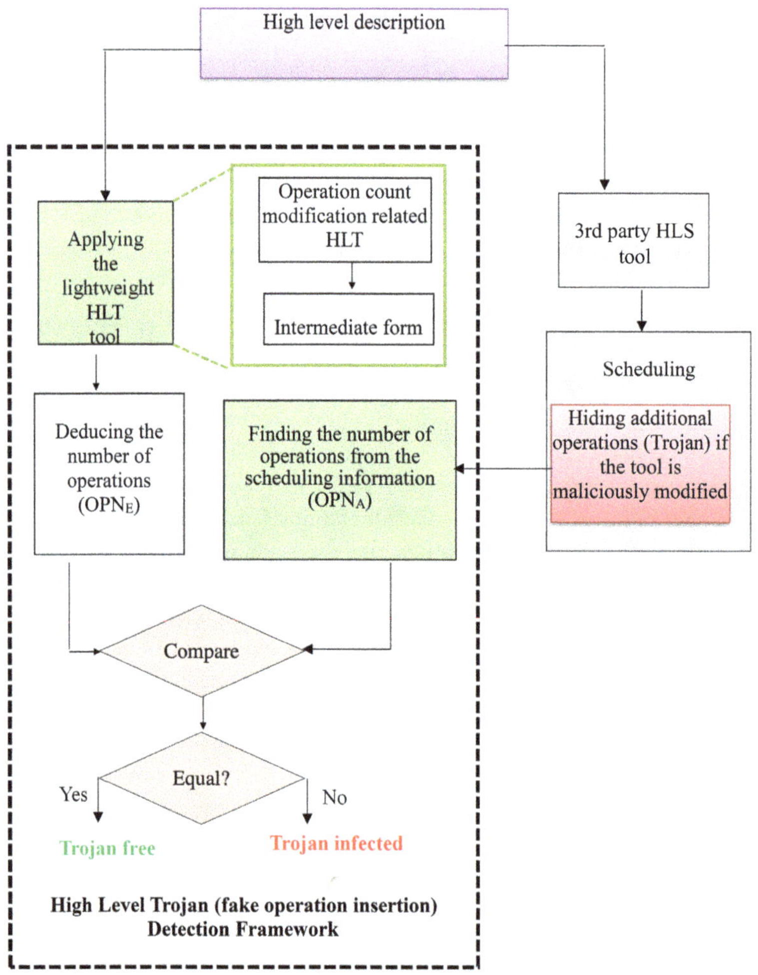

Figure 11.2 The Trojan detection flow using the LC-HLTD approach (Rathor and Sengupta, 2023)

LT, and LICM are applicable). Thus, a modified intermediate representation is generated. Further, the total number of operations, let's say OPN_E is deduced from this modified representation. The procured HLS tool is expected to produce the same operation count even after the scheduling is performed. Therefore, the OPN_E is the expected operation count from a genuine HLS tool.

2. **Finding the number of operations from the scheduling generated by the HLS tool (potentially compromised):**
   The third-party HLS tool is also fed with the same high-level description (as used in phase 1). It is to be noted that the LC-HLTD approach uses the assumption of enabling the same operation count modification-related HLT directives as performed in phase 1. If the third-party HLS tool is malicious (capable of performing battery exhaustion attack), it places the fake operations during the scheduling process. Hence, it produces scheduling with a higher number of operations compared to the expected operation count from a Trojan-free scheduling. Let's say the number of extracted operations from the scheduling generated by an HLS tool (suspected to be compromised) is OPN_A.
3. **Comparing the operation count obtained from the phase (1) and (2):**
   Post obtaining the operation count from phase 1 and phase 2 (i.e., OPN_E and OPN_A), a comparison is performed to find the presence of Trojan. If OPN_E and OPN_A are found to be equal, then any additional (pseudo) operations have not been inserted by the third-party HLS tool and hence the design is not affected by the battery exhaustion Trojan attack. However, if OPN_E and OPN_A are not equal, then there is a possibility that some additional operations have been placed by the HLS tool, hence exposing its maliciousness.

11.4.5 LC-HLTD approach for the applications where operation count modification-related transformations are not applicable (Rathor and Sengupta, 2023)

To discuss the LC-HLTD approach for this case, let us consider a high-level description shown in Figure 11.3(a) (Basu *et al.*, 2019; Rathor and Sengupta, 2023). Originally, this description contains three multiplication and six addition operations. When this description is subjected to a black-hat HLS process as input, the maliciously modified scheduling process of the black-hat HLS framework introduces additional/fake operations into the scheduling as shown in Figure 11.3(b), where scheduling is performed with two multipliers and two adders. As evident, the given high-level description cannot undergo any operation count modification-related transformation. In order to apply the LC-HLTD approach for Trojan detection for this case, the below steps are followed:

- Firstly, we deduce the expected operations count OPN_E directly from the high-level description given in Figure 11.3(a). It is deduced to be $OPN_E = 9$ (i.e., six additions and three multiplications).

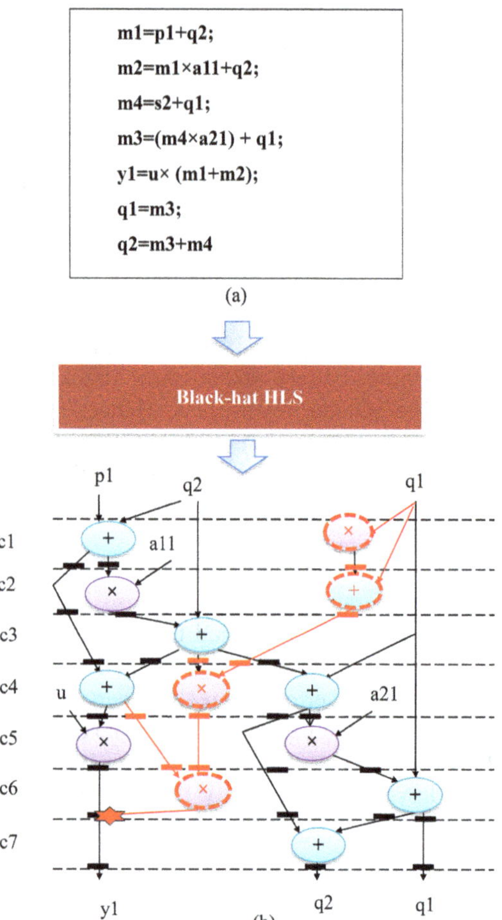

Figure 11.3 Insertion of additional operations by a malicious HLS, where fake operations are denoted using red dotted circles (Rathor and Sengupta, 2023). (a) Sequence of operations in a sample high-level description and (b) corresponding scheduled data flow graph after fake operations insertion.

- A potential black-hat HLS places additional one addition and three multiplication operations using the maliciously modified scheduling phase. Thus, a Trojan-infected scheduling is produced. It is to be noted that the insertion of fake operations does not result in any violation of the resource constraints.
- Further, the operation count OPN_A is obtained from the Trojan-infected scheduling. It is obtained to be 13.
- As the operation count extracted from the high-level description is not the same as the operation count deduced from the Trojan infected scheduling ($OPN_E \neq OPN_A$), therefore the presence of fake operations (Trojan) is guaranteed.

11.4.6 LC-HLTD approach for the applications wherein the operation count modification-related transformations can be applied (Rathor and Sengupta, 2023)

To discuss the LC-HLTD approach for this case, let us consider a high-level description shown in Figure 11.4(a) (Rathor and Sengupta, 2023). Originally, this

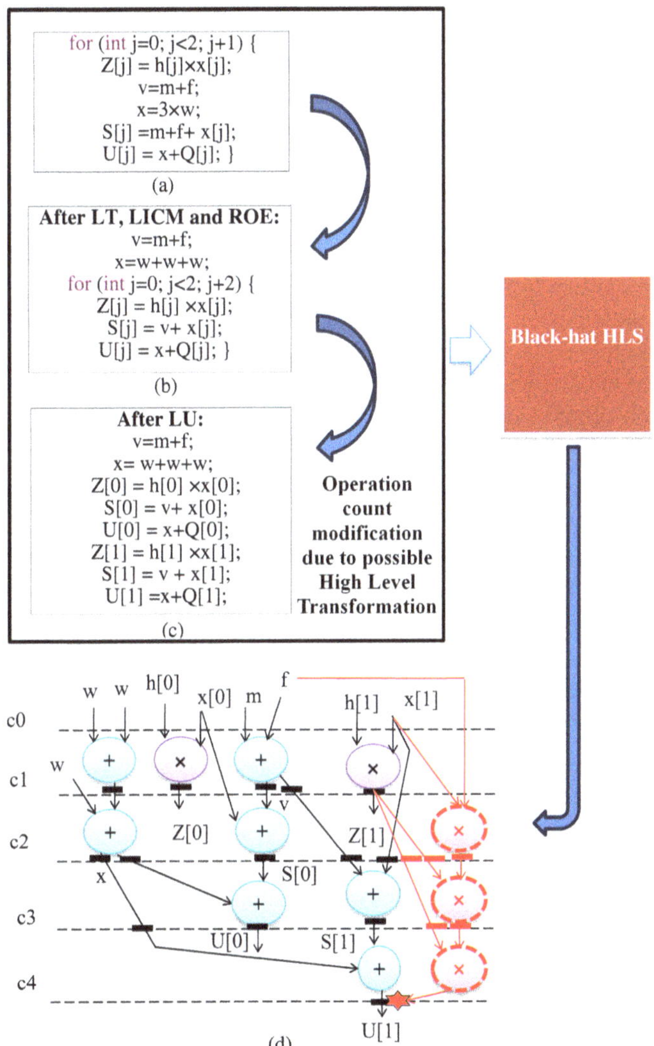

Figure 11.4 *Insertion of additional operations during scheduling while considering the HLT directives applied by the malicious HLS (Rathor and Sengupta, 2023). (a) A sample high-level description of a loop based application, (b) LT, LICM and ROE based high-level transformations, (c) LU based high-level transformation, and (d) insertion of fake operations in the scheduled data flow graph post high-level transformations.*

description contains two multiplications and five additions operations. In this case, the operation count modification-related HLTs are possible. As discussed earlier, the high-level transformations such as ROE, LICM, LT, and LU can modify the overall operation count of the application being considered for the scheduling process. The possible operation count modification related high-level transformation to be applied are discussed below:

- The LT can be applied to the expression "x = 3 × w" where it can be logically converted to "x = w + w + w", affecting the overall operation count.
- The operation "m + f" is redundant in the original expression "S[j] = m + f + x [j]" shown in Figure 11.4(a), as the operation "v = m + f" already exists. Hence, ROE can be performed to eliminate the redundant execution of "m + f" in S[j] = m + f + x[j]. It can directly use the result of computation obtained from v = m + f to execute the expression S[j] = m + f + x[j]. After ROE, the expression is modified to S[j] = v + x[j] (reducing the number of operations from two to one in this particular expression) as shown in Figure 11.4(b).
- The expressions "x = w + w + w" and "v = m + f" are loop invariant and therefore are shifted outside the loop as presented in Figure 11.4(b).
- The loop in the high-level description is fully unrolled which also modifies the operation count as depicted in Figure 11.4(c).

After applying the relevant HLTs, an intermediate representation/description is obtained which is shown in Figure 11.4(c). This intermediate description has an operation count 9. When this description is subjected to a black-hat HLS process as input, the maliciously modified scheduling process of the black-hat HLS framework introduces additional/fake operations into the scheduling as shown in Figure 11.4(d), where scheduling is performed with two multipliers and two adders.

In order to apply the LC-HLTD approach for the Trojan detection for this case, the below steps are followed:

- Firstly, we deduce the expected operations count OPN_E directly from the high-level description given in Figure 11.4(a). It is deduced to be $OPN_E = 7$ (i.e., five additions and two multiplications).
- The same (original) high-level description, shown in Figure 11.4(a), is subjected to the third-party HLS tool while enabling the same HLS directives (i.e., ROE, LICM, LT, and LU) for the high-level transformation phase. The potential black-hat HLS places additional three multiplication operations in the transformed (based on LU, ROE, LT, and LICM) high-level description using the maliciously modified scheduling phase. Thus, a Trojan-infected scheduling is produced.
- Further, the operation count OPN_A is obtained from the Trojan-infected scheduling. It is obtained to be 12 as shown in Figure 11.4(d).
- As the operation count extracted from the high-level description is not the same as the operation count deduced from the Trojan infected scheduling ($OPN_E \neq OPN_A$), the presence of fake operations (Trojan) is guaranteed.

11.5 Analysis of case studies (Rathor and Sengupta, 2023)

The case studies have been performed on a number of digital signal processing (DSP) applications such as 8-tap finite impulse response (FIR) filter, discrete wavelet transform (DWT), autoregression filter (ARF), etc. In the studied applications, the operation count-related HLTs were performed as discussed in Section 11.4. The HLS process is compromised to hide the fake operations during the scheduling of the case studies. The insertion of fake operations is performed such that a minimum single additional operation is placed in a control step where the resources have not been reused, to increase the reusing of functional resources.

The results are studied in terms of the following:

- Count of additional operations placed by the malicious HLS
- The Trojan detection time
- The false positive and false negative
- The cost of Trojan insertion in terms of area and delay

Table 11.1 reports the effect of the applied HLTs on operation count. As shown, the operation count is modified after applying the HLTs for different target applications. Further, Table 11.2 reports the following information: (i) the

Table 11.1 Operation count (Oc) affected due to HLTs (Rathor and Sengupta, 2023)

Applications	**Test-sample**	**JPEG**	**ARF**	**8-tap FIR**
LU	✓	—	—	✓
ROE	✓	✓	✓	—
LT	✓	—	—	—
LICM	✓	—	—	—
O_c (Before HLTs)	8	112	28	4
O_c (after HLTs)	9	95	23	15

Table 11.2 Analysis of fake operations insertion by a compromised HLS for Trojan detection (Rathor and Sengupta, 2023)

Applications	**Test-sample**	**Mesa Horner**	**JPEG IDCT**	**MPEG**	**DWT**	**ARF**	**8-tap FIR**
Resource constraints	2+,2*	1+,2*	12 +,12*	2+,3*	1+,2*	2 +,4*	1+,4*
Expected operation count (OPN_E)	9	14	95	28	17	23	15
Extracted operation count (OPN_A) from the scheduling	12	16	103	33	22	30	22
Trojan detection	**Yes**	**Yes**	**Yes**	**Yes**	**Yes**	**Yes**	**Yes**

Note: Bold indicates that the Trojan detection is performed successfully.

operation count extracted post applying the related possible HLTs, (ii) the operation count deduced from the scheduling performed by the malicious HLS framework. As evident, both operation counts are different due to the hiding of extra operations during the scheduling phase by the malicious HLS tool. The table shows that if the fake operations are inserted then the operation count is certainly modified. Therefore, fake operations are found in all the tested applications scheduled using the compromised HLS. Hence, this shows the ability of LC-HLTD approach to correctly identify the Trojan attack (i.e., fake operations). The feature of having the false positive and false negative as zero is also evident as reported in Table 11.3. The Trojan detection time of the LC-HLTD approach is also reported in Table 11.3. As evident, the Trojan detection time is significantly lower.

Further, the effect of hiding the additional operations on design area and delay are also discussed as reported in Table 11.4. The following two cases may exist:

1. Existing multiplexers/demultiplexers may be used to bind the extra operations with the existing FUs. This situation may occur if, for example, an m×1 multiplexer needs to multiplex between less than m inputs (before Trojan insertion). In this case, the remaining unused inputs of the multiplexer are sufficient to bind the extra operations with the existing FUs. Therefore, area and delay overhead do not incur in this scenario as shown in Table 11.4 for some benchmarks.
2. The currently used multiplexers/demultiplexers are not sufficient to bind the extra operations with the existing FUs. This situation may occur if, for

Table 11.3 Analysis of the false positive, false negative, and detection time (Rathor and Sengupta, 2023)

False positive	0
False Negative	0
Trojan detection time (Average and estimated)	~50 milliseconds

Table 11.4 Design overhead analysis due to fake operation insertion by the compromised HLS (Rathor and Sengupta, 2023)

Applications	**Area (μm²) (without Trojan)**	**Area (μm²) (affected due to Trojan)**	**Delay (ps) (without Trojan)**	**Delay (ps) (affected due to Trojan)**	**%Area overhead**	**%Delay overhead**
Mesa Horner	198.62	198.62	2191.54	2191.54	0.0	0.0
MPEG	323.67	323.67	2278.29	2278.29	0.0	0.0
8-tap FIR	341.95	353.45	1488.37	1539.37	3.4	3.4
ARF	381.91	389.58	1885.83	1936.83	2.0	2.7
DWT	213.96	217.79	2849.25	2874.75	1.8	0.9
JPEG IDCT	1316.50	1316.50	2477.02	2477.02	0.0	0.0
Test-sample	200.25	207.91	667.69	667.69	3.8	0.0

example, an m×1 multiplexer is multiplexing between exactly m inputs. Hence, multiplexer/demultiplexers of relatively bigger size are required to bind the extra operations with the existing FUs. In this case, slight area and delay overhead incur as shown in Table 11.4.

11.6 Conclusion

Battery exhaustion attack is a serious threat to modern battery-powered systems. A black-hat HLS process can enable the battery exhaustion attack by hiding the additional/pseudo operations in the design during the scheduling phase of the HLS process. The maliciously modified scheduling process hides the pseudo operations in such a fashion that the resource constraints are not violated and simultaneously the Trojan remains undetectable. The black-hat HLS process can be exposed by carefully examining the scheduling output of the malicious HLS tool. If the expected number of operations does not match with the number of operations that are extracted from the scheduling generated by the HLS tool, then it guarantees the insertion of fake operations by the HLS tool, hence exposing its maliciousness. This chapter has discussed a lightweight and efficient countermeasure to detect the insertion of fake operations by a malicious HLS tool, in detail.

At the end of this chapter, a reader is expected to gain the following insights:

1. Idea of Black hat HLS process
2. Pseudo-operations insertion and battery exhaustion attack
3. Preventive countermeasure against battery exhaustion attack and its limitations
4. Lightweight detective countermeasure for high-level Trojan (fake operations insertion) detection
5. Use of high-level transformation tool to aid the high-level Trojan detection
6. Concept of modification of operation count due to the high-level transformation
7. Detection of Trojan using LC-HLTD approach for the applications where operation count does not modify due to high-level transformations
8. Detection of Trojan using LC-HLTD approach for the applications where operation count is modified due to high-level transformations
9. Analysis of LC-HLTD approach on a number of use cases

11.7 Questions and exercise

1. How does a CAD tool raise the issue of untrustworthiness?
2. What is black hat HLS?
3. What is a preventive control measure against fake operation insertion-based Trojan attack?
4. What are the salient features of the LC-HLTD approach?
5. How does a high-level transformation tool aid to detect high-level Trojan attack?

6. Explain how a loop unrolling transformation modifies the operation count.
7. Explain how a logic transformation modifies the operation count.
8. Explain how a loop invariant code motion transformation modifies the operation count.
9. Explain how a redundant operation elimination transformation modifies the operation count.
10. Explain the Trojan detection flow of the LC-HLTD approach.
11. List out some applications where operation-related high-level transformations are applicable.

References

M. Abderehman, R. Gupta, R. R. Theegala and C. Karfa, (2022) "BLAST: Belling the Black-Hat High-Level Synthesis Tool," *IEEE Transactions on Computer-Aided Design of Integrated Circuits and Systems*, vol. 41, no. 11, pp. 3661–3672.

K. Basu, S. Mohamed Saeed, C. Pilato, *et al.*, (2019) "CAD-Base: An Attack Vector into the Electronics Supply Chain," *ACM Transactions on Design Automation of Electronic Systems (TODAES),* vol. 24, no. 4, Article 38, 30 pp.

J. S. Hurtarte, E. A. Wolsheimer and L. M. Tafoya, (2007) *Understanding Fabless IC Technology*, Amsterdam: Elsevier.

C. Pilato and F. Ferrandi, (2013) "Bambu: A Modular Framework for the High-Level Synthesis of Memory-Intensive Applications," in *Proceedings of the FPL*, pp. 1–4.

C. Pilato, K. Basu, F. Regazzoni and R. Karri, (2019) "Black-Hat High-Level Synthesis: Myth or Reality?" *IEEE Transactions on Very Large Scale Integration (VLSI) Systems*, vol. 27, no. 4, pp. 913–926.

I. Polian, G. T. Becker and F. Regazzoni, (2016) "Trojans in Early Design Steps—An Emerging Threat," in *Proceedings of TRUDEVICE*, 2016, pp. 55:1–55:6.

M. Potkonjak, (2010) "Synthesis of Trustable ICs Using Untrusted CAD Tools," in *Proceedings of the DAC*, New York, NY, USA, pp. 633–634.

M. Rathor and A. Sengupta, (2023) "Revisiting Black-Hat HLS: A Lightweight Countermeasure to HLS-Aided Trojan Attack," in *IEEE Embedded Systems Letters*, vol. 16, no. 2, pp. 170–173, doi:10.1109/LES.2023.3327793.

A. Sengupta and D. Roy, (2017) "Protecting an Intellectual Property Core during Architectural Synthesis Using High-Level Transformation Based Obfuscation," *IET Electronics Letters*, vol. 53, no. 13, pp. 849–851.

A. Sengupta, D. Roy, S. P. Mohanty and P. Corcoran, (2017) "DSP Design Protection in CE through Algorithmic Transformation Based Structural Obfuscation," *IEEE Transactions on Consumer Electronics*, vol. 63, no. 4, pp. 467–476.

Chapter 12
Conclusion

Anirban Sengupta[1]

12.1 Concepts learned through the book

Research on HLS has been ongoing for the last two decades by the scientific community. However, HLS exploited for security, has become the focus in recent years. It has been an active area of interest for the security community all over the globe, especially in the last few years due to manifold benefits offered by HLS such as lower overhead, flexibility in architecture exploration, and design automation. Security and trust in hardware have become more important than ever before due to the emergence of several points of vulnerabilities and the growing sophistication of cyber-attacks.

Through this book, a reader gained valuable insights on how efficiently the HLS framework can be leveraged to offer several benefits for hardware security. This is particularly important when designing specialized hardware (or IPs/modules) for data-intensive/power-hungry applications such as convolutional layer in CNN, signal filtering through digital filters, image processing through image processing filters, multimedia compression/decompression using JPEG CODECs, as well as other digital signal processing applications, etc. The hardware IPs of such algorithms/applications are typically designed using HLS instead of starting from RTL (or lower levels) due to the complexity involved. Therefore, it is natural, that the security of such hardware IPs should be dealt with by exploiting the HLS framework such that it results in lower overhead, lesser design time, and greater flexibility for the designer. Hardware security of smaller applications with lesser complexity such as traditional combinational/sequential circuit blocks can be handled at the RTL or gate level. However, it's not pragmatic to do so for the aforementioned complex applications.

This book, therefore, provided an important guide for readers to understand how to design secure specialized hardware IPs using HLS. The book covers how HLS flow can be used to address critical security threats (that can potentially affect reusable hardware IPs) such as IP piracy, IP ownership abuse, watermark collision,

[1]Department of Computer Science and Engineering, Indian Institute of Technology Indore, India

ghost insertion search attack, tampering attack, forgery attack, reverse engineering, malicious RTL alteration, SCA, etc.

12.2 Summary

In recent years, several emerging techniques for HLS-based hardware security have surfaced. This book presented state-of-the-art high-level synthesis methodologies for hardware security and trust. No such book exists in the market yet that covers the state-of-the-art contents/topics presented in this book. The book provided a one-stop reading for readers who wish to learn the state of the art on "HLS solutions for hardware security, trust and IP protection." Through the proposed book, the readers have learned important topics such as HLS-based security for IP piracy countermeasure, employing protein molecular biometric, retinal biometric, and facial biometric during HLS for hardware security, detecting black-hat HLS-aided Trojan attack, crypto-chain driven HLS security, structural obfuscation in HLS, and exploiting voice biometric as watermark during HLS security.

Three key features of the proposed book:

- Detective countermeasure against IP piracy threat and false ownership claim using high-level synthesis: protein molecular biometric signature, retina biometric-based, and facial biometric encryption-based HLS security techniques.
- HLS-based lightweight countermeasure to Black-HAT Trojan attack.
- Multi-modal biometric and multi-cut-based structural obfuscation during HLS for securing IP cores.

The book covered the following important topics through different chapters along with questions and exercises for students.

Chapter 1: Introduction to hardware security, trust, and IPP;
Chapters 2–6 on recent HLS-based watermarking approaches;
Chapter 7 on recent HLS-based fingerprinting approach;
Chapters 8–9 on HLS-based hardware obfuscation approaches;
Chapters 10–11 on HLS-based hardware Trojans and their security countermeasure;
Chapter 12—concludes the book with an important take-away from a reader's perspective.

- **Chapter 1—Introduction to hardware security and trust and high-level synthesis**: This chapter introduced high-level synthesis-based hardware security and trust techniques (as well as IP protection approaches). It discussed the threat models, taxonomy of different threats, and threat actors. It also covered a summary of different hardware attacks/threats and their security countermeasures such as IP (hardware) watermarking, IP steganography, digital signature, hardware metering, hardware fingerprinting, computational forensic engineering, biometric-based watermarking, structural obfuscation, functional obfuscation/logic locking.

- **Chapter 2—High-level synthesis-based watermarking using protein molecular biometric with facial biometric encryption:** This chapter introduced a state of the art high-level synthesis-based protein molecular biometric watermarking with facial biometric encryption technique for securing hardware accelerators. It explained the key concepts of the specific cyber threat, defense mechanism, and its principles along with several illustrative examples.
- **Chapter 3—High-level synthesis-based watermarking using retinal biometric**: This chapter introduced a state-of-the-art high-level synthesis-based retina biometric watermarking technique for securing hardware IPs. It explained the key concepts of the specific cyber threat, defense mechanism, and its principles along with several illustrative examples.
- **Chapter 4—HLS-based mathematical watermarks for hardware security and trust:** This chapter introduced a state-of-the-art high-level synthesis-based mathematical watermarks for hardware security and trust using dispersion matrix, variance, covariance, and Eigen decomposition frameworks.
- **Chapter 5—High-level synthesis-based watermarking using multi-modal biometric:** This chapter introduced a state-of-the-art high-level synthesis-based multi-modal biometric watermarking solution for protecting hardware IP cores. It explained the key concepts of the specific cyber threat, defense mechanism, and its principles along with several illustrative examples.
- **Chapter 6—High-level synthesis-based watermarking using crypto-chain signature framework:** This chapter presented a state-of-the-art high-level synthesis-based crypto-chain watermarking technique for securing hardware IPs. It explained the key concepts of the specific cyber threat, defense mechanism, and its principles along with several illustrative examples.
- **Chapter 7—HLS-based fingerprinting**: This chapter explained the utility of IP fingerprinting in the context of IP buyer in the IP business model. The chapter covered different HLS-based IP fingerprint techniques including multivariable crypto-fingerprinting approach and symmetrical IP protection techniques along with examples.
- **Chapter 8—Hardware obfuscation—high-level synthesis-based structural obfuscation for hardware security and trust:** This chapter introduced a state-of-the-art high-level synthesis-based structural obfuscation technique along with physical biometric for security against reverse engineering of integrated circuits or IPs. It explains the key concepts of the specific cyber threat, defense mechanism, and its principles along with several illustrative examples.
- **Chapter 9—Hardware obfuscation—algorithmic transformation-based obfuscation for secure floorplan-driven high-level synthesis:** This chapter discussed designing hardware obfuscated IPs using key-controlled algorithmic transformation while generating secure floorplan. The description of the approach, along with specific examples from secure HLS to floorplan was discussed.
- **Chapter 10—Fundamentals on HLS-based hardware Trojan:** This chapter presented HLS-based Trojans including various types of threats possible by

exploiting the HLS framework such as battery exhaustion Trojan attack, degradation Trojan attack, downgrade attack, and hardware Trojan in 3PIP module.

- **Chapter 11—Hardware Trojans—detective countermeasure against HLS-based hardware Trojan attack:** This chapter discussed a lightweight detective countermeasure technique against HLS-based Trojan attack. The description of the approach along with the security analysis is also provided.
- **Chapter 12—Conclusion:** This chapter highlighted the key take-away for a reader along with the concepts learned from this book.

Interested readers can also read the following books on hardware security and IPP for additional learning:

- "*Physical Biometrics for Hardware Security of DSP and Machine Learning Coprocessors*," The Institute of Engineering and Technology (IET), 2023
- "*Secured Hardware Accelerators for DSP and Image Processing Applications,*" The Institute of Engineering and Technology (IET), 2021
- "*Frontiers in Securing IP Cores – Forensic detective control and obfuscation techniques,*" The Institute of Engineering and Technology (IET), 2020
- "*IP Core Protection and Hardware-Assisted Security for Consumer Electronics,*" The Institute of Engineering and Technology (IET), 2019

Index

www.ingramcontent.com/pod-product-compliance
Lightning Source LLC
Chambersburg PA
CBHW050510190326
41458CB00005B/1485
9781837241170